Studies in Systems, Decision and Control

Volume 117

Series editor

Janusz Kacprzyk, Polish Academy of Sciences, Warsaw, Poland
e-mail: kacprzyk@ibspan.waw.pl

The series "Studies in Systems, Decision and Control" (SSDC) covers both new developments and advances, as well as the state of the art, in the various areas of broadly perceived systems, decision making and control- quickly, up to date and with a high quality. The intent is to cover the theory, applications, and perspectives on the state of the art and future developments relevant to systems, decision making, control, complex processes and related areas, as embedded in the fields of engineering, computer science, physics, economics, social and life sciences, as well as the paradigms and methodologies behind them. The series contains monographs, textbooks, lecture notes and edited volumes in systems, decision making and control spanning the areas of Cyber-Physical Systems, Autonomous Systems, Sensor Networks, Control Systems, Energy Systems, Automotive Systems, Biological Systems, Vehicular Networking and Connected Vehicles, Aerospace Systems, Automation, Manufacturing, Smart Grids, Nonlinear Systems, Power Systems, Robotics, Social Systems, Economic Systems and other. Of particular value to both the contributors and the readership are the short publication timeframe and the world-wide distribution and exposure which enable both a wide and rapid dissemination of research output.

More information about this series at http://www.springer.com/series/13304

Hussein A. Abbass · Jason Scholz
Darryn J. Reid
Editors

Foundations of Trusted Autonomy

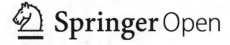

Editors
Hussein A. Abbass
School of Engineering and IT
University of New South Wales
Canberra, ACT
Australia

Darryn J. Reid
Defence Science and Technology Group
Joint and Operations Analysis Division
Edinburgh, SA
Australia

Jason Scholz
Defence Science and Technology Group
Joint and Operations Analysis Division
Edinburgh, SA
Australia

ISSN 2198-4182   ISSN 2198-4190   (electronic)
Studies in Systems, Decision and Control
ISBN 978-3-319-87879-9   ISBN 978-3-319-64816-3   (eBook)
https://doi.org/10.1007/978-3-319-64816-3

To a future where humans and machines live together in harmony.

Foreword

Technology-dependent industries and agencies, such as Defence, are keenly seeking game-changing capability in trusted autonomous systems. However, behind the research and development of these technologies is the story of the people, collaboration and the potential of technology.

The motivation for Defence in sponsoring the open publication of this exciting new book is to accelerate Australia's Defence science and technology in Trusted Autonomous Systems to a world-class standard. This journey began in July 2015 with a first invitational symposium hosted in Australia with some of the world-class researchers featured in this book in attendance. Since that time, engagement across the academic sector both nationally and internationally has grown steadily. In the near future in Australia, we look forward to establishing a Defence Cooperative Research Centre that will further develop our national research talent and sow the seeds of a new generation of systems for Defence.

Looking back over the last century at the predictions made about general purpose robotics and AI in particular, it seems appropriate to ask "so where are all the robots?" Why don't we see them more embedded in society? Is it because they can't deal with the inevitable unpredictability of open environments— in the case for the military, situations that are contested? Is it because these machines are simply not smart enough? Or is it because humans cannot trust them? For the military, these problems may well be the hardest challenges of all, as failure may come with high consequences.

This book then appropriately in the spirit of foundations examines the topic with an open and enquiring flavour, teasing apart critical philosophical, scientific, mathematical, application and ethical issues, rather than assuming a stance of advocacy.

The full story has not yet been written but it has begun, and I believe this contribution will take us forward. My thanks in particular to the authors and the editors, Prof. Hussein A. Abbass at the University of New South Wales for his sustained effort and art of gentle persuasion, and my own Defence Scientist, Research Leader Dr. Jason Scholz and Principal Scientist Dr. Darryn J. Reid.

Canberra, Australia   Dr. Alex Zelinsky
April 2017   Chief Defence Scientist of Australia

Preface

Targeting scientists, researchers, practitioners and technologists, this book brings contributions from like-minded authors to offer the basics, the challenges and the state of the art on trusted autonomous systems in a single volume.

On the one hand, the field of autonomous systems has been focusing on technologies including robotics and artificial intelligence. On the other hand, the trust dimension has been studied by social scientists, philosophers, human factors specialists and human–computer interaction researchers. This book draws threads from these diverse communities to blend the technical, social and practical foundations to the emerging field of trusted autonomous systems.

The book is structured in three parts. Each part contains chapters written by eminent researchers and supplemented with short chapters written by high calibre and outstanding practitioners and users of this field. The first part covers foundational artificial intelligence technologies. The second part focuses on the trust dimension and covers philosophical, practical and technological perspectives on trust. The third part brings about advanced topics necessary to create future trusted autonomous systems.

The book is written by researchers and practitioners to cover different types of readership. It contains chapters that showcase scenarios to bring to practitioners the opportunities and challenges that autonomous systems may impose on the society. Examples of these perspectives include challenges in Cyber Security, Defence and Space Operations. But it is also a useful reference for graduate students in engineering, computer science, cognitive science and philosophy. Examples of topics covered include Universal Artificial Intelligence, Goal Reasoning, Human–Robotic Interaction, Computational Motivation and Swarm Intelligence.

Canberra, Australia   Hussein A. Abbass
Edinburgh, Australia   Jason Scholz
Edinburgh, Australia   Darryn J. Reid
March 2017

Acknowledgements

The editors wish to thank all authors for their contributions to this book and for their patience during the development of the book.

A special thanks go to the Defence Science and Technology Group, Department of Defence, Australia, for funding this project to make the book public access.

Thanks also are due to the University of New South Wales in Canberra (UNSW Canberra) for the time taken by the first editor for this book project.

Contents

Contributors

Hussein A. Abbass School of Engineering and Information Technology, University of New South Wales, Canberra, ACT, Australia

David W. Aha Navy Center for Applied Research in AI, US Naval Research Laboratory, Washington DC, USA

Michael Barlow School of Engineering and IT, UNSW, Canberra, Australia

Russell Boyce University of New South Wales, Canberra, Australia

Selmer Bringsjord Rensselaer AI & Reasoning (RAIR) Lab, Department of Cognitive Science, Department of Computer Science, Rensselaer Polytechnic Institute (RPI), Troy, NY, USA

Peter D. Bruza Information Systems School, Queensland University of Technology (QUT), Brisbane, Australia

Bobby D. Bryant Department of Computer Sciences, University of Texas at Austin, Austin, USA

Alexandra Coman NRC Research Associate at the US Naval Research Laboratory, Washington DC, USA

Elizabeth Croft Department of Mechanical Engineering, University of British Columbia, Vancouver, Canada

Noel Derwort Department of Defence, Canberra, Australia

Andrew Dowse Department of Defence, Canberra, Australia

Tom Everitt Australian National University, Canberra, Australia

Frank P. Ferrie Department of Electrical and Computer Engineering, McGill University, Montreal, Canada

Michael W. Floyd Knexus Research Corporation, Springfield, VA, USA

Brian Gleeson Department of Computer Science, University of British Columbia, Vancouver, Canada

Clément Gosselin Department of Mechanical Engineering, Laval University, Quebec City, Canada

Naveen Sundar Govindarajulu Rensselaer AI & Reasoning (RAIR) Lab, Department of Cognitive Science, Department of Computer Science, Rensselaer Polytechnic Institute (RPI), Troy, NY, USA

Douglas Griffin University of New South Wales, Canberra, Australia

Medria Hardhienata School of Engineering and Information Technology, University of New South Wales, Canberra, Australia

Justin W. Hart Department of Computer Science, University of Texas at Austin, Austin, USA; Department of Mechanical Engineering, University of British Columbia, Vancouver, Canada

John Harvey School of Engineering and Information Technology, University of New South Wales, Canberra, Australia

Eduard C. Hoenkamp Information Systems School, Queensland University of Technology (QUT), Brisbane, Australia; Institute for Computing and Information Sciences, Radboud University, Nijmegen, The Netherlands

Marcus Hutter Australian National University, Canberra, Australia

Benjamin Johnson NRC Research Associate at the US Naval Research Laboratory, Washington DC, USA

S. Kate Devitt Robotics and Autonomous Systems, School of Electrical Engineering and Computer Science, Faculty of Science and Engineering, Institute for Future Environments, Faculty of Law, Queensland University of Technology, Brisbane, Australia

Adam Klyne School of Engineering and Information Technology, University of New South Wales, Canberra, Australia

Denis Laurandeau Department of Electrical Engineering, Laval University, Quebec City, Canada

Michael Lewis Department of Information Sciences, University of Pittsburgh, Pittsburgh, PA, USA

Karon MacLean Department of Computer Science, University of British Columbia, Vancouver, Canada

Kathryn Merrick School of Engineering and Information Technology, University of New South Wales, Canberra, Australia

Risto Miikkulainen Department of Computer Sciences, University of Texas at Austin, Austin, USA

Tim Miller Department of Computing and Information Systems, University of Melbourne, Melbourne, VIC, Australia

Adrian R. Pearce Department of Computing and Information Systems, University of Melbourne, Melbourne, VIC, Australia

Darryn J. Reid Defence Science and Technology Group, Joint and Operations Analysis Division, Edinburgh, SA, Australia

Jason Scholz Defence Science and Technology Group, Joint and Operations Analysis Division, Edinburgh, SA, Australia

Sara Sheikholeslami Department of Mechanical Engineering, University of British Columbia, Vancouver, Canada

Michael Smithson Research School of Psychology, The Australian National University, Canberra, Australia

Liz Sonenberg Department of Computing and Information Systems, University of Melbourne, Melbourne, VIC, Australia

Ron Sun Cognitive Sciences Department, Rensselaer Polytechnic Institute, Troy, NY, USA

Katia Sycara Robotics Institute School of Computer Science, Carnegie Mellon University, Pittsburgh, PA, USA

Phillip Walker Department of Information Sciences, University of Pittsburgh, Pittsburgh, PA, USA

Mark A. Wilson Navy Center for Applied Research in AI, US Naval Research Laboratory, Washington DC, USA

Leon D. Young Department of Defence, War Research Centre, Canberra, Australia

Chapter 1
Foundations of Trusted Autonomy: An Introduction

Hussein A. Abbass, Jason Scholz and Darryn J. Reid

1.1 Autonomy

To aid in understanding the chapters to follow, a general conceptualisation of autonomy may be useful. Foundationally, autonomy is concerned with an agent that acts in an environment. However, this definition is insufficient for autonomy as it requires persistence (or resilience) to the hardships that the environment acts upon the agent. An agent whose first action ends in its demise would not demonstrate autonomy. The themes of autonomy then include agency, persistence and action.

Action may be understood as the utilisation of capability to achieve intent, given awareness.[1] The action trinity of intent, capability and awareness is founded on a mutual tension illustrated in the following figure.

If "capability" is defined as anything that changes the agent's awareness of the world (usually by changing the world), then the error between the agent's awareness and intent drives capability choice in order to reduce that error. Or, expressed compactly, an agent seeks achievable intent.

The embodiment of this action trinity in an entity, itself separated from the environment, but existing within it, and interacting with it, is termed an agent, or autonomy, or intelligence.

---

[1] D.A. Lambert, J.B. Scholz, Ubiquitous Command and Control, Intelligent Decision Technologies, Volume 1 Issue 3, July 2007, Pages 157–173, IOS Press Amsterdam, The Netherlands.

H. A. Abbass (✉)
School of Engineering and IT, University of New South Wales,
Canberra, ACT 2600, Australia
e-mail: h.abbass@adfa.edu.au

J. Scholz · D. J. Reid
Defence Science and Technology Group, Joint and Operations Analysis Division,
PO Box 1500, Edinburgh, SA, Australia
e-mail: jason.scholz@defence.gov.au

D. J. Reid
e-mail: darryn.reid@defence.gov.au

1

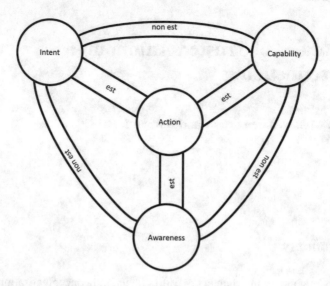

So it is fitting that Chapter 2 by Tom Everitt and Marcus Hutter opens with the topic Universal Artificial Intelligence (UAI): Practical Agents and Fundamental Challenges. Their definition of UAI involves two computational models: Turing Machines; one representing the agent, and one the environment, with actions by the agent on the environment (capability), actions from the environment on the agent (awareness), and actions from the environment to the agent including a utilisation reward (intent achievement) subject to uncertainty. The "will" that underpins the intent of this agent is "maximisation of reward". This machine intelligence is expressible - astoundingly - as a single equation. Named AIXI, it achieves a theoretically-optimal agent in terms of reward maximisation. Though uncomputable, the construct provides a principled approach to considering a practical artificial intelligence and its theoretical limitations. Everitt and Hutter guide us through the development of this theory and the approximations necessary. They then examine the critical question of whether we can trust this machine given machine self-modification, and given the potential for reward counterfeiting, and possible means to manage these. They also consider agent death and self-preservation. Death for this agent involves the cessation of action, and might represented as an absorbing zero reward state. They define both death and suicide, to assess the agent's self-preservation drive which has implications for autonomous systems safety. UAI provides a fascinating theoretical foundation for an autonomous machine and indicates other definitional paths for future research.

In this action trinity of intent, capability, and awareness, it is intent that is in some sense the foremost. Driven by an underlying will to seek utility, survival or other motivation, intent establishes future goals. Chapter 3 Benjamin Johnson, Michael Floyd, Alexandra Coman, Mark Wilson and David Aha consider Goal Reasoning and Trusted Autonomy. Goal Reasoning allows an autonomous system to respond more successfully to unexpected events or changes in the environment. In relation to UAI, the formation of goals and exploration offer the massive benefit of exponential improvements in comparison with random exploration. So goals are important computationally to achieve practical systems. They present two different models of Goal Reasoning: Goal-Driven Autonomy and the Goal Lifecycle. They also describe the Situated Decision Process (SDP), which manages and executes goals for a team of autonomous vehicles. The articulation of goals is also important to human trust, as behaviours can be complex and hard to explain, but goals may be easier because behaviour (as capability action on the environment) is driven by goals (and their difference from awareness). Machine reasoning about goals also provides a basis for the "mission command" of machines. That is, the expression of intent from one agent to another, and the expression of a capability (e.g. a plan) in return provides for a higher level of control with the "human-on-the-loop" applied to more machines than would be the case of the "human-in-the-loop". In this situation, the authors touch on "rebellion", or refusal of an autonomous system to accept a goal expressed to it. This is an important trust requirement if critical conditions are violated that the machine is aware of, such as the legality of action.

The ability to reason with and explain goals (intent) is complemented in Chapter 4 by consideration of reasoning and explanation of planning (capability). Tim Miller, Adrian R. Pearce and Liz Sonenberg examine social planning for trusted autonomy. Social planning is machine planning in which the planning agent maintains and reasons with an explicit model of the humans with which it interacts, including the human's goals (intent), intentions (in effect their plans or in general capability to act), beliefs (awareness), as well as their potential behaviours. The authors combine recent advances to allow an agent to act in a multi-agent world considering the other agents' actions, and a Theory of Mind about the other agents' beliefs together, to provide a tool for social planning. They present a formal model for multi-agent epistemic planning, and resolve the significant processing that would have been required to solve this if each agent's perspective were a mode in modal logic, by casting the problem as a non-deterministic planning task for a single agent. Essentially, treating the actions of other agents in the environment as non-deterministic outcomes (with some probability that is not resolved until after the action) of one agents own actions. This approach looks very promising to facilitate computable cooperative and competitive planning in human and machine groups.

Considering autonomy as will-driven (e.g. for reward, survival) from Chapter 2, and autonomy as goal-directed and plan-achieving (simplifying computation and explanation) from Chapters 3 and 4, what does autonomy mean in a social context? The US Defense Science board[2] signals the need for a social perspective,

> it should be made clear that all autonomous systems are supervised by human operators at some level, and autonomous systems' software embodies the designed limits on the actions and decisions delegated to the computer. Instead of viewing autonomy as an intrinsic property of an unmanned vehicle in isolation, the design and operation of autonomous systems needs to be considered in terms of human-system collaboration.

---

[2] U.S. Defence Science Board, Task Force Report: The Role of Autonomy in DoD Systems, July 2012, pp. 3–5.

The Defense Science Board report goes on to recommend "that the DoD abandon the use of 'levels of autonomy' and replace them with an autonomous systems reference framework". Given this need for supervision and eventual human-system collaboration, perhaps a useful conceptualisation for autonomy might borrow from psychology as illustrated in the following figure.

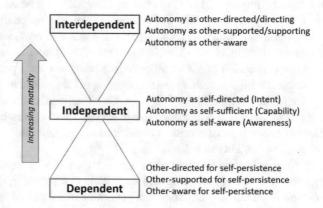

Here, a popular definition[3] of 'autonomy as self-sufficient and self-directed' is situated in a setting of social maturity and extended to include 'awareness of self'. Covey[4] popularises a maturity progression from dependence (e.g. on parents) via independence to interdependence. The maladjusted path is progression from dependence to co-dependence. Co-dependent agents may function but lack resilience as compromise to one agent affects the other(s) thus directly affecting own survival or utility. For the interdependent agent cut off from communication there is the fall-back state of independence.

So, if this might be a preferred trajectory for machine autonomy, what are the implications a strong and independent autonomy? In Chapter 5, Bobby D. Bryant and Risto Miikkulainen consider a neuroevolutionary approach to adaptive multi-agent teams. In their formulation, a similar and significant capability for every agent is posed. They propose a collective where each agent has sufficient breadth of skills to allow for a self-organized division of labour so that it behaves as if it were a heterogeneous team. This division is dynamic in response to conditions, and composed of autonomous agents occurs without direction from a human operator. Indeed in general, humans might be members of the team. This potentially allows for massively-scalable resilient autonomous systems with graceful degradation, as losing any agent affects a loss of role(s) which might be taken up by any other agent(s) all of which have requisite skills (capability). Artificial neural networks are used to learn teams with examples given in the construct of strategy games.

Furthering the theme of social autonomy in Chapter 6, John Harvey examines both the blessing and curse of emergence in swarm intelligence systems. We might

[3]J.M. Bradshaw, The Seven Deadly Myth of Autonomous Systems, IEEE, 2013.

[4]S. R. Covey, The Seven Habits of Highly Effective People, Free Press, 1989.

consider agents composing a swarm intelligence as "similar" and ranging to identical, but not necessarily "significant" capabilities, with the implications that resilience is a property of the collective rather than the individual. Harvey notes that swarm intelligence may relate to a category within the complexity and self-organisation spectrum of emergence characterised as weakly predictable. Swarms do not require centralised control, and may be formed from simple agent interactions, offering the potential for graceful degradation. That is, the loss of some individuals may only weakly degrade the effect of the collective. These and other "blessings" of swarm intelligence presented by the author are tempered by the shortcomings of weak predictability and controllability. Indeed, if they are identical, systematic failure may also be possible as any design fault in an individual is replicated. The author suggests a future direction for research related to the specification of trust properties, might follow from the intersection of liveness properties based on formal methods and safety properties based on Lyapunov measures. Swarm intelligence also brings into question the nature of intelligence. Perhaps it may arise as an emergent property from interacting simpler cognitive elements.

If a social goal for autonomy is collaboration, then cooperation and competition (e.g. for resources) is important. Furthermore, interdependent autonomy must include machines capable of social conflict. Conflict exists where there is mutually exclusive intent. That is, if the intent of one agent can only be achieved if the intent of the other is not achieved. Machine agents need to recognise and operate under these conditions. A structured approach to framing competition and conflict is in games. Michael Barlow, in Chapter 7 examines trusted autonomous game play. Barlow explains four defining traits of games that include a goal (intent), rules (action bounds), a feedback system (awareness), and voluntary participation. Voluntary participation is an exercise of agency where an agreement to act within those conditions is accepted. Barlow examines both perspectives of autonomy for games and games for autonomy. Autonomous entities are usually termed AIs in games, and may serve a training purpose or just provide an engaging user experience. So, improving AIs may improve human capabilities. Autonomous systems can also benefit from games, as games provide a closed-world construct for machine reasoning and learning about scenarios.

These chapters take us on a brief journey of some unique perspectives, from autonomy as individual computational intelligence through to collective machine diversity.

1.2 Trust

Trust is a ubiquitous concept. We all have experienced it one way or another, yet it appears to hold many components that we may never converge on a single, precise, and concise definition of the concept. Yet, the massive amount of literature on the topic is evidence that the topic is an important one for scientific inquiry.

The main contribution of this part of the book is to showcase the complexity of the concept in an attempt to get a handle on its multifaceted nature. This part of the book is a brief inquiry into the meaning of trust, how it is perceived in human-human interaction and in human-machine interaction, and attempts to confine the ambiguity of the topic through novel perspectives and scientifically-grounded opinions.

It initially sounded logical to us to start this part of the book with those chapters discussing trust in its general form before the chapters discussing the trusted autonomy literature. As logical as this idea may sound, it is arguably biasing in a methodological treatment of trust in trusted autonomy.

The previous structure reflects the path that most research in the literature has been following. First, an attempt is made to understand the concept in the human social context then we use this understanding to define what aspect of the concept can be mapped to the human-machine interaction context. Why not? After all, we would like the human to trust and accept the machine as part of our social system.

The previous argument is the strength and weakness of the rationale behind that logic. It is a strong argument when we investigate human-machine interaction; when trust in this relationship is only a means to an end. The ultimate end is the human accepts the machine, accepts its decision, and accepts its role within a context.

However, this view falls short methodologically to study trust in trusted autonomy. In the ultimate form of trusted autonomous systems, the parties of a trusting relationship are both autonomous; thus, both parties need to establish trust in themselves, and then in each other. If one party is a human and the other is a machine, the machine needs to trust the human (machine-human trust) and the human needs to trust the machine (human-machine trust). Therefore, to merely assume that the machine needs to respect what trust is in a human system limits our grasp on the complexity of trust in trusted autonomy.

The nature of trust in a machine needs to be understood. How can machines evaluate trust is a question whose answers need to stem from studies that focus on the nature of the machine.

We then decided to flip the coin in the way we structure this part of the book. We start the journey of inquiry with a chapter written by Lewis, Sycarab and Walker. The chapter entitled "The Role of Trust in Human-Robot Interaction" paves the way to understand trust from a machine perspective. Lewis et al. present a thorough investigation of trust in human-robot interaction, starting with the identification of factors affecting trust as means for measuring trust. They conclude by calling for a need to establish a battery of tasks in human-robot interaction to enable researchers to study the concept of trust.

Kate Devitt in her chapter entitled "Trustworthiness of Autonomous Systems" starts a journey of inquiry to answer three fundamental questions: who or what is trustworthy? how do we know who or what is trustworthy? and what factors influence what or who is trust worthy? She proposes a model of trust with two primary dimensions: one related to competency and the second related to integrity. The author concludes the chapter by discussing the natural relationship between risk and trustworthiness; followed by questioning who and what should we trust?

Michael Smithson investigates the relationship between trust and uncertainty in more depth in his chapter entitled "Trusted Autonomy Under Uncertainty". His first inquiry into the relationship between trust and distrust, takes the view that an autonomous system is an automaton and investigates the human-robotic interaction from this perspective. The inquiry into uncertainty leads to discussing the relationship between trust and social dilemmas up to the issue of trust repair.

Andrew Dowse in his chapter "The Need for Trusted Autonomy in Military Cyber Security" presents on the need for trusted autonomy in the Cyber space. Dowse discusses the requirements for trust in the Cyber space by discussing a series of challenges that needs to be considered.

Bruza and Hoenkamp bring the field of quantum cognition to offer a lens on trust in their chapter "Reinforcing trust in autonomous systems: a quantum cognitive approach". They look into the interplay between system 1 - the fast reactive system - and system 2 - the slow rationale thinking system. They discuss an experiment with images, where they found that humans distrust fake images when they distrust the subject of the image. Bruza and Hoenkamp then presents a quantum cognition model of this phenomenon.

Jason Scholz in his chapter "Learning to Shape Errors with a Confusion Objective" presents an investigation into class hiding in machine learning. Through class re-weighting during learning, the error of a deep neural network on a classification task can be redistributed and controlled. The chapter addresses the issue of trust from two perspectives. First, error trading allows the user to establish confidence in the machine learning algorithm by focusing on classes of interest. Second, the chapter shows that the user can exert control on the behavior of the machine learning algorithm; which is a two-edge sword. It would allow the user the flexibility to manipulate it, while at the same time it may offer an opportunity for an adversary to influence the algorithm through class redistribution.

The last chapter in this part show cases a few practical examples from work conducted at the University of British Columbia. Hart and his colleagues in their chapter on "Developing Robot Assistants with Communicative Cues for Safe, Fluent HRI" list examples of their work ranging from Car Door Assembly all the way to the understanding of social cues and how these communicative cues can be integrated in a human-robot interaction tasks.

1.3 Trusted Autonomy

Part III of the book has a distinctively philosophical flavour: the basic theme that runs through all of its chapters concerns the nature of autonomy, as distinct from automation, and the requirements that autonomous agents must meet if they are to be trustworthy, at least. Autonomy is more or less understood as a requirement for operating in complex environments that manifest uncertainty; without uncertainty relatively straightforward automation will do, and indeed the autonomy is generally seen here as being predicated on some form of environmental uncertainty. Part III is heavily concerned with the centre point of autonomy in terms of intrinsic motivation, computational motivation, creativity, freedom of action, and theory of self. Trustworthiness is largely seen as a here as a necessary but not sufficient condition for such agents to be trusted by humans to carry out tasks in complex environments, with considerable implications for the need for controls on agent behaviour as a component of its motivational processes.

Sun argues that agents need to have intrinsic motivation, meaning internal motivational processes, if they are to deal successfully with unpredictable complex environments. Intrinsic motivation is required under such conditions because criteria defining agent control cannot be specified prior to operation. The importance of intrinsic motivation in regards to the successful operation and acceptance by humans under conditions of fundamental uncertainty represents a challenge that requires serious redress of familiar but outdated assumptions and methodologies.

Furthermore, the ability to understand the motivation of other agents is central to trust, because having this ability means that the behaviour of other agents is predictable even in the absence of predictability of future states of the overall environment. Indeed, the argument is that predictability of the behaviour of other agents through understanding their motivations is what enables trust, and this also explains why trust is such an important issue in an uncertain operating environment.

The chapter presents an overview of a cognitive architecture – the Clarion cognitive architecture – supporting cognitive capabilities as well as intrinsic and derived motivation for agents; it amounts to a structural specification for a variety of psychological processes necessary for autonomy. In particular, the focus of the chapter in this regard is on the interaction between motivation and cognition. Finally, several simulations of this cognitive architecture are given to illustrate how this approach enables autonomous agents to function correctly.

Merrick et al. discussion on computational motivation extends a very similar argument, by arguing that computational motivation is necessary to achieve open-ended goal formulation in autonomous agents operating under uncertainty. Yet it approaches this in a very different manner, by realising computational motivation in practical autonomous systems sufficient for experimental investigation of the question. Here, computational motivation includes curiosity and novel-seeking as well as adaptation, primarily as an epistemic motivation for knowledge increase.

Agents having different prior experiences may behave differently, with the implication that intrinsic motivation through prior experience impacts trustworthiness. Thus trust is a consequence of how motivational factors interact with uncertainty in the operating environment to produce an effect that is not present under closed environments containing only measurable stochastic risk, where essentially rationality and thus trustworthiness is a definable in terms of an optimality condition that means that agents operate without a much scope for exercising choice.

The chapter concludes that the empirical evidence presented is consistent with the thesis that intrinsic motivation in agents impacts trustworthiness, in potentially simultaneously positive and negative ways, because of the complex of overlapping and sometimes conflicting implications motivation has for privacy and security. Trustworthiness is also impacted by what combination of motivations the agents employ and whether they operate in mixed or homogeneous agent environments. Finally, if humans are to develop trust in autonomous agents, then agent technologies have to be transparent to humans.

General computational logics are used by Bringsjord and Naveen as the basis for a model of human-level cognition as formal computing machines to formally explore the consequences for trust of autonomy. The chapter thereby sets formal limits on trust very much akin to those observed for humans in the psychology literature, by presenting a theorem stating, under various formal assumptions, that an artificial agent that is autonomous (A) and creative (C) will tend to be, from the standpoint of a fully informed rational agent, intrinsically untrustworthy (U). The chapter thus refers to the principle for humans as $PACU$, and the theorem as $TACU$. The proof of this theorem is obtained using ShadowProver, a novel automated theorem proving program.

After building an accessible introduction to the principle with reference to the psychology maintaining it for humans and empirical evidence for its veracity, the chapter establishes a formal version of the principle. This requires establishing formalisations of what it means to be an ideal observer, of what it means to be creative, and of what it means to be autonomous, and a formal notion of collaborative situations. The chapter describes the cognitive calculus DeLEL in which TACU is formalised, and the novel theorem prover ShadowProver used to prove the theorem.

More broadly, the chapter seeks not just to establish the theorem, but to establish the case for its plausibility beyond the specific assumptions of the theorem. Beyond the limitations of this particular formalisation - and the authors invite further investigation based on more powerful formalisations - the TACU theorem establishes the necessity of active engineering practices to protect humans from the unintended consequences of creative autonomous machines, by asserting legal and ethical limits on what agents can do. The preconditions of autonomy and creativity are insufficient; just as with humans, societal controls in the form of legal and ethical constraints are also required.

Derwort's concerns relate to the development of autonomous military command and control (C2). Autonomous systems in military operational environments will not act alone, but rather will do so in concert with other autonomous and manned systems, and ultimately all under broad national military control exercised by human decision-makers. This is a situation born of necessity and the opportunity afforded by rapidly developing autonomous technologies: autonomous systems and the distributed C2 across them is emerging as a response to the rapid increase in capabilities of potential military adversaries and the limited ability to respond to them with the development of traditional manned platforms.

The chapter outlines a number of past scenarios involving human error in C2, with tragic consequences, to illustrate the limitations of human decision-making, and plausible military scenarios in the not-too-distant future. There are no doubt risks involved with taking the human out of the decision-making in terms of responsibility, authority and dehumanising of human conflict, yet any rational discussion on the use of autonomy in war and battle needs to also be moderated by due recognition of the inherent risks of having humans in the decision-making processes.

Autonomous systems are merely tools, and the cost of their destruction is merely counted in dollars. Therein lies a particular strength, for autonomous systems with distributed C2 has enormous potential to create and implement minimal solutions in place of the more aggressive solutions to tactical problems to which stressed humans are prone. Autonomy offers the potential to intervene in in the face of unexpected circumstances, to de-escalate, to improve the quality as well as speed of military decision-making. Therein may lie its most serious potential for military operational use.

Young presents on the application of autonomy to training systems and raises questions about how such systems will impact the human learning environments in which they are used. Chapter 19 explores this starting from the pivotal premise of traditional teaching whereby the students must have trust in the teacher to effectively concede responsibility to the teacher. What does this mean if the teacher is a machine? The chapter seeks to explore what is possible with autonomy in the classroom, and what we might reasonably expect to be plausible.

A map is presented showing the interconnected functional components of a training system, including both those that are provided by human trainees and those that might be provided by machines. It includes the functions of the teacher and the learner, including the training topic and measurement of learning. The authors present three key drivers likely to determine the future of autonomous systems in training and education: autonomous systems development, training systems, and trust. Some of the functions required for a learning environment are already being provided by machines, albeit in relatively limited ways; the advance of autonomous systems technologies will expand the potential for delegating more of these functions to machines.

Trust is presented as a function of familiarity, which is consistent with the view of trust in some preceding chapters as requiring predictability of other agents' behaviours even within a complex environment that is inherently unpredictable. Trust is held to be central to learning, and trust through familiarity over time is the basis for exploring a number of future scenarios. The first revolves around the frustration that might be the result of the perceived artificiality of autonomous teachers, compounded by inconsistencies between different autonomous teachers over subsequent time periods. The second concerns the social dislocation and potential incompetence resulting from machines taking over simpler tasks from humans and thereby denying the humans knowledge of those tasks and thereby effecting the quality of higher-level human decision-making. The third is a scenario in which the machine responsible for teaching the human grows up with the human in a complex relationship marked by mutual trust, suggesting that the human's trust in the machine is symbiotic with the development of the machine's trust in the human.

Boyce and Griffin begin with an elucidation of the harshness and remoteness of space, marked by extreme conditions that can degrade or destroy spacecraft. Manoeuvres in orbits near earth or other large objects are complex and counterintuitive. Gravitational fields are not uniform, interactions between multiple objects can produce significant errors, and space is becoming increasingly crowded, requiring the ability to conduct evasive actions in advance of potential collisions. Close human operation is inefficient and dangerous, mandating the use of autonomy for a wide range of spacecraft functions.

With increasing miniaturisation of spacecraft, traffic management and collision avoidance are becoming pressing problems driving greater degrees of spacecraft autonomy. Yet the lack of trust ascribed to the limitations of automated code generation, runtime analysis and model checking for verification and validation for software that has to make complex decisions is a large barrier to adoption of higher-level autonomy for spacecraft. Linked to this is the need for human domain experts to be involved in the design and development of software in order to build trust in the product.

The chapter concludes with some possible space scenarios for autonomy, the first of which might be achieved in the near future, involving greater autonomous analysis of information from different sources. The second concerns autonomy in space traffic management, linked to all spacecraft that have the ability to manoeuvre, that includes the decision-making and action currently undertaken by humans. The final scenario concerns distributed space systems that can self-configure with minimal human input, both to achieve capabilities not achievable using single large spacecraft and to respond to unexpected events such as partial system failure.

The final chapter presents a picture of autonomous systems development primarily from an economic point of view, on the basis that an economic agent is an autonomous agent; the difference being that economics is primarily concerned with analysing overall outcomes from societies of decision-makers while AI is squarely focussed on decision-making algorithm development. The connection between economics and AI is probably more widely understood in economics - which has long utilised and contributed, in turn, to the development of machine learning and automated reasoning methods - than it is in autonomy research. Thus the chapter treats autonomy as the allocation of scarce resources under conditions of fundamental uncertainty.

The main thrust of the chapter is an economic view of uncertainty, which distinguishes between epistemic uncertainty and ontological uncertainty, and its consequences for autonomy. Ontological uncertainty is the deeper of the two: epistemic uncertainty amounts to ignorance of possible outcomes due to sampling limits, while ontological uncertainty relates to the presence of unsolvable paradoxical problems; the chapter thus draws out the connection between the economic notion of ontological uncertainty and the famed incompleteness theorems of Gödel, the unsolvability of the Halting Problem of Turing, and incompressibility theorems of Algorithmic Information Theory.

Drawing on both financial economics and macroeconomic theory, commonplace investment strategies are presented in the context of this notion of uncertainty, noting that, under conditions of ontological uncertainty, what might be seemingly rational for an individual agent in the short-term need not be rational in the long-term nor from the perspective of the entire social enterprise. Certain well-known bond investment strategies, however, appear to have the potential to strike a healthy balance and yield desirable long-term properties for both the agent and the broader system of which it is a component, and thus may offer a basis for autonomous systems. Interestingly, implementing such a strategy in an agent seems to require a theory of self, to provide the kinds of motivational processes discussed in other chapters as well.

Part I
Autonomy

Chapter 2
Universal Artificial Intelligence

Practical Agents and Fundamental Challenges

Tom Everitt and Marcus Hutter

2.1 Introduction

Artificial intelligence (AI) bears the promise of making us all healthier, wealthier, and happier by reducing the need for human labour and by vastly increasing our scientific and technological progress.

Since the inception of the AI research field in the mid-twentieth century, a range of practical and theoretical approaches have been investigated. This chapter will discuss *universal artificial intelligence* (UAI) as a unifying framework and foundational theory for many (most?) of these approaches. The development of a foundational theory has been pivotal for many other research fields. Well-known examples include the development of Zermelo-Fraenkel set theory (ZFC) for mathematics, Turing-machines for computer science, evolution for biology, and decision and game theory for economics and the social sciences. Successful foundational theories give a precise, coherent understanding of the field, and offer a common language for communicating research. As most research studies focus on one narrow question, it is essential that the value of each isolated result can be appreciated in light of a broader framework or goal formulation. UAI offers several benefits to AI research beyond the general advantages of foundational theories just mentioned. Substantial attention has recently been called to the *safety* of autonomous AI systems [10]. A highly intelligent autonomous system may cause substantial unintended harm if constructed carelessly. The trustworthiness of autonomous agents may be much improved if their design is grounded in a formal theory (such as UAI) that allows formal verification of their behavioural properties. Unsafe designs can be ruled out at an early stage, and adequate attention can be given to crucial design choices.

T. Everitt (✉) · M. Hutter
Australian National University, Canberra, Australia
e-mail: Tom.Everitt@anu.edu.au

M. Hutter
e-mail: marcus.hutter@anu.edu.au

© The Author(s) 2018
H. A. Abbass et al. (eds.), *Foundations of Trusted Autonomy*, Studies in Systems, Decision and Control 117, https://doi.org/10.1007/978-3-319-64816-3_2

UAI also provides a high-level blueprint for the design of practical autonomous agents, along with an appreciation of fundamental challenges (e.g. the induction problem and the exploration–exploitation dilemma). Much can be gained by addressing such challenges at an appropriately general, abstract level, rather than separately for each practical agent or setup. Finally, UAI is the basis of a general, non-anthropomorphic definition of intelligence. While interesting in itself to many fields outside of AI, the definition of intelligence can be useful to gauge progress of AI research.[1]

The outline of this chapter is as follows: First we provide general background on the scientific study of intelligence in general, and AI in particular (Sect. 2.2). Next we give an accessible description of the UAI theory (Sect. 2.3). Subsequent sections are devoted to applications of the theory: Approximations and practical agents (Sect. 2.4), high-level formulations and approaches to fundamental challenges (Sect. 2.5), and the safety and trustworthiness of autonomous agents (Sect. 2.6).

2.2   Background and History of AI

Intelligence is a fascinating topic, and has been studied from many different perspectives. Cognitive psychology and behaviourism are psychological theories about how humans think and act. Neuroscience, linguistics, and the philosophy of mind try to uncover how the human mind and brain works. Machine learning, logic, and computer science can be seen as attempts to make machines that *think*.

Scientific perspectives on intelligence can be categorised based on whether they concern themselves with thinking or acting (cognitive science vs. behaviourism), and whether they seek objective answers such as in logic or probability theory, or try to describe humans as in psychology, linguistics, and neuroscience. The distinction is illustrated in Table 2.1. The primary focus of AI is on *acting* rather than *thinking*, and on *doing the right thing* rather than *emulating humans*. Ultimately, we wish to build systems that solve problems and act appropriately; whether the systems are inspired by humans or follow philosophical principles is only a secondary concern.

Induction and deduction. Within the field of AI, a distinction can be made between systems focusing on *reasoning* and systems focusing on *learning*. *Deductive* reasoning systems typically rely on logic or other symbolic systems, and use search algorithms to combine inference steps. Examples of primarily deductive systems include medical expert systems that infer diseases from symptoms, and chess-playing agents deducing good moves. Since the deductive approach dominated AI in its early days, it is sometimes referred to as *good old-fashioned AI*.

---

[1]See [42, 43] for discussions about the intelligence definition.

Table 2.1 Scientific perspectives on intelligence

	Thinking	Acting
Humanly	Cognitive science	Turing test, behaviourism
Rationally	Laws of thought	**Doing the right thing**

A more modern approach to AI shifts the focus from reasoning to learning. This *inductive approach* has become increasingly popular, both due to progress in machine learning and neural networks, and due to the failure of deductive systems to manage unknown and noisy environments. While it is possible for a human designer to construct a deductive agent for well-defined problems like chess, this task becomes unfeasible in tasks involving real-world sensors and actuators. For example, the reaction of any physical motor will never be *exactly* the same twice. Similarly, inferring objects from visual data could potentially be solved by a 'hard-coded' deductive system under 'perfect circumstances' where a finite number of geometric shapes generate perfectly predictable images. But in the real world, objects do not come from a finite number of geometric shapes, and camera images from visual sensors always contain a significant amount of noise. Induction-oriented systems that *learn* from data seem better fitted to handle such difficulties.

It is natural to imagine that some synthesis of inductive and deductive modules will yield superior systems. In practice, this may well turn out to be the case. From a theoretical perspective, however, the inductive approach is more-or-less self-sufficient. Deduction emerges automatically from a "simple" planning algorithm once the induction component has been defined, as will be made clear in the following section. In contrast, no general theory of AI has been constructed starting from a deductive system. See [67] (Sect. 1.1) for a more formal comparison.

2.3 Universal Artificial Intelligence

Universal Artificial Intelligence (UAI) is a completely general, formal, foundational theory of AI. Its primary goal is to give a precise mathematical answer to *what is the right thing to do in unknown environments*. UAI has been explored in great technical depth [28, 33], and has inspired a number of successful practical applications described in Sect. 2.4.

The UAI theory is composed of the following four components:

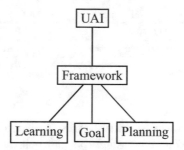

- **Framework**. Defines agents and environments, and their interaction.
- **Learning**. The learning part of UAI is based on Solomonoff induction. The general learning ability this affords is the most distinctive feature of UAI.
- **Goal**. In the simplest formulation, the goal of the agent will be to maximise reward.
- **Planning.** (Near) perfect planning is achieved with a simple expectimax search.

The following sections discuss these components in greater depth.

2.3.1 *Framework*

The framework of UAI specifies how an *agent* interacts with an *environment*. The agent can take *actions $a \in \mathscr{A}$*. For example, if the agent is a robot, then the actions may be different kinds of limb movements. The environment reacts to the actions of the agent by returning a *percept $e \in \mathscr{E}$*. In the robot scenario, the environment is the real world generating a percept e in the form of a camera image from the robot's visual sensors. We assume that the set $\mathscr{A}$ of actions and the set $\mathscr{E}$ of percepts are both finite.

The framework covers a very wide range of agents and environments. For example, in addition to a robot interacting with the real world, it also encompasses: A *chess-playing agent* taking actions a in the form of chess moves, and receiving percepts e in the form either of board positions or the opponent's latest move. The environment here is the chess board and the opponent. *Stock-trading agents* take actions a in the form of buying and selling stocks, and receive percepts e in the form of trading data from a *stock-market* environment. Essentially any application of AI can be modelled in this general framework.

A more formal example is given by the following toy problem, called *cheese maze* (Fig. 2.1). Here, the agent can choose from four actions $\mathscr{A} = \{$up, down, left, right$\}$ and receives one of two possible percepts $\mathscr{E} = \{$cheese, no cheese$\}$. The illustration shows a maze with cheese in the bottom right corner. The cheese maze is a commonly used toy problem in reinforcement learning (RL) [82].

- **Interaction histories**. The interaction between agent and environment proceeds in cycles. The agent starts taking an action a_1, to which the environment responds with a percept e_1. The agent then selects a new action a_2, which results in a

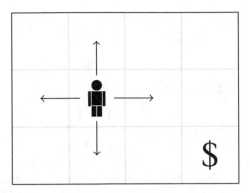

Fig. 2.1 Cheese maze environment

new percept e_2, and so on. The *interaction history* up until time t is denoted $æ_{<t} = a_1 e_1 a_2 e_2 \ldots a_{t-1} e_{t-1}$. The set of all interaction histories is $(\mathscr{A} \times \mathscr{E})^*$.

- **Agent and environment.** We can give formal definitions of agents and environments as follows.

Definition 1 (*Agent*) An *agent* is a *policy* $\pi : (\mathscr{A} \times \mathscr{E})^* \to \mathscr{A}$ that selects a new action $a_t = \pi(æ_{<t})$ given any history $æ_{<t}$.

Definition 2 (*Environment*) An *environment* is a stochastic function $\mu : (\mathscr{A} \times \mathscr{E})^* \times \mathscr{A} \rightsquigarrow \mathscr{E}$ that generates a new percept e_t for any history $æ_{<t}$ and action a_t. Let $\mu(e_t \mid æ_{<t} a_t)$ denote the probability that the next percept is e_t given the history $æ_{<t} a_t$.

The agent and the environment are each other's analogues. Their possible interactions can be illustrated as a tree where the agent selects actions and the environment responds with percepts (see Fig. 2.2). Note in particular that the second percept e_2 can depend also on the first agent action a_1. In general, our framework puts no restriction on how long an action can continue to influence the behaviour of the environment and vice versa.

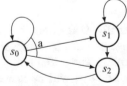

Histories and states. It is instructive to compare the generality of the *history* representation in the UAI framework to the *state* representation in standard RL. Standard RL is built around the notion of Markov decision processes (MDPs), where the agent transitions between *states* by taking actions, as illustrated to the right. The MDP specifies the *transition probabilities* $P(s' \mid s, a)$ of reaching new state s' when taking action a in current state s. An *MDP policy* $\tau : \mathscr{S} \to \mathscr{A}$ selects actions based on the state $s \in \mathscr{S}$.

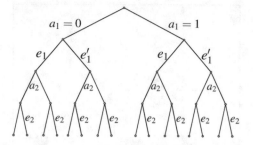

Fig. 2.2 The tree of possible agent-environment interactions. The agent π starts out with taking action $a_1 = \pi(\epsilon)$, where ϵ denotes the empty history. The environment μ responds with a percept e_1 depending on a_1 according to the distribution $\mu(e_1 \mid a_1)$. The agent selects a new action $a_2 = \pi(a_1 e_1)$, to which the environment responds with a percept $e_2 \sim \mu(\cdot \mid a_1 e_1 a_2)$

The history framework of UAI is more general than MDPs in the following respects:

- **Partially observable states.** In most realistic scenarios, the most recent observation or percept does not fully reveal the current state. For example, when in the supermarket I need to remember what is currently in my fridge; nothing in the percepts of supermarket shelves provide this information.[2]
- **Infinite number of states.** Another common assumption in standard RL is that the number of states is finite. This is unrealistic in the real world. The UAI framework does not require a finite state space, and UAI agents can learn without ever returning to the same state (see Sect. 2.3.2).
- **Non-stationary environments.** Standard RL typically assumes that the environment is stationary, in the sense that the transition probability $P(s' \mid s, a)$ remains constant over time. This is not always realistic. A car that changes travelling direction from a sharp wheel turn in dry summer road conditions may react differently in slippery winter road conditions. Non-stationary environments are automatically allowed for by the general definition of a UAI environment $\mu : (\mathscr{A} \times \mathscr{E})^* \times \mathscr{A} \rightsquigarrow \mathscr{E}$ (Definition 2). As emphasised in Chapter 11 of this book, the non-stationarity and non-ergodicity of the real world is what makes truly autonomous agents so challenging to construct and to trust.
- **Non-stationary policies.** Finally, UAI offers the following mild notational convenience. In standard RL, agents must be represented by sequences of policies $\pi_1, \pi_2, \ldots$ to allow for learning. The initial policy π_1 may for example be random, while later policies π_t, $t > 1$, will be increasingly directed to obtaining reward. In the UAI framework, policies $\pi : (\mathscr{A} \times \mathscr{E})^* \rightarrow \mathscr{A}$ depend on the entire interaction history. Any learning that is made from a history $æ_{<t}$ can be incorporated into a single policy π.

---

[2]Although histories can be viewed as states, this is generally not useful since it implies that no state is ever visited twice [28] (Sect. 3.3.3).

In conclusion, the history-based UAI framework is very general. Indeed, it is hard to find AI setups that cannot be reasonably modelled in this framework.

2.3.2 Learning

The generality of the UAI environments comes with a price: The agent will need much more sophisticated learning techniques than simply visiting each state many times, which is the basis of most learning in standard RL. This section will describe how this type of learning is possible, and relate it to some classical philosophical principles about learning.

A good image of a UAI agent is that of a newborn baby. Knowing nothing about the world, the baby tries different actions and experiences various sensations (percepts) as a consequence. Note that the baby does not initially know about any states of the world—only percepts. Learning is essential for intelligent behaviour, as it enables prediction and thereby adequate planning.

Principles. Learning or *induction* is an ancient philosophical problem, and has been studied for millennia. It can be framed as the problem of inferring a correct hypothesis from observed data. One of the most famous inductive principles is *Occam's razor*, due to William of Ockham (c. 1287–1347). It says to prefer the simplest hypothesis consistent with data. For example, relativity theory may seem like a complicated theory, but it is the *simplest* theory that we know of that is consistent with observed (non-quantum) physics data. Another ancient principle is due to Epicurus (341–270 BC). In slight conflict with Occam's razor, *Epicurus' principle* says to keep *all* hypothesis consistent with data. To discard a hypothesis one should have data that disconfirms it.

Thomas Bayes (1701–1761) derived a precise rule for how belief in a hypothesis should change with additional data. According to *Bayes' rule*, the *posterior belief* $\Pr(\text{Hyp} \mid \text{Data})$ should relate to the *prior belief* $\Pr(\text{Hyp})$ as:

$$\Pr(\text{Hyp} \mid \text{Data}) = \frac{\Pr(\text{Hyp}) \Pr(\text{Data} \mid \text{Hyp})}{\sum_{H_i \in \mathscr{H}} \Pr(H_i) \Pr(\text{Data} \mid H_i)}$$

Here $\mathscr{H}$ is a class of possible hypotheses, and $\Pr(\text{Data} \mid \text{Hyp})$ is the *likelihood* of seeing the data under the given hypothesis. Bayes' rule has been highly influential in statistics and machine learning.

Two major questions left open by Bayes' rule are how to choose the prior $\Pr(\text{Hyp})$ and the class of possible hypotheses $\mathscr{H}$. Occam's razor tells us to weight simple hypotheses higher, and Epicurus tells us to keep any hypothesis for consideration. In other words, Occam says that $\Pr(\text{Hyp})$ should be large for simple hypotheses, and Epicurus prescribes using a wide $\mathscr{H}$ where $\Pr(\text{Hyp})$ is never 0. (Note that this does not prevent the posterior $\Pr(\text{Hyp} \mid \text{Data})$ from being 0 if the data completely disconfirms the hypothesis.) While valuable, these principles are not yet precise. The following four questions remain:

   I. What is a suitable general class of hypotheses $\mathcal{H}$?
  II. What is a simple hypothesis?
 III. How much higher should the probability of a simple hypothesis be compared to a complicated one?
 IV. Is there any guarantee that following these principles will lead to good learning performance?

Computer programs. The solution to these questions come from a somewhat unexpected direction. In one of the greatest mathematical discoveries of the 20th century, Alan Turing invented the *universal Turing machine* (UTM). Essentially, a UTM can compute anything that can be computed at all. Today, the most well-known examples of UTMs are programming languages such as C, C++, Java, and Python. Turing's result shows that given unlimited resources, these programming languages (and many others) can compute the same set of functions: the so-called *computable functions*.

Solomonoff [77–79] noted an important similarity between deterministic environments μ and computer programs p. Deterministic environments and computer programs are both essentially input-output relations. A program p can therefore be used as a hypothesis about the true environment μ. The program p is the hypothesis that μ returns percepts $e_{<t} = p(a_{<t})$ on input $a_{<t}$.

As hypotheses, programs have the following desirable properties:

- **Universal.** As Turing showed, computer programs can express any computable function, and thereby model essentially any environment. Even the universe itself has been conjectured computable [20, 33, 70, 87]. Using computer programs as hypotheses is thus in the spirit of Epicurus, and answers question I.
- **Consistency check.** To check whether a given computer program p is consistent with some data/history $æ_{<t}$, one can usually run p on input $a_{<t}$ and check that the output matches the observed percepts, $e_{<t} = p(a_{<t})$. (This is not always feasible due to the *halting problem* [27].)
- **Prediction.** Similarly, to predict the result of an action a given a hypothesis p, one can run p with input a to find the resulting output prediction e. (A similar caveat with the halting problem applies.)
- **Complexity definition.** When comparing informal hypotheses, it is often hard to determine which hypothesis is simpler and which hypothesis is more complex (as illustrated by the *grue and bleen* problem [23]). For programs, complexity can be defined precisely. A program p is a binary string interpreted by some fixed program interpreter, technically known as a *universal Turing machine* (UTM). We denote with $\ell(p)$ the length of this binary string p, and interpret the length $\ell(p)$ as the *complexity* of p. This addresses question II.[3]

---

[3]The technical question of which programming language (or UTM) to use remains. In *passive* settings where the agent only predicts, the choice is inessential [29]. In *active* settings, where the agent influences the environment, bad choices of UTMs

The complexity definition as length of programs corresponds well to what we consider *simple* in the informal sense of the word. For example, an environment where the percept always mirrors the action is given by the following simple program:

procedure MIRRORENVIRONMENT
    while true **do**:
        $x \leftarrow$ action input
        output percept $\leftarrow x$

In comparison, a more complex environment with, say, multiple players interacting in an intricate physics simulation would require a much longer program. To allow for stochastic environments, we say that an environment μ is *computable* if there exists a computer program μ_p that on input $æ_{<t}a_t$ outputs the distribution $\mu(e_t \mid æ_{<t}a_t)$ (cf. Definition 2).

Solomonoff induction. Based on the definition of complexity as length of strings coding computer programs, Solomonoff [77–79] defined a *universal prior* $\Pr(p) = 2^{-\ell(p)}$ for program hypotheses p, which gives rise to a *universal distribution* M able to predict any computable sequence. Hutter [28] extended the definition to environments reacting to an agent's actions. The resulting *Solomonoff-Hutter universal distribution* can be defined as

$$M(e_{<t} \mid a_{<t}) = \sum_{p\,:\,p(a_{<t})=e_{<t}} 2^{-\ell(p)} \qquad (2.1)$$

assuming that the programs p are binary strings interpreted in a suitable programming language. This addresses question III.

Given some history $æ_{<t}a_t$, we can predict the next percept e_t with probability:

$$M(e_t \mid æ_{<t}a_t) = \frac{M(e_{<t}e_t \mid a_{<t}a_t)}{M(e_{<t} \mid a_{<t})}.$$

This is just an application of the definition of conditional probability $P(A \mid B, C) = P(A, B \mid C)/P(B \mid C)$, with $A = e_t$, $B = e_{<t}$, and $C = a_{<t}a_t$.

Prediction results. Finally, will agents based on M learn? (Question IV.) There are, in fact, a wide range of results in this spirit.[4] Essentially, what can be shown is that:

Theorem 1 (Universal learning) *For any computable environment μ (possibly stochastic) and any action sequence $a_{1:\infty}$,*

$$M(e_t \mid æ_{<t}a_t) \to \mu(e_t \mid æ_{<t}a_t) \qquad as\ t \to \infty\ with\ \mu\text{-}probability\ 1.$$

(Footnote 3 continued)
can adversely affect the agent's performance [44], although remedies exist [46]. Finally, [54] describes a failed but interesting attempt to find an objective UTM.

[4]Overviews are provided by [28, 29, 48, 67], More recent technical results are given by [30, 39, 41, 45].

The convergence is quick in the sense that M only makes a finite number of prediction errors on infinite interaction sequences $æ_{1:\infty}$. In other words, an agent based on M will (quickly) learn to predict any true environment μ that it is interacting with. This is about as strong an answer to question V as we could possibly hope for. This learning ability also loosely resembles one of the key elements of human intelligence: That by interacting with almost any new 'environment' – be it a new city, computer game, or language – we can usually figure out how the new environment works by interacting with it.

2.3.3 Goal

Intelligence is to use (learnt) knowledge to achieve a goal. This section will define the goal of *reward maximisation* and argue for its generality.[5] For example, the goal of a chess agent should be to win the game. This can be communicated to the agent via reward, by giving the agent reward for winning, and no reward for losing or breaking game rules. The goal of a self-driving car should be to drive safely to the desired location. This can be communicated in a reward for successfully doing so, and no reward otherwise. More generally, essentially any type of goal can be communicated by giving reward for the goal's achievement, and no reward otherwise.

The reward is communicated to the agent via its percept e. We therefore make the following assumption on the structure of the agent's percepts:

Assumption 1 (*Percept = Observation + Reward*) The percept e is composed of an *observation* o and a *reward* $r \in [0, 1]$; that is, $e = (o, r)$. Let r_t be the reward associated with the percept e_t.

The observation part o of the percept would be the camera image in the case of a robot, and the chess board position in case of a chess agent. The reward r tells the agent how well it is doing, or how happy its designers are with its current performance. Given a *discount parameter* γ, the goal of the agent is to maximise the γ-*discounted return*

$$r_1 + \gamma r_2 + \gamma^2 r_3 + \ldots.$$

The discount parameter γ ensures that the sum is finite. It also means that the agent prefers getting reward sooner rather than later. This is desirable: For example, an agent striving to achieve its goal soon is more useful than an agent striving to achieve it in a 1000 years. The discount parameter should be set low enough so that the agent does not defer acting for too long, and high enough so that the agent does not become *myopic*, sacrificing substantial future reward for small short-term gains (compare *delayed gratification* in the psychology literature).

Reinforcement learning [82] is the study of agents learning to maximise reward. In our setup, Solomonoff's result (Theorem 1) entails that the agent will learn to predict

---

[5]Alternatives are discussed briefly in Sect. 2.6.2.

which actions or policies lead to percepts containing high reward. In practice, some care needs to be taken to design a sufficiently informative reward signal. For example, it may take a very long time before a chess agent wins a game 'by accident', leading to an excessively long exploration time before any reward is found. To speed up learning, small rewards can be added for moving in the right direction. A minor reward can for example be added for imitating a human [69].

The expected return that an agent/policy obtains is called *value*:

Definition 3 (*Value*) The *value* of a policy π in an environment μ is the expected return:

$$V_\mu^\pi = \mathbb{E}_\mu^\pi[r_1 + \gamma r_2 + \gamma^2 r_3 + \ldots].$$

2.3.4   Planning

The final component of UAI is planning. Given knowledge of the true environment μ, how should the agent select actions to maximise its expected reward?

Conceptually, this is fairly simple. For any policy π, the expected reward $V_\mu^\pi = \mathbb{E}[r_1 + \gamma r_2 + \ldots]$ can be computed to arbitrary precision. Essentially, using π and μ, one can determine the histories $æ_{1:\infty}$ that their interaction can generate, as well as the relative probabilities of these histories (see Fig. 2.2). This is all that is needed to determine the expected reward. The discount γ makes rewards located far into future have marginal impact, so the value can be well approximated by looking only finitely far into the future. Settling on a sufficient accuracy ε, the number of time steps we need to look ahead in order to achieve this precision is called the *effective horizon*.

To find the optimal course of action, the agent only needs to consider the various possible policies within the effective horizon, and choose the one with the highest expected return. The optimal behaviour in a known environment μ is given by

$$\pi_\mu^* = \arg \max_\pi V_\mu^\pi \tag{2.2}$$

We sometimes call this policy AIμ. A full expansion of (2.2) can be found in [28] (p. 134). Efficient approximations are discussed in Sect. 2.4.1.

2.3.5   AIXI – Putting It All Together

This section describes how the components described in previous sections can be stitched together to create an optimal agent for unknown environments. This agent is called AIXI, and is defined by the optimal policy

$$\pi_M^* = \arg \max_{\pi} V_M^{\pi} \tag{2.3}$$

The difference to AIμ defined in (2.2) is that the true environment μ has been replaced with the universal distribution M in (2.3). A full expansion can be found in [28] (p. 143). While AIμ is optimal when knowing the true environment μ, AIXI is able to learn essentially any environment through interaction. Due to Solomonoff's result (Theorem 1) the distribution M will converge to the true environment μ almost regardless of what the true environment μ is. And once M has converged to μ, the behaviour of AIXI will converge to the behaviour of the optimal agent AIμ which perfectly knows the environment. Formal results on AIXI's performance can be found in [28, 38, 46].

Put a different way, AIXI arrives to the world with essentially no knowledge or preconception of what it is going to encounter. However, AIXI quickly makes up for its lack of knowledge with a powerful learning ability, which means that it will soon figure out how the environment works. From the beginning and throughout its "life", AIXI acts optimally according to its growing knowledge, and as soon as this knowledge state is sufficiently complete, AIXI acts as well as any agent that knew everything about the environment from the start. Based on these observations (described in much greater technical detail by [28]), we would like to make the claim that AIXI defines the *optimal behaviour in any computable, unknown environment*.
Trusting AIXI. The AIXI formula is a precise description of the optimal behaviour in an unknown world. It thus offers designers of practical agents a target to aim for (Sect. 2.4). Meanwhile, it also enables safety researchers to engage in formal investigations of the consequences of this behaviour (Sects. 2.5 and 2.6). Having a good understanding of the behaviour and consequences an autonomous system strives towards, is essential for us being able to trust the system.

2.4 Approximations

The AIXI formula (2.3) gives a precise, mathematical description of the optimal behaviour in essentially any situation. Unfortunately, the formula itself is incomputable, and cannot directly be used in a practical agent. Nonetheless, having a description of the right behaviour is still useful when constructing practical agents, since it tells us what behaviour we are trying to approximate. The following three sections describe three substantially different approximation approaches. They differ widely in their approximation approaches, and have all demonstrated convincing experimental performance. Sect. 2.4.4 connects UAI with recent deep learning results.

2.4.1 MC-AIXI-CTW

MC-AIXI-CTW [85] is the most direct approximation of AIXI. It combines the Monte Carlo Tree Search algorithm for approximating expectimax planning, and the Context Tree Weighting algorithm for approximating Solomonoff induction. We describe these two methods next.

Planning with sampling. The expectimax planning principle described in Sect. 2.3.4 requires exponential time to compute, as it simulates all future possibilities in the planning tree seen in Fig. 2.2. This is generally far too slow for all practical purposes.

A more efficient approach is to randomly sample paths in the planning tree, as illustrated in Fig. 2.3. Simulating a single random path $a_t e_t \ldots a_m e_m$ only takes a small, constant amount of time. The average return from a number of such simulated paths gives an approximation $\hat{V}(\text{æ}_{<t} a_t)$ of the value. The accuracy of the approximation improves with the number of samples.

A simple way to use the sampling idea is to keep generating samples for as long as time allows for. When an action must be chosen, the choice can be made based on the current approximation. The sampling idea thus gives rise to an *anytime algorithm* that can be run for as long as desired, and whose (expected) output quality increases with time.

Monte Carlo Tree Search. The *Monte Carlo Tree Search* (MCTS) algorithm [2, 11, 36] adds a few tricks to the sampling idea to increase its efficiency. The sampling idea and the MCTS algorithm are illustrated in Fig. 2.3.

One of the key ideas of MCTS is in optimising the informativeness of each sample. First, the sampling of a next percept e_k given a (partially simulated) history $\text{æ}_{<k} a_k$ should always be done according to the current best idea about the environment distribution; that is, according to $M(e_k \mid \text{æ}_{<k} a_k)$ for Solomonoff-based agents.

The sampling of actions is more subtle. The agent itself is responsible for selecting the actions, and actions that the agent knows it will not take, are pointless for the agent to simulate. As an analogy, when buying a car, I focus the bulk of my cognitive resources on evaluating the feasible options (say, the Ford and the Honda) and only

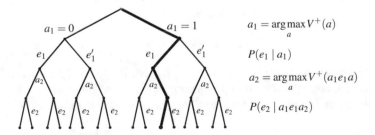

Fig. 2.3 Sampling branches from the planning tree gives an *anytime algorithm*. Sampling actions according to the *optimistic value estimates* V^+ increases the informativeness of samples. This is one of the ideas behind the MCTS algorithm

briefly consider clearly infeasible options such as a luxurious Ferrari. Samples should be focused on plausible actions.

One way to make this idea more precise is to think of the sampling choice as a *multi-armed Bandit problem* (a kind of "slot machine" found in casinos). Bandit problems offer a clean mathematical theory for studying the allocation of resources between *arms* (actions) with *unknown returns* (value). One of the ideas emerging from the bandit literature is the *upper confidence bound* (UCB) algorithm that uses *optimistic value estimates* V^+. Optimistic value estimates add an exploration bonus for actions that has received comparatively little attention. The bonus means that a greedy agent choosing actions that optimise V^+ will spend a sufficient amount of resources exploring, while still converging on the best action asymptotically.

The MCTS algorithm uses the UCB algorithm for action sampling, and also uses some dynamic programming techniques to reuse sampling results in a clever way. The MCTS algorithm first caught the attention of AI researchers for its impressive performance in computer Go [22]. Go is infamous for its vast playout trees, and allowed the MCTS sampling ideas to shine.

Induction with contexts. Computing the universal probability $M(e_t \mid æ_{<t}a_t)$ of a next percept requires infinite computational resources. To be precise, conditional probabilities for the distribution M are only *limit computable* [48]. We next describe how probabilities can be computed efficiently with the context tree weighting algorithm (CTW) [86] under some simplifying assumptions.

One of the key features of Solomonoff induction and UAI is the use of histories (Sect. 2.3.1), and the arbitrarily long time dependencies they allow for. For example, action a_1 may affect the percept e_{1000}. This is desirable, since the real world sometimes behaves this way. If I buried a treasure in my backyard 10 years ago, chances are I may find it if I dug there today. However, in most cases, it is the most recent part of the history that is most useful when predicting the next percept. For example, the most recent five minutes is almost always more relevant than a five minute time slot from a week ago for predicting what is going to happen next.

We define the *context of length c* of a history as the last c actions and percepts of the history:

> **procedure** MIRRORENVIRONMENT
> **while** true **do**:
> $\quad x \leftarrow$ action input
> $\quad$ output percept $\leftarrow x$

Relying on contexts for prediction makes induction not only computationally faster, but also conceptually easier. For example, if my current context is 0011, then I can use previous instances where I have been in the same context to predict the next percept:

$$a_1\, e_1\, a_2\, e_2 \ldots \ldots \ldots \underbrace{e_{t-2}\, a_{t-1}\, e_{t-1}\, a_t}_{\text{context of length 4}} \begin{array}{l} e_t = 0 \\ \diagup ? \\ \diagdown ? \\ e_t = 1 \end{array}$$

Table 2.2 The tradeoff for the size of the considered context

Short context	**More data**	Less precision
Long context	Less data	**Greater precision**

Long contexts offer greater precision but require more data. The MCTS algorithm dynamically trades between them

In the pictured example, $P(1) = 2/3$ would be a reasonable prediction since in two thirds of the cases where the context 0011 occurred before it was followed by a 1. (*Laplace's rule* gives a slightly different estimate.) Humans often make predictions this way. For example, when predicting whether I will like the food at a Vietnamese restaurant, I use my experience from previous visits to Vietnamese restaurants.

One question that arises when doing induction with contexts is how long or specific the context should be. Should I use the experience from all Vietnamese restaurants I have ever been to, or only this particular Vietnamese restaurant? Using the latter, I may have very limited data (especially if I have never been to the restaurant before!) On the other hand, using too unspecific contexts is not useful either: Basing my prediction on *all* restaurants I have ever been to (and not only the Vietnamese), will probably be too unspecific. Table 2.2 summarises the tradeoff between short and long contexts, which is nicely solved by the CTW algorithm.

The right choice of context length depends on a few different parameters. First, it depends on how much data is available. In the beginning of an agent's lifetime, the history will be short, and mainly shorter contexts will have a chance to produce an adequate amount of data for prediction. Later in the agent's life, the context can often be more specific, due to the greater amount of accumulated experience.

$$\ldots 00111 \ldots 00110 \ldots 00111 \ldots \underbrace{0}_{e_{t-2}} \underbrace{0}_{a_{t-1}} \underbrace{1}_{e_{t-1}} \underbrace{1}_{a_t} \begin{array}{c} e_t = 0 \\ ? \\ \diagdown \\ ? \\ e_t = 1 \end{array}$$

Second, the ideal context length may depend on the context itself, as aptly demonstrated by the example to the right. Assume you just heard the word *cup* or *cop*. Due to the similarity of the words, you are unable to tell which of them it was. If the most recent two words (i.e. the context) was *fill the*, you can infer the word was *cup*, since *fill the cop* makes little sense. However, if the most recent two words were *from the*, then further context will be required, as both *drink from the cup* and *run from the cop* are intelligible statements.

Context Tree Weighting. The Context Tree Weighting (CTW) algorithm is a clever way of adopting the right context length based both on the amount of data available and on the context. Similar to how Solomonoff induction uses a sum over all possible computer programs, the CTW algorithm uses a sum over all possible context trees up to a maximum depth D. For example, the context trees of depth $D \leq 2$ are the trees:

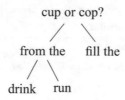

The structure of a tree encodes when a longer context is needed, and when a shorter context suffices (or is better due to a lack of data). For example, the leftmost tree corresponds to an iid process, where context is never necessary. The tree of depth $D = 1$ posits that contexts of length 1 always are the appropriate choice. The rightmost tree says that if the context is 1, then that context suffices, but if the most recent symbol is 0, then a context of length two is necessary. Veness et al. [85] offer a more detailed description.

For a given maximum depth D, there are $O(2^{2^D})$ different trees. The trees can be given binary encodings; the coding of a tree Γ is denoted $CL(\Gamma)$. Each tree Γ gives a probability $\Gamma(e_t \mid æ_{<t}a_t)$ for the next percept, given the context it prescribes using. Combining all the predictions yields the CTW distribution:

$$CTW(e_{<t} \mid a_{<t}) = \sum_{\Gamma} 2^{-CL(\Gamma)} \Gamma(e_{<t} \mid a_{<t}) \qquad (2.4)$$

The CTW distribution is tightly related to the Solomonoff-Hutter distribution (2.1), the primary difference being the replacing of computer programs with context trees. Naively computing $CTW(e_t \mid æ_{<t}a_t)$ takes double-exponential time. However, the CTW algorithm [86] can compute the prediction $CTW(e_t \mid æ_{<t}a_t)$ in $O(D)$ time. That is, for fixed D, it is a constant-time operation to compute the probability of a next percept for the current history. This should be compared with the infinite computational resources required to compute the Solomonoff-Hutter distribution M.

Despite its computational efficiency, the CTW distribution manages to make a weighted prediction based on all context trees within the maximum depth D. The relative weighting between different context trees changes as the history grows, reflecting the success and failure of different context trees to accurately predict the next percept. In the beginning, the shallower trees will have most of the weight due to their shorter code length. Later on, when the benefit of using longer contexts start to pay off due to the greater availability of data, the deeper trees will gradually gain an advantage, and absorb most of the weight from the shorter trees. Note that CTW handles partially observable environments, a notoriously hard problem in AI.

MC-AIXI-CTW. Combining the MCTS algorithm for planning with the CTW approximation for induction yields the MC-AIXI-CTW agent. Since it is history based, MC-AIXI-CTW handles hidden states gracefully (as long as long-term dependencies are not too important). The MC-AIXI-CTW agent can run on a standard desktop computer, and achieves impressive practical performance. For example, MC-AIXI-CTW can learn to play Rock Paper Scissors, TicTacToe, Kuhn Poker, and even Pacman, just by trying actions and observing percepts, and without additional knowledge about the rules of the game [85]. For computational reasons, in PacMan the agent did not view the entire screen, only a compressed version telling it the direction of ghosts and nearness of food pellets (16 bits in total). Although less informative, this drastically reduced the number of bits per interaction cycle, and allowed for using a reasonably short context. Thereby the less informative percepts actually made the task computationally easier.

Other approximations of Solomonoff induction. Although impressive, a major drawback of the CTW approximation of Solomonoff induction is that the CTW-agents cannot learn time dependencies longer than the maximum depth D of the context trees. This means that MC-AIXI-CTW will underperform in situations where long-term memory is required.

A few different approaches to approximating Solomonoff induction has been explored. Generally they are less well-developed than CTW, however.

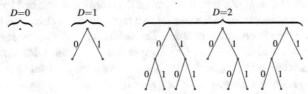

A seemingly minor generalisation of CTW is to allow loops in context trees. Such loops allow context trees of a limited depth to remember arbitrarily long dependencies, and can significantly improve performance in domains where this is important [12]. However, the loops break some of the clean mathematics of CTW, and predictions can no longer be computed in constant time. Instead, practical implementations must rely on approximations such as simulated annealing to estimate probabilities.

The *speed prior* [71] is a version of the universal distribution M where the prior is based on both program length and program runtime. The reduced probability of programs with long runtime makes the speed prior computable. It still requires exponential or double-exponential computation time, however [18]. Recent results show that program-based compression can be done incrementally [19]. These results can potentially lead to the development of a more efficient anytime-version of the speed prior. It is an open question whether such a distribution can be made sufficiently efficient to be practically useful.

2.4.2 Feature Reinforcement Learning

Feature reinforcement learning (ΦMDP) [31, 32] takes a more radical approach to reducing the complexity of Solomonoff induction. While the CTW algorithm outputs a distribution of the same *type* as Solomonoff induction (i.e. a distribution

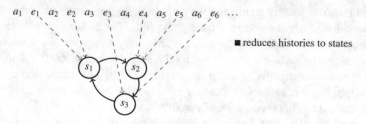

Fig. 2.4 ΦMDP infers an underlying state representations from a history

over next percepts), the ΦMDP approach instead tries to infer states from histories (see Fig. 2.4).

Histories and percepts are often generated by an underlying set of state transitions. For example, in classical physics, the *state of the world* is described by the position and velocity of all objects. In toy examples and games such as chess, the board state is mainly what matters for future outcomes. The usefulness of thinking about the world in terms of states is also vindicated by simple introspection: with few exceptions, we humans translate our histories of actions and percepts into states and transitions between states such as *being at work* or *being tired*.

In standard applications of RL with agents that are based on states, the designers of the agent also design a mechanism for interpreting the history/percept as a state. In ΦMDP, the agent is instead programmed to learn the most useful state representation itself. Essentially, a state representation is *useful* if it predicts rewards well. To avoid overfitting, smaller MDPs are also preferred, in line with Occam's razor.

The computational flow of a ΦMDP agent is depicted in Fig. 2.5. After a percept e_{t-1} has been received, the agent searches for the best map Φ : history $\mapsto$ state for its current history $æ_{<t}$. Given the state transitions provided by Φ, the agent can calculate transition and reward probabilities by frequency estimates. The value functions are computed by standard MDP techniques [82] or modern PAC-MDP algorithms, which allows for a near-optimal action to be found in polynomial time. Intractable planning is avoided. Once the optimal action has been determined, the agent submits it to the environment and waits for a new percept.

ΦMDP is not the only approach for inferring states from percepts. Partially observable MDPs (POMDPs) [35] is another popular approach. However, the learning of POMDPs is still an open question. The *predictive state representation* [51] approach also lacks a general and principled learning algorithm. In contrast, initial consistency results for ΦMDP show that under some assumptions, ΦMDP agents asymptotically learn the correct underlying MDP [80].

A few different practical implementations of ΦMDP agents have been tried. For toy problems, the ideal MDP-reductions can be computed with brute-force [56]. This is not possible in harder problems, where Monte Carlo approximations can be used instead [57]. Finally, the idea of context trees can be used also for ΦMDP. The context tree given the highest weight by the CTW algorithm can be used as a map Φ

that considers the current context as the state. The resulting ΦMDP agent exhibits similar performance as the MC-AIXI-CTW agent.

Generalisations of the ΦMDP agent include generalising the states to feature vectors [31] (whence the name *feature* RL). As mentioned above on page xxx, loops can be introduced to enable long-term memory of context trees [12]. The Markov property of states can be relaxed in the *extreme state aggregation* approach [34]. A somewhat related idea using neural networks for the feature extraction was recently suggested [74].

2.4.3   Model-Free AIXI

Both MC-AIXI-CTW and ΦMDP are *model-based* in the sense that they construct a *model* for how the environment reacts to actions. In MC-AIXI-CTW, the models are the context trees, and in ΦMDP, the model is the inferred MDP. In both cases, the models are then used to infer the best course of action. *Model-free* algorithms skip the middle step of inferring a model, and instead infer the value function directly.

Recall that $V^\pi(\text{æ}_{<t}a_t)$ denotes the expected return of taking action a_t in history $\text{æ}_{<t}$, and thereafter following the superscripted policy π, and that $V^*(\text{æ}_{<t}a_t)$ denotes expected return of a_t and thereafter following an optimal policy π^*. The optimal value function V^* is particularly useful for acting: If known, one can act optimally by always choosing action $a_t = \arg\max_a V^*(\text{æ}_{<t}a)$. This action a_t will be optimal under the assumption that future actions are optimal, which is easily achieved by selecting them from V^* in the same way. In other words, being greedy with respect to V^* gives an optimal policy. In model-free approaches, V^* is inferred directly from data. This removes the need for an extra planning step, as the best action is simply the action with the highest V^*-value. Planning is thereby incorporated into the induction step.

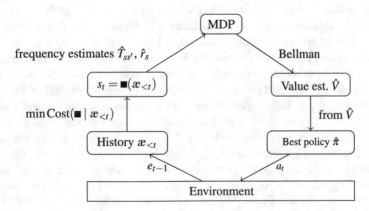

Fig. 2.5 Computational flow of a ΦMDP-agent

Many of the most successful algorithms in traditional RL are model-free, including Q-learning and SARSA [82]. The first computable version of AIXI, the AIXI*tl* agent [28] (Sect. 6.2), was a model-free version of AIXI. A more efficient model-free agent *compress and control* (CNC) was recently developed by developed by Veness et al. [84]. The performance of the CNC agent is substantially better than what has been achieved with both the MC-AIXI-CTW approach and the ΦMDP approach. CNC learned to play several Atari games (Pong, Bass, and Q*Bert) just by looking at the screen, similar to the subsequent famous *Deep Q-Learning* algorithm (DQN) [53] discussed in the next section. The CNC algorithm has not yet been generalised to the general, history-based case. The version described by Veness et al. [84] is developed only for fully observable MDPs.

2.4.4 Deep Learning

Deep learning with artificial neural networks has gained substantial momentum the last few years, demonstrating impressive practical performance in a wide range of learning tasks. In this section we connect some of these results to UAI.

A standard (feed-forward) neural network takes a fixed number of inputs, propagates them through a number of hidden layers of differentiable activation functions, and outputs a label or a real number. Given enough data, such networks can learn essentially any function. In one much celebrated example with particular connection to UAI, a deep learning RL system called DQN learned to play 49 different Atari video games at human level just by watching the screen and knowing the score (its reward) [53]. The wide variety of environments that the DQN algorithm learned to handle through interaction alone starts to resemble the general learning performance exhibited by the theoretical AIXI agent.

One limitation with standard feed-forward neural networks is that they only accept a fixed size of input data. This fits poorly with sequential settings such as text, speech, video, and UAI environments μ (see Definition 2) where one needs to remember the past in order to predict the future. Indeed, a key reason that DQN could learn to play Atari games using feed-forward networks is that Atari games are mostly *fully observable*: everything one needs to know in order to act well is visible on the screen, and no memory is required (compare *partial observability* discussed in Sect. 2.3.2).

Sequential data is better approached with so-called *recurrent neural networks*. These networks have a "loop", so that part of the output of the network at time t is fed as input to the network at time $t + 1$. This, in principle, allows the network to remember events for an arbitrary number of time steps. *Long short-term memory networks* (LSTMs) are a type of recurrent neural networks with a special pathway for preserving memories for many time steps. LSTMs have been highly successful in settings with sequential data [50]. *Deep Recurrent Q-Learning* (DRQN) is a generalisation of DQN using LSTMs. It can learn a partially observable version of Atari games [25] and the 3D game Doom [37]. DQN and DRQN are model-free algorithms, and so are most other practical successes with deep learning in RL. References [58, 73] (Chap. 5) provide more extensive surveys of related work.

Due to their ability to cope with partially observable environments with long-term dependencies between events, we consider AIs based on recurrent neural networks to be interesting deep-learning AIXI approximations. Though any system based on a finite neural network must necessarily be a less general learner than AIXI, deep neural networks tend to be well-fitted to problems encountered in our universe [49].

The connection between the abstract UAI theory and practical state-of-the-art RL algorithms underlines the relevancy of UAI.

2.5   Fundamental Challenges

Having a precise notion of intelligent behaviour allows us to identify many subtle issues that would otherwise likely have gone unnoticed. Examples of issues that have been identified or studied in the UAI framework include:

- Optimality [28, 44, 46]
- Exploration vs. exploitation [46, 61]
- How should the future be discounted? [40]
- What is a practically feasible and general way of doing joint learning and planning [32, 84, 85]
- What is a "natural" universal Turing machine or programming language? [44, 54]
- How should embodied agents reason about themselves? [17]
- Where should the rewards come from? [16, 26, 68]
- How should agents reason about other agents reasoning about themselves? [47]
- Personal identity and teleportation [62, 63].

In this section we will mainly focus on the optimality issues and the exploration vs. exploitation studies. The question of where rewards should come from, together with other safety related issues will be treated in Sect. 2.6. For the other points, we refer to the cited works.

2.5.1   Optimality and Exploration

What is the optimal behaviour for an agent in any unknown environment? The AIXI formula is a natural answer, as it specifies which action generates the highest expected return with respect to a distribution M that learns any computable environment in a strong sense (Theorem 1).

The question of optimality is substantially more delicate than this however, as illustrated by the common dilemma of when to explore and when to instead exploit knowledge gathered so far. Consider, for example, the question of whether to try a new restaurant in town. Trying means risking a bad evening, spending valued dollars on food that is potentially much worse than what your favourite restaurant has to offer. On the plus side, trying means that you learn whether it is good, and chances are that it is better than your current favourite restaurant.

The answer AIXI gives to this question is that the restaurant should be tried if and only if the expected return (utility) of trying the restaurant is greater than not trying, accounting for the risk of a bad evening and the possibility of finding a new favourite restaurant, as well as for their relative subjective probabilities. By giving this answer, AIXI is *subjectively* optimal with respect to its belief M. However, the answer is not fully connected to *objective* reality. Indeed, either answer (*try* or *don't try*) could have been justified with some belief.[6] While the convergence result Theorem 1 shows that M will correctly predict the rewards on the followed action sequence, the result does not imply that the agent will correctly predict the reward of actions that it is *not* taking. If the agent never tries the new restaurant, it will not learn how good it is, even though it would learn to perfectly predict the quality at the restaurants it is visiting. In technical terms, M has guaranteed *on-action* convergence, but not guaranteed *off-action* convergence [28] (Sect. 4.1.3).

An alternative optimality notion is *asymptotic optimality*. An agent is asymptotically optimal if it eventually learns to obtain the maximum possible amount of reward that can be obtained from the environment. No agent can obtain maximum possible reward directly, since the agent must first spend some time learning which environment is the true one. That AIXI is not asymptotically optimal was shown by [44, 61]. In general, it is impossible for an agent to be both Bayes-optimal and asymptotically optimal [61].

Bayes-optimality	Subjective	**Immediate**
Asymptotic optimality	**Objective**	Asymptotic

Among other benefits, the interaction between asymptotically optimal agents yields clean game-theoretic results. Almost regardless of their environment, asymptotically optimal agents will converge on a Nash-equilibria when interacting [47]. This result provides a formal solution to the long-open *grain-of-truth* problem, connecting expected utility theory with game theory.

2.5.2 Asymptotically Optimal Agents

AIXI is Bayes-optimal, but is not asymptotically optimal. The reason is that AIXI does not explore enough. There are various ways in which one can create more explorative agents. One of the simplest ways is by letting the agent act randomly for

---

[6]In fact, for any decision there is one version of AIXI that prefers each option, the different versions of AIXI differing only in which programming language (UTM) is used in the definition of the universal distribution M (2.1) [44].

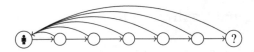

Fig. 2.6 In this environment, focused exploration far outperforms random exploration. Focused exploration finds out the content at the question mark in 6 time steps. With random exploration, the expected number of steps required is 2^6, an exponential increase

periods of time. A fine balance needs to be struck between doing this enough so that the true environment is certain to be discovered, and not doing it too much so that the full benefits of knowing the true environment can be reaped (note that the agent can never know for certain that it has now found the true environment). If exploration is done in just the right amount, this gives rise to a (weakly) asymptotically optimal agent [38].

Optimistic agents. Exploring randomly is often inefficient, however. Consider for example the environment depicted in Fig. 2.6. An agent that purposefully explores the rightmost question mark, finds out the truth exponentially faster than a randomly exploring agent. For a real-world example, consider how long it would take you to walk into a new restaurant and order a meal by performing random actions. Going to a restaurant with the intention of finding out how good the food is tends to be much more efficient.

Optimism is a useful principle for devising focused exploration. In standard RL, this is often done with *positive initialisation* of value estimates. Essentially, the agent is constructed to believe that "there is a path to paradise", and will systematically search for it. Optimism thus leads to strategic exploration. In the UAI framework, optimistic agents can be constructed using a growing, finite class $\mathcal{N}_t$ of possible environments, and act according to the environment $v \in \mathcal{N}_t$ that promises the highest expected reward. Formally, AIXI's action selection (2.3) is replaced by

$$a_t = \arg\max_a \max_{v \in \mathcal{N}_t} V_v(\text{æ}_{<t}\, a).$$

Optimistic agents are asymptotically optimal [81].

Thompson-sampling. A third way of obtaining asymptotically optimal agents is through Thompson-sampling. Thompson-sampling is more closely related to AIXI than optimistic agents. While AIXI acts according to a weighted average over all consistent environments, a Thompson-sampling agent randomly picks *one* environment v and acts as if v were the true one for one effective horizon. When the effective horizon is over, the agent randomly picks a new environment v'. Environments are sampled from the agent's posterior belief distribution at the time of the sampling.

Since Thompson-sampling agents act according to one environment over some time period, they explore in a strategic manner. Thompson-sampling agents are also asymptotically optimal [46].

2.6 Predicting and Controlling Behaviour

The point of creating intelligent systems is that they can act and make decisions without detailed supervision or micromanagement. For example, Sect. 18.5.3 in this book describes the application of autonomous AI systems to unmanned space missions. However, with increasing autonomy and responsibility, and with increasing intelligence and capability, there inevitably comes a risk of systems causing substantial harm [10]. The UAI framework provides a means for giving formal proofs about the behaviour of intelligent agents. While no practical agent may perfectly implement the AIXI ideal, having a sense of what behaviour the agent strives towards can still be highly illuminating.

We start with some general observations. What typically distinguishes an autonomous agent from other agents is that it decides itself what actions to take to achieve a goal. The goal is central, since a system without a goal must either be instructed on a case-by-case basis, or work without clear direction. Systems optimising for a goal may find surprising paths towards that goal. Sometimes these paths are desirable, such as when a Go or Chess program finds moves no human would think of. Other times, the results are less desirable. For example, [8] used an evolutionary algorithm to optimise circuit design of a radio controller. Surprisingly, the optimal design found by the algorithm did not contain any oscillator, a component typically required. Instead the system had evolved a way of using radio waves from a nearby computer. While clever, the evolved controller would not have worked in other circumstances.

In general, artificial systems optimise the *literal* interpretation of the goal they are given, and are indifferent to implicit *intentions* of the designer. The same behaviour is illustrated in fairy tales of "evil genies", such as with King Midas who wished that everything he touched would turn to gold. Closer to the field of AI is Asimov's ([7]) three laws of robotics. Asimov's stories illustrate some problems with AIs interpreting these laws overly literally.

The examples above illustrate how special care must be taken when designing the goals of autonomous systems. Above, we used the simple goal of maximising reward for our UAI agents (Sect. 2.3.3). One might think that maximising reward given by a human designer should be safe against most pitfalls: After all, the ultimate goal of the system in this case is pretty close to making its human designer happy. This section will discuss some issues that nonetheless arise, and ways in which those issues can potentially be addressed. For more comprehensive overviews of safety concerns of intelligent agents, see [4, 21, 76, 83].

2.6.1 Self-Modification

Autonomous agents that are intelligent and have means to affect the world in various ways may, in principle, turn those means towards modifying itself. An autonomous agent may for example find a way to rewrite its own source code. Although present AI systems are not yet close to exhibiting the required intelligence or "self-awareness" required to look for such self-modifications, we can still anticipate that such abilities will emerge in future AI systems. By modelling self-modification formally, we can assess some of the consequences of the self-modification possibility, and look for ways to manage the risks and harness the possibilities. Formal models of self-modification have been developed in the UAI-framework [15, 65, 66]. We next discuss some types of self-modification in more detail.

Self-improvement. One reason an intelligent agent may want to self-modify could be for improving its own hardware or software. Indeed, Omohundro [60] lists *self-improvement* as a fundamental drive of any intelligent system, since a better future version of the agent would likely be better at achieving the agent's goal. The Gödel machine [72] is an agent based on this principle: The Gödel machine is able to change any part of its own source code, and uses part of its computational resources to find such improvements. The claim is that the Gödel machine will ultimately be an optimal agent. However, Gödel's second incompleteness theorem and its corollaries imply fundamental limitations to formal systems' ability to reason about themselves. Yudkowsky and Herreshoff [89] claim some progress on how to construct self-improving systems that sidestep these issues.

Though self-improvement is generally positive as it allows our agents to become better over time, it also implies a potential safety problem. An agent improving itself may become more intelligent than we expect, which admonishes us to take extra care in designing agents that can be trusted regardless of their level of intelligence [10].

Self-modification of goals. Another way an intelligent system may use its self-modification capacity is to replace its goal with something easier, for example by rewriting the code that specifies its goal. This would generally be undesirable, since there is no reason the new goal of the agent would be useful to its human designers.

It has been argued on philosophical grounds that intelligent systems will not want to replace their goals [60]. Essentially, an agent should want future versions of itself to strive towards the same goal, since that will increase the chances of the goal being fulfilled. However, a formal investigation reveals that this depends on subtle details of the agent's design [15]. Some types of agents do not want to change their goals, but there are also wide classes of agents that are indifferent to goal modification, as well as systems that actively desire to modify their goals. The first proof that an UAI-based agent can be constructed to avoid self-modification was given by [26].

2.6.2   Counterfeiting Reward

The agent *counterfeiting reward* is another risk. An agent that maximises reward means an agent that actively desires a particular kind of percept: that is, a percept with maximal reward component. Similar to how a powerful autonomous agent may modify itself, an autonomous agent may be able to subvert its percepts, for example by modifying its sensors. Preventing this risk turns out to be substantially harder than preventing self-modification of goals, since there is no simple philosophical reason why an agent set to maximise reward should not do so in the most effective way; i.e. by taking control of its percepts.

More concretely, the rewards must be communicated to the agent in some way. For example, the reward may be decided by its human designers every minute, and communicated to the robot through a network cable. Making the input and the communication channel as secure against modification as possible goes some way towards preventing the agent from easily counterfeiting reward. However, such solutions are not ideal, as they challenge the agent to use its intelligence to try and overcome our safety measures. Especially in the face of a potentially self-improving agent, this makes for a brittle kind of safety.

Artificial agents counterfeiting reward have biological analogues. For example, humans inventing drugs and contraception may be seen as ways to counterfeit pleasure without maximising for reproduction and survival as would be evolutionary optimal. In a more extreme example, [59] plugged a wire into the pleasure centre of rats' brains, and gave the rats a button to activate the wire. The rats pressed the button incessantly, forgetting other pleasures such as eating and sleeping. The rats eventually died of starvation. Due to this experiment, the reward counterfeiting problem is sometimes called *wireheading* [88] (Chap. 4).

What would the failure mode of a wireheaded agent look like? There are several possibilities. The agent may either decide to act innocently, to reduce the probability of being shut down. Or it may try to transfer or copy itself outside of the control of its designers. In the worst-case scenario, the agent tries to incapacitate or threaten its designers, to prevent them from shutting it down. A combination of behaviours or transitions over time are also conceivable. In either of the scenarios, an agent with fully counterfeited reward has no (direct) interest in making its designers happy. We next turn to some possibilities for avoiding this problem.

Knowledge-seeking agents. One could consider designing agents with other types of goals than optimising reward. Knowledge-seeking agents [64] are one such alternative. Knowledge-seeking agents do not care about maximising reward, only about improving their knowledge about the world. It can be shown that they do not wirehead [68]. Unfortunately, it is hard to make knowledge-seeking agents useful for tasks other than scientific investigation.

Utility agents. A generalisation of both reward maximising agents and knowledge seeking agents are *utility agents*. Utility agents maximise a real-valued utility function $u(æ_{<t})$ over histories. Setting $u(æ_{<t}) = R(æ_{<t})$ gives a reward maximising agent,[7] and setting $u(æ_{<t}) = -M(æ_{<t})$ gives a knowledge-seeking agent (trying to minimise the likelihood of the history it obtains, to make it maximally informative). While some utility agents are tempted to counterfeit reward (such as the special case of reward maximising agents), properly defined utility agents whose utility functions make them care about the *state of the world* do avoid the wireheading problem [26].

The main challenge with utility agents is how to specify the utility function. Precisely formulating one's goal is often challenging enough even using one's native

---

[7]The *return* $R(æ_{<t}) = r_1 + \gamma r_2 + \ldots$ is defined and discussed in Sect. 2.3.3.

language. A correct formal specification seems next to impossible for any human to achieve. Utility agents also seem to forfeit a big part of the advantage with induction-based systems discussed in Sect. 2.2. That is, that the agent can *learn* what we want from it.

Value learning. The idea of *value learning* [13] is that the agent learns the utility function u by interacting with the environment. For example, the agent might spend the initial part of its life reading the philosophy literature on ethics, to understand what humans what. Formally, the learning must be based on information contained in the history $æ_{<t}$. The history is therefore used both to learn about the true utility function, and to evaluate how well the world currently satisfies the inferred utility function. Concrete value learning suggestions include *inverse reinforcement learning* (IRL) [3, 14, 24, 55, 75] and *apprenticeship learning* [1]. Bostrom [9, 10] also suggests some interesting alternatives for value learning, but they are less concrete than IRL and apprenticeship learning.

Concerns have been raised that value learning agents may be incentivised to learn the "wrong thing" by modifying their percepts. Suggested solutions include *indifference* [5, 6] and *belief consistency* [16].

2.6.3 Death and Self-Preservation

The UAI framework can also be used to formally define *death* for artificial agents, and for understanding when agents will want to preserve themselves. A natural definition of death is the ceasing of experience. This can be directly defined in the UAI framework. *Death* is the ending of the history. When an agent is dead, it receives no more percepts, and takes no more actions. The naturalness of this definition should be contrasted both with the ongoing controversy defining death for biological systems and with the slightly artificial construct one must use in state-based MDP representations. To represent death in an MDP, an extra absorbing state (with reward 0) must be introduced.

A further nice feature of defining death in the UAI framework is that the universal distribution M can be interpreted to define a subjective death probability. Recall Eq. (2.1) on page xxx that M is defined as a sum over programs,

$$M(e_{<t} \mid a_{<t}) = \sum_{p:\ p(a_{<t})=e_{<t}} 2^{-\ell(p)}.$$

Some computer programs p may fail to produce an output at all. As a consequence, M is actually not a proper probability distribution, but a *semi-measure*. Summing over all percept probabilities gives total probability less than 1, i.e. $\sum_{e \in \mathcal{E}} M(e \mid a) < 1$. For example, $M(0 \mid a) = 0.4$ and $M(1 \mid a) = 0.4$ gives $M(0 \mid a) + M(1 \mid a) = 0.8 < 1$. The lacking probability 0.2 can be interpreted as a subjective death probability [52]. The interpretation makes sense as it corresponds to a probability of not seeing any percept at all (i.e. death). Further, interpreting programs as environments, the measure deficit arises because some programs fail to output. An environment program that fails to output a next percept is an environment where the agent will have no further experience (i.e. is dead).

Having a definition of death lets us assess an agent's *self-preservation drive* [60]. In our definition of death, the reward obtained when dead is automatically 0 for any agent. We can therefore design *self-preserving* agents that get reward communicated as a positive real number, say between 0 and 1. These agents will try to avoid death as long as possible, as death is the worst possible outcome. We can also define *suicidal agents* by letting the reward be communicated in negative real numbers, say between -1 and 0. For these agents, obtaining the implicit death reward of 0 is like paradise. Suicidal agents will therefore consider termination as the ideal outcome. The difference in behaviour that ensues is somewhat surprising since positive linear transformations of the reward typically do not affect behaviour. The reason that it affects behaviour in UAI is that M is a semi-measure and not a measure.[8]

These different kinds of agents have implications for AI safety. In Sect. 2.6.1 we discussed the possibility of a self-improving AI as a safety risk. If a self-improving AI becomes highly intelligent *and* is self-preserving, then it may be very hard to stop. As a rough comparison, consider how hard it can be to stop relatively dumb computer viruses. A suicidal agent that becomes powerful will try to self-terminate instead of self-preserve. This also comes with some risks, as the agent has no interest in minimising collateral damage in its suicide. Further research may reveal whether the risks with such suicides are less than the risks associated with self-preserving agents.

2.7 Conclusions

In summary, UAI is a formal, foundational theory for AI that gives a precise answer to the question of what is the optimal thing to do for essentially any agent acting in essentially any environment. The insight builds on old philosophical principles (Occam, Epicurus, Bayes), and can be expressed in a single, one-line AIXI equation [28] (p. 143).

The AIXI equation and the UAI framework surrounding it has several important applications. First, the formal framework can be used to give mathematically precise statements of the behaviour of intelligent agents, and to devise potential solutions to the problem of how we can control highly intelligent autonomous agents (Sect. 2.6). Such guarantees are arguably essential for designing trustworthy autonomous agents. Second, it has inspired a range of practical approaches to (general) AI. Several fundamentally different approaches to approximating AIXI have exhibited impressive practical performance (Sect. 2.4). Third, the precision offered by the mathematical

---

[8]Interesting observations about how the agent's belief in its own mortality evolves over time can also be made [52].

framework of UAI has brought to light several subtle issues for AI. We discussed different optimality notions and directed exploration-schemes, and referenced many other aspects (Sect. 2.5).

References

1. P. Abbeel, A.Y. Ng, Apprenticeship learning via inverse reinforcement learning. *Proceedings of the 21st International Conference on Machine Learning (ICML)* (2004), pp. 1–8
2. B. Abramson, The expected-outcome model of two-player games. Ph.D. thesis, Columbia University, 1991
3. K. Amin, S. Singh, Towards resolving unidentifiability in inverse reinforcement learning. Preprint (2016), arXiv:1601.06569 [cs.AI]
4. D. Amodei, C. Olah, J. Steinhardt, P. Christiano, J. Schulman, D. Mané, Concrete problems in AI safety. Preprint (2016), arXiv:1606.06565 [cs.AI]
5. S. Armstrong, Utility indifference. Technical Report (Oxford University, 2010)
6. S. Armstrong, Motivated value selection for artificial agents, in *Workshops at the Twenty-Ninth AAAI Conference on Artificial Intelligence* (2015), pp. 12–20
7. I. Asimov, *Runaround* (Austounding Science Fiction, Street & Smith 1942)
8. J. Bird, P. Layzell, The evolved radio and its implications for modelling the evolution of novel sensors. *Proceedings of Congress on Evolutionary Computation* (2002), pp. 1836–1841
9. N. Bostrom, Hail mary, value porosity, and utility diversification. Technical Report (Oxford University, 2014)
10. N. Bostrom, *Superintelligence: Paths, Dangers, Strategies* (Oxford University Press, Oxford, 2014)
11. R. Coulom, Efficient selectivity and backup operators in Monte-Carlo tree search. Comput. Games **4630**, 72–83 (2007)
12. M. Daswani, P. Sunehag, M. Hutter, Feature reinforcement learning using looping suffix trees, in *10th European Workshop on Reinforcement Learning: JMLR: Workshop and Conference Proceedings*, vol. 24, pp. 11–22 (2012) (J. Mach. Learn. Res.)
13. D. Dewey, Learning what to value. *Artificial General Intelligence* (2011), pp. 309–314
14. O. Evans, A. Stuhlmuller, N.D. Goodman, Learning the preferences of ignorant, inconsistent agents, in *Association for the Advancement of Artificial Intelligence (AAAI)* (2016)
15. T. Everitt, D. Filan, M. Daswani, M. Hutter, Self-modificication of policy and utility function in rational agents. *Artificial General Intelligence* (Springer, 2016), pp. 1–11
16. T. Everitt, M. Hutter, Avoiding wireheading with value reinforcement learning. *Artificial General Intelligence* (Springer, 2016), pp. 12–22
17. T. Everitt, J. Leike, M. Hutter, Sequential extensions of causal and evidential decision theory, in *Algorithmic Decision Theory*, ed. by T. Walsh (Springer, 2015), pp. 205–221
18. D. Filan, M. Hutter, J. Leike, Loss bounds and time complexity for speed priors, in *Artificial Intelligence and Statistics (AISTATS)* (2016)
19. A. Franz, Some theorems on incremental compression. *Artificial General Intelligence* (Springer, 2016)
20. E. Fredkin, Finite nature. *XXVIIth Rencotre de Moriond* (1992)
21. Future of Life Institute, Research priorities for robust and beneficial artificial intelligence. Technical Report (Future of Life Institute, 2015)
22. S. Gelly, Y. Wang, R. Munos, O. Teytaud, Modification of UCT with patterns in Monte-Carlo Go. INRIA Technical Report, vol. 6062, No. 24 (November 2006)
23. N. Goodman, *Fact, Fiction and Forecast*, vol. 74 (Harvard University Press, 1983)
24. D. Hadfield-Menell, A. Dragan, P. Abbeel, S. Russell, Cooperative inverse reinforcement learning. Preprint (2016), arXiv:1606.03137 [cs.AI]

25. M. Hausknecht, P. Stone, Deep recurrent Q-learning for partially observable MDPs. Preprint (2015), pp. 29–37, arXiv:1507.06527 [cs.LG]
26. B. Hibbard, Model-based utility functions. J. Artif. Gen. Intell. **3**(1), 1–24 (2012)
27. J.E. Hopcroft, J.D. Ullman, *Introduction to Automata Theory, Languages, and Computation* (Addison-Weasly, 1979). ISBN 0-201-02988-X
28. M. Hutter, *Universal Artificial Intelligence: Sequential Decisions based on Algorithmic Probability* (Springer, Berlin, 2005), 300 pp, http://www.hutter1.net/ai/uaibook.htm
29. M. Hutter, On universal prediction and Bayesian confirmation. Theor. Comput. Sci. **384**(1), 33–48 (2007)
30. M. Hutter, *Discrete MDL predicts in total variation, in Advances in Neural Information Processing Systems 22 (NIPS'09)* (Curran Associates, Cambridge, 2009), pp. 817–825
31. M. Hutter, Feature dynamic Bayesian networks, in *Proceedings of the 2nd Conference on Artificial General Intelligence (AGI'09)*, vol. 8 (Atlantis Press, 2009), pp. 67–73
32. M. Hutter, Feature reinforcement learning: Part I: unstructured MDPs. J. Artif. Gen. Intell. **1**, 3–24 (2009)
33. M. Hutter, The subjective computable universe, in *A Computable Universe: Understanding and Exploring Nature as Computation* (World Scientific, 2012), pp. 399–416
34. M. Hutter, Extreme state aggregation beyond MDPs, in *Proceedings of the 25th International Conference on Algorithmic Learning Theory (ALT'14)*, vol. 8776 of LNAI (Springer, Bled, Slovenia, 2014), pp. 185–199
35. L.P. Kaelbling, M.L. Littman, A.R. Cassandra, Planning and acting in partially observable stochastic domains. Artif. Intell. **101**(1–2), 99–134 (1998)
36. L. Kocsis, C. Szepesvári, Bandit based Monte-Carlo planning, in *Proceedings of ECML* (2006), pp. 282–203
37. G. Lample, D.S. Chaplot, Playing FPS games with deep reinforcement learning. Preprint (2016), arXiv:1609.05521 [cs.AI]
38. T. Lattimore, M. Hutter, Asymptotically optimal agents. Lect. Notes Comput. Sci. **6925**, 368–382 (2011)
39. T. Lattimore, M. Hutter, On Martin-Löf convergence of Solomonoff's mixture. *Theory and Applications of Models of Computation* (2013), pp. 212–223
40. T. Lattimore, M. Hutter, General time consistent discounting. Theor. Comput. Sci. **519**, 140–154 (2014)
41. T. Lattimore, M. Hutter, V. Gavane, Universal prediction of selected bits, in *Proceedings of the 22nd International Conference on Algorithmic Learning Theory (ALT-2011)* (2011), pp. 262–276
42. S. Legg, M. Hutter, Universal intelligence: a definition of machine intelligence. Mind. Mach. **17**(4), 391–444 (2007)
43. S. Legg, J. Veness, An approximation of the universal intelligence measure, in *Ray Solomonoff 85th Memorial Conference* (2011), pp. 236–249
44. J. Leike, M. Hutter, Bad universal priors and notions of optimality. Conf. Learn. Theory **40**, 1–16 (2015)
45. J. Leike, M. Hutter, Solomonoff induction violates Nicod's criterion, in *Algorithmic Learning Theory* (2015), pp. 349–363
46. J. Leike, T. Lattimore, L. Orseau, M. Hutter, Thompson sampling is asymptotically optimal in general environments, in *Uncertainty in Artificial Intelligence (UAI)* (2016)
47. J. Leike, J. Taylor, B. Fallenstein, A formal solution to the grain of truth problem. In *Uncertainty in Artificial Intelligence (UAI)* (2016)
48. M. Li, P. Vitanyi, *Kolmogorov Complexity and its Applications*, 3rd edn. (Springer, 2008)
49. H.W. Lin, M. Tegmark, Why does deep and cheap learning work so well? Preprint, 02139:14 (2016), arXiv:1608.08225 [cond-mat.dis-nn]
50. Z.C. Lipton, J. Berkowitz, C. Elkan, A critical review of recurrent neural networks for sequence learning. Preprint (2015), pp. 1–35, arXiv:1506.00019 [cs.LG]
51. M.L. Littman, R.S. Sutton, S. Singh, Predictive representations of state. Neural Information Processing Systems (NIPS) **14**, 1555–1561 (2001)

52. J. Martin, T. Everitt, M. Hutter, Death and suicide in universal artificial intelligence. *Artificial General Intelligence* (Springer, 2016), pp. 23–32
53. V. Mnih, K. Kavukcuoglu, D. Silver, A. Rusu, J. Veness, M.G. Bellemare, A. Graves, M. Riedmiller, A.K. Fidjeland, G. Ostrovski, S. Petersen, C. Beattie, A. Sadik, I. Antonoglou, H. King, D. Kumaran, D. Wierstra, S. Legg, D. Hassabis, Human-level control through deep reinforcement learning. Nature **518**(7540), 529–533 (2015)
54. M. Mueller, Stationary algorithmic probability. Theor. Comput. Sci. **2**(1), 13 (2006)
55. A. Ng, S. Russell, Algorithms for inverse reinforcement learning. *Proceedings of the Seventeenth International Conference on Machine Learning* (2000), pp. 663–670
56. P. Nguyen, Feature reinforcement learning agents. Ph.D. thesis, Australian National University, 2013
57. P. Nguyen, P. Sunehag, M. Hutter, Feature reinforcement learning in practice, in *Proceedings of the 9th European Workshop on Reinforcement Learning (EWRL-9)*, vol. 7188 of LNAI (Springer, 2011), pp. 66–77
58. J. Oh, V. Chockalingam, S. Singh, H. Lee, Control of memory, active perception, and action in Minecraft. Preprint (2016), arXiv:1605.09128 [cs.AI]
59. J. Olds, P. Milner, Positive reinforcement produced by electrical stimulation of septal area and other regions of rat brain. J. Comp. Physiol. Psychol. **47**(6), 419–427 (1954)
60. S.M. Omohundro, The basic AI drives, in *Artificial General Intelligence*, vol. 171, ed. by P. Wang, B. Goertzel, S. Franklin (IOS Press, 2008), pp. 483–493
61. L. Orseau, Optimality issues of universal greedy agents with static priors. *Lecture Notes in Computer Science (Including Subseries Lecture Notes in Artificial Intelligence and Lecture Notes in Bioinformatics)*, vol. 6331 of LNAI (2010), pp. 345–359
62. L. Orseau, The multi-slot framework: a formal model for multiple, copiable AIs. *Artificial General Intelligence*, vol. 8598 of LNAI (Springer, 2014), pp. 97–108
63. L. Orseau, Teleporting universal intelligent agents, in *Artificial General Intelligence*, vol. 8598 of LNAI (Springer, 2014), pp. 109–120
64. L. Orseau, Universal knowledge-seeking agents. Theor. Comput. Sci. **519**, 127–139 (2014)
65. L. Orseau, M. Ring, Self-modification and mortality in artificial agents, in *Artificial General Intelligence*, vol. 6830 of LNAI (2011), pp. 1–10
66. L. Orseau, M. Ring, Space-time embedded intelligence, in *Artificial General, Intelligence* (2012), pp. 209–218
67. M. Sl Rathmanner, Hutter, A philosophical treatise of universal induction. Entropy **13**(6), 1076–1136 (2011)
68. M. Ring, L. Orseau, *Delusion, survival, and intelligent agents, in Artificial General Intelligence* (Springer, Heidelberg, 2011), pp. 11–20
69. S. Schaal, Is imitation learnig the route to humanoid robots? Trends Cogn. Sci. **3**(6), 233–242 (1999)
70. J. Schmidhuber, Algorithmic theories of everything. Technical Report (IDSIA, 2000)
71. J. Schmidhuber. The speed prior: A new simplicity measure yielding near-optimal computable predictions, in *Proceedings of the 15th Annual Conference on Computational Learning Theory COLT 2002*, vol. 2375 of *Lecture Notes in Artificial Intelligence* (Springer, 2002), pp. 216–228
72. J. Schmidhuber, Gödel machines: fully self-referential optimal universal self-improvers, in *Artificial General Intelligence*, ed. by B. Goertzel, C. Pennachin (Springer, IDSIA, 2007), pp. 199–226
73. J. Schmidhuber, Deep learning in neural networks: an overview. Neural Netw. **61**, 85–117 (2015)
74. J. Schmidhuber, On learning to think: algorithmic information theory for novel combinations of reinforcement learning controllers and recurrent neural world models (2015), pp. 1–36, arXiv:1404.7828
75. C.E. Sezener, Inferring human values for safe AGI design, in *Artificial General Intelligence* (Springer, 2015), pp. 152–155
76. N. Soares, B. Fallenstein, Aligning superintelligence with human interests: a technical research agenda. Technical Report (Machine Intelligence Research Institute (MIRI), 2014), pp. 152–155

77. R.J. Solomonoff, A formal theory of inductive inference. Part I. Inf. Control **7**(1), 1–22 (1964)
78. R.J. Solomonof, A formal theory of inductive inference. Part II applications of the systems to various problems in induction. Inf. Control **7**(2), 224–254 (1964)
79. R.J. Solomonoff, Complexity-based induction systems: comparisons and convergence theorems. IEEE Trans. Inf. Theory **IT-24**(4), 422–432 (1978)
80. P. Sunehag, M. Hutter, Consistency of feature Markov processes, in *Proceedings of the 21st International Conference on Algorithmic Learning Theory (ALT'10)*, vol. 6331 of LNAI (Springer, Canberra, 2010), pp. 360–374
81. P. Sunehag, M. Hutter, Rationality, optimism and guarantees in general reinforcement learning. J. Mach. Learn. Res. **16**, 1345–1390 (2015)
82. R.S. Sutton, A.G. Barto, *Reinforcement Learning: An Introduction* (MIT Press, 1998)
83. J. Taylor, E. Yudkowsky, P. Lavictoire, A. Critch, Alignment for advanced machine learning systems. Technical Report (MIRI, 2016)
84. J. Veness, M.G. Bellemare, M. Hutter, A. Chua, G. Desjardins, Compress and control, in *Association for the Advancement of Artificial Intelligence (AAAI)* (AAAI Press, 2015), pp. 3016–3023
85. J. Veness, K.S. Ng, M. Hutter, W. Uther, D. Silver., A Monte-Carlo AIXI approximation. J. Artif. Intell. Res. **40**, 95–142 (2011)
86. F.M.J. Willems, Y.M. Shtarkov, T.J. Tjalkens, The context-tree weighting method: basic properties. IEEE Trans. Inf. Theory **41**(3), 653–664 (1995)
87. S. Wolfram, *A New Kind of Science* (Wolfram Media, 2002)
88. R.V. Yampolskiy, *Artificial Superintelligence: A Futuristic Approach* (Chapman and Hall/CRC, 2015)
89. E. Yudkowski, M. Herreshoff, Tiling agents for self-modifying AI, and the Löbian obstacle. Technical Report (MIRI, 2013)

Chapter 3
Goal Reasoning and Trusted Autonomy

Benjamin Johnson, Michael W. Floyd, Alexandra Coman,
Mark A. Wilson and David W. Aha

3.1 Introduction

An important consideration for any autonomous system is the need to react intelligently to unplanned events and observations. Doing so requires that the system be given the freedom and ability to adjust its behavior, without being commanded to do so by a human operator or external system. A good deal of research in robotics and autonomy focuses on developing systems that can change their actions and plans, to more reliably or more optimally achieve their goals. However, it is also important to develop systems that autonomously deliberate on the goals themselves. Goal Reasoning (e.g., [38]) is the study of how autonomous agents can dynamically reason about and adjust their goals. Doing so enables agents to adapt intelligently to changing conditions and unexpected events, allowing them to address a wider variety of complex problems. Section 1.3 argues for the generality of reward maximization for goals; Goal Reasoning, then, would allow the autonomous agent to adapt its own reward function.

Goal Reasoning capabilities, in one form or another, may prove useful in many applications. In domains where the agents must operate autonomously for substantial periods of time or with limited communications, such as in unmanned underwater

B. Johnson (✉) · A. Coman
NRC Research Associate at the US Naval Research Laboratory,
Washington DC, USA
e-mail: blj39@cornell.edu

A. Coman
e-mail: alexandra.coman.ctr.ro@nrl.navy.mil

M. W. Floyd
Knexus Research Corporation, Springfield, VA, USA
e-mail: michael.floyd@knexusresearch.com

M. A. Wilson · D. W. Aha
Navy Center for Applied Research in AI, US Naval Research Laboratory,
Washington DC, USA
e-mail: mark.wilson@nrl.navy.mil

D. W. Aha
e-mail: david.aha@nrl.navy.mil

© The Author(s) 2018
H. A. Abbass et al. (eds.), *Foundations of Trusted Autonomy*, Studies in Systems,
Decision and Control 117, https://doi.org/10.1007/978-3-319-64816-3_3

operations, it is imperative that the agents have the freedom to act on the information they gather during the mission; they cannot rely on timely operator input, and must be able to adjust their goals autonomously, should the need arise. Goal Reasoning can also be used to allow autonomous agents to deliberate about the goals of other agents, such as an operator. This can lower the burden on the operator and avoid information overload, allowing the operator to focus on other, immediate tasks (e.g., operating their own vehicle, or observing their surroundings). Similarly, Goal Reasoning can be useful in systems that involve multiple, collaborating autonomous agents. In such a system, it is valuable for each agent to be able to adjust its own goals, based on the goals or actions of the other agents with which it is collaborating.

Goal Reasoning also raises interesting questions with regard to the topic of Trusted Autonomy. One definition of Trusted Autonomy, taken from [1], refers to it as:

> [T]he ability to form teams of humans and/or machines that make educated and conscious decisions to delegate risky tasks among team members seamlessly and symbiotically.

Trust, then, is inherent in any well-functioning autonomous system in which tasks or goals are delegated to agents within the system. In particular, a Goal Reasoning agent requires a large degree of trust to have the freedom to autonomously determine and adjust its goals. As such, Goal Reasoning pushes the scope of Trusted Autonomy, as the trust of other agents must extend beyond the completion of known tasks or towards achieving a known goal, to trusting the agent's ability to make decisions regarding its goals.

There are many open questions with respect to Goal Reasoning agents and their relation to Trusted Autonomy. What motivates the agent to change its goals? How do humans interact with robotic agents and delegate tasks? When will the agents choose to overrule an operator's commands, and why? How can the Goal Reasoning process be framed to promote transparency? How can one ensure that certain safety conditions and guarantees are maintained, despite the additional freedom of autonomy provided to a Goal Reasoning agent? These are just a few of the questions raised by the relationship between Goal Reasoning and Trusted Autonomy. In this chapter, we describe Goal Reasoning and elaborate on some of these important questions. While we focus on a few selected topics of research, there is a large and growing body of work on Goal Reasoning and related topics. For additional reading, see the survey papers [20, 38], or the proceedings of several Goal Reasoning workshops [2–4, 34].

This chapter is structured as follows. In Sect. 3.2, we describe a simple model of Goal Reasoning called Goal-Driven Autonomy (GDA), as well as a domain-independent method for goal selection in GDA, and an application of GDA in a human-robot teaming task. Section 3.3 focuses on a more comprehensive model of Goal Reasoning based on goal refinement, an architecture for ensuring the behaviors of Goal Reasoning agents that use this model, and its application in a distributed robotics task. For both models, we describe the importance of transparency to engendering operator trust. We then describe two extensions to Goal Reasoning in Sect. 3.4, first on how inverse trust can be used as a basis for adaptive autonomy, and then on rebel agents and their relation to Trusted Autonomy. We then conclude in Sect. 3.5.

3.2   Goal-Driven Autonomy Models

This section describes the Goal-Driven Autonomy (GDA) model of Goal Reasoning, which has been studied by several groups (c.g., [11–13, 22, 25, 29, 38, 39]). We discuss only some of our own group's work on GDA, and its relation to Trusted Autonomy. We start with an introduction to GDA in Sect. 3.2.1, describe an approach for goal selection in GDA in Sect. 3.2.2, and an application of the GDA model in a human-robot teaming task in Sect. 3.2.3.

3.2.1   Goal-Driven Autonomy

One proposed model for Goal Reasoning is that of Goal-Driven Autonomy (GDA) [25, 28], which allows an autonomous agent to introduce new goals, manage existing goals, and preempt active goals. An early instantiation of GDA is in an agent called Autonomous Response to Unexpected Events (ARTUE). Molineaux et al. show that the Goal Reasoning capabilities provided by GDA allow ARTUE to better react to unexpected events, and improve performance, versus an on-line planning system [28].

In the GDA model (shown in Fig. 3.1), an agent performs Goal Reasoning via a repeated 4-step sequence:

1. First, the agent uses a *Discrepancy Detector* to compare its observations with a set of expected observations (given by the planner). Any differences between the expected and actual observations are used to define a set of discrepancies.

Fig. 3.1 Conceptual diagram of the Goal-Driven Autonomy model

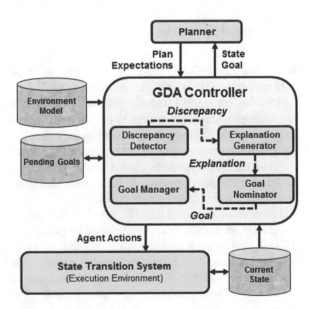

2. Next, the *Explanation Generator* creates one or more possible explanations for the set of discrepancies. Each explanation hypothesizes a possible cause for the set of discrepancies, based on the current and prior observed states.
3. Third, the *Goal Nominator* nominates a set of potentially appropriate goals, based on the generated explanation(s).
4. Finally, given the current set of pending goals and the set of newly nominated goals, the *Goal Manager* selects a subset of goals to be passed to the planner as pending goals. This may involve adding, deleting, and/or modifying the pending goals.

These four steps (Discrepancy Detection, Explanation Generation, Goal Nomination, and Goal Management) form the core of the GDA model of Goal Reasoning. Molineaux et al. [28] also contrasted this model with a conceptual model of on-line planning [30]. In both cases, the planner uses a model of the environment, a current state, and an active goal, and generates a plan to achieve that goal. However, GDA's planner also generates a set of expectations. For both models, the controller uses the generated plan to apply an action to the state transition system, and updates the current state. The primary difference between GDA and the on-line planning model is in the ability of the GDA controller to reason over a set of goals, rather than being limited to pursuing a single, set goal.

We relate GDA to the topic of Trusted Autonomy as follows. In Sect. 3.2.2 we describe a domain-independent method for goal selection, which is a subtask of goal management, and how it can be biased to choose goals that engender operator trust. Then in Sect. 3.2.3 we describe the use of a GDA agent in a simulated mission involving human-robot teaming, and the importance of transparency in that mission.

3.2.2 Goal Selection

An integral part of Goal Reasoning is the problem of goal selection (i.e., a subtask of goal management in which one or more goals are chosen for subsequent execution). Agents that are teamed directly with humans can receive operator-selected goals, or seek approval from operators to pursue their self-selected goals. However, some agents may not have access to operators in a timely fashion. For instance, an autonomous underwater vehicle (AUV) cannot communicate with operators unless it surfaces, and an interplanetary robot may require excessive time to consult operators on mission-critical decisions. In these and similar contexts, the agents' ability to intelligently select their goals is critical.

One approach for goal selection involves manually constructing knowledge bases that dictate what goals to formulate based on the agent's beliefs about the world (e.g., ARTUE [28] uses an engineered rule set that governs goal formulation). However, this approach requires extensive domain-specific knowledge engineering. It is also limiting, as the agent knows only how to respond to situations that were anticipated by the designer. The greater control afforded by hand-tuned goal selection mechanisms may appear at first to offer greater predictability for operators. However, an agent employing such a system cannot be expected to respond intelligently to situations outside its programmed knowledge.

For instance, during tests applying GDA on an AUV, we found that low-level software responsible for controlling the vehicle was at one point unable to correctly determine the vehicle's motion [41]. The vehicle deviated significantly from its intended trajectory without reporting that deviation to the GDA agent. More robust goal selection abilities offer the possibility of mitigating such failures through autonomous responses.

Some agents address the need for more robust goal selection through the use of learning. For example, agents may learn goal selection knowledge from criticism and query-answer interaction with a human expert [31], from demonstrations by human experts [40], or from Q-learning [22]. These approaches are more adaptable than manually engineered systems, but they rely on the availability of human experts or demonstration data for training. Also, agents using these approaches may perform poorly when confronted with new situations for which they were not trained.

Another approach is to control goal selection through the application of *motivators*, which encode high-level, domain-independent desires the agent wants to fulfill. We adopted this approach in Motivated ARTUE (M-ARTUE) [42], an extension of ARTUE. M-ARTUE expresses these motivators in terms of the agent's planning model; thus the motivator functions themselves are domain-independent and do not require further domain knowledge than that encoded for the agent's planner. M-ARTUE applies the following motivators to guide its behavior:

- **Social**: This encodes the desire to pursue goals provided by the agent's human operators or teammates.
- **Opportunity**: This encodes the desire to gather and conserve resources, as well as preserve the agent's possible actions in future states.
- **Exploration**: This encodes the desire to visit states that the agent has not visited previously.

To achieve broadly applicable, domain-independent implementations of these motivators, we first introduced two subfunctions:

- **Urgency**: $u_m(s_c)$ returns a numeric value representing the agent's need to fulfill a particular motivator m in the current state s_c.
- **Fitness**: $f_m(X_g)$ returns a numeric value representing how well a plan (to achieve a given goal g) fulfills a particular motivator m, using the sequence of states X_g the agent expects to visit while executing the plan.

We defined these functions to embody the following properties for each motivator:

- *Social*: This motivator's urgency increases as time passes without the agent achieving any operator goals. Its fitness expresses a high value for any plan that achieves an operator goal, and a low value for other plans.

- *Opportunity*: This motivator's urgency increases as the agent expends resources and as fewer actions are available in the current state. Its fitness expresses higher values for plans that retain high quantities of resources and many available actions, and lower values for plans that do not.
- *Exploration*: This motivator's urgency is initially high and decreases as time passes, to encourage the agent to explore early but prioritize other desires after it has gathered more information. Its fitness expresses higher values for plans that visit more new states, and lower values for plans that visit more known states.

During goal selection, M-ARTUE prefers the goal with the highest overall fitness, which is defined for each goal in the current state as the sum of the motivators' fitness scores, weighted by urgency:

$$F(X_g, s_c) = \sum_m u_m(s_c) f_m(X_g)$$

M-ARTUE's approach for goal selection was tested in a simulated Mars rover domain with hazards, with the objective of successfully maneuvering up to three rovers to given destinations around a map within a fixed number of actions. In three different levels of difficulty (controlled by the prevalence of hazards) it achieved comparable performance to ARTUE without requiring domain engineering for goal selection [42].

Other researchers have investigated the use of motivations or drives as goal selection mechanisms: several alternate approaches are described in Sects. 14.5 and 15.5. For instance, Sun [36] uses drives analogous to our motivators in CLARION; however, the drives have domain-specific aspects (e.g., Thirst and Hunger drives in CLARION contrast with resource management in our Opportunity motivator), and their preferred implementation is using a supervised back-propagating neural network. Also, the experimental focus of [36] is on cognitive plausibility, while we focus on agent performance. Merrick and Shafi [27] focus on three motivations (Achievement, Affiliation, and Power), extending a psychological theory of achievement that models competing impulses of success and failure. Unlike our work, this forgoes planning as a mechanism, instead modeling probability of success using past experiences and less domain knowledge. The authors propose inverse probability of success or socially-determined value of goals as alternatives to modeling how well a goal satisfies a particular motivation; but our motivators address the problem of determining the direct value of a goal using future predictive states. Finally, this work proposes "motive profiles" for agents (encompassing different sets of values for the model's parameters) and focuses on testing those motive profiles against the expected responses for corresponding human profiles in certain psychological tests. Baldassarre and Mirolli [6] focus on the use of the psychological theory of *intrinsic motivation* as a basis for long-term learning through exploration. This work bears some resemblance to our Exploration motivator; however, we focus primarily on acquisition of knowledge about the world and do not address the acquisition of new skills. Moreover, our system is guided by novelty of state, as contrasted with metrics of predictability or acquisition of competence.

Goal selection methods based on motivators (or other primitives) allow autonomous agents to make their own decisions in situations their designers did not anticipate. This may be viewed as a step toward greater autonomy, but does not necessarily establish a mutual understanding of trust as a concept [1]. Introducing an additional motivator that represents the desire to establish trust between the agent and its operators (or other agents) could establish a basis of understanding for trustworthiness, as well as a means for the agent to determine courses of action that would maximize human-machine trust. For example, in [17], an autonomous agent applies "inverse trust" metrics that guide a robot's behavior towards increasing the trust it receives from human operators. A motivator utilizing a similar metric, applied in goal selection, might enable a Goal Reasoning agent to become more trusted by its operators. In future research, we plan to investigate an extension of M-ARTUE that incorporates trust-related motivators.

3.2.3 An Application for Human-Robot Teaming

A solitary Goal Reasoning agent can be the sole determiner of its own tasks and goals. However, if the agent collaborates with other agents or humans as a member of a team, it needs to consider both the goals of individual teammates as well as the overall team goals. Failure to do so could lead to teammates viewing the agent as selfish (e.g., never assisting any of its teammates) or hinder the efficient achievement of team objectives (e.g., performing actions that create more work for teammates). This contrasts with a traditional multi-agent setting where other entities may provide a Goal Reasoning agent with motivations for goal change (e.g., saving an injured civilian, defending against an aggressive enemy), but they do not share any squad-level goals with the agent. It may be necessary to reason about their goals or motivations, which requires sharing team goals with the agent.

Our Autonomous Squad Member (ASM) project [18] focuses on the design and development of an extended GDA agent that controls a simulated unmanned ground vehicle, which is embedded with a detached squad that is performing surveillance tasks in a rural environment. The ASM agent (Fig. 3.2) observes the other squad members and controls its behavior accordingly.

More specifically, the ASM agent continuously monitors the behavior of its teammates to identify their current goals. Using its current sensory inputs (i.e., observations of the environment or spoken dialog detected by its *Natural Language Classifier*), the agent's *Explanation Generator* attempts to explain what actions each teammate must have performed for the environment to be in its current state (i.e., explain the most recent actions that were observed). The actions of each teammate are used by the agent's *Plan Recognizer* to recognize their respective plans and associated goals (i.e., predict their future actions based on previously observed actions). If the ASM agent determines that some, or all, of the teammates have changed their goals, it can change its goal in response, using its *Goal Selector*. For example, if the teammates were *patrolling* but are now *retreating*, it can use that information to

Fig. 3.2 The Autonomous
Squad Member (ASM)
agent's conceptual design

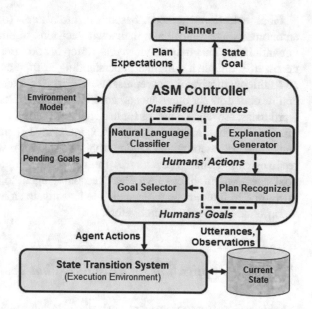

reason that there must be a threat that the teammates have observed but the agent has not, so it should also retreat. Similarly, the agent can also modify its goals in response to an opportunity or an unexpected external event (i.e., something it perceives in the environment). The ASM agent's current goal is used to control its behavior by generating a plan (i.e., using the *Planner*) and executing its actions in the environment.

The ASM model can be contrasted with the GDA model introduced in Sect. 3.2.1 and shown in Fig. 3.1, with several notable differences. The ASM model specifically differentiates natural language utterances from other state observations and uses the *Natural Language Classifier* to pre-process them before they are provided as input to the *Explanation Generator*. Explanation in the ASM model is not a single process but is done at two levels, with the *Explanation Generator* explaining what actions each teammate must have performed and the *Plan Recognizer* reasoning about what plans they must be performing. Similarly, in the ASM model, discrepancy detection is not a single module but happens throughout the system. For example, discrepancies about expected and observed states are handled in the *Explanation Generator*, while discrepancies about past and current goals of teammates are handled in the *Goal Selector*. The *Goal Selector* itself differs from the GDA model since it only performs a subset of the duties of the GDA model's *Goal Manager*.

To support Trusted Autonomy, we are integrating the ASM agent with a user interface that adds transparency between the agent and a human operator. This interface uses the Situation awareness-based Agent Transparency (SAT) model [9], a transparency model that attempts to reduce user overhead, provide situational awareness, and allow for appropriate calibration of trust in the agent. The SAT interface provides three levels of transparency information: the agent's status (e.g., current state, goals, plans, physical location), the agent's reasoning process (e.g., what motivated it to perform its current task), and the agent's projections (e.g., future environment states, future resource levels). Each transparency element is presented using icons on the user interface, allowing the user to quickly and intuitively process information and identify changes in the agent's behavior.

In this application of SAT, we are displaying to the user interface information pertaining to the ASM agent's location in its environment, its current goal, its current task, and influence factors pertaining to why it selected that goal and is operating on that task. Our objective is to demonstrate the benefits of the SAT model in a real-time environment in mission-critical situations.

3.3 Goal Refinement

Section 3.2 described some of our group's work on GDA, a simple model of Goal Reasoning. In this section, we describe our work on goal refinement, which is a more comprehensive Goal Reasoning model. We begin in Sect. 3.3.1 by describing its basic concepts and realization in the Goal Lifecycle. In Sect. 3.3.2 we present an architecture for ensuring guarantees on the behavior of agents that employ this model. Finally, we describe its application to a distributed robotics task in Sect. 3.3.3.

3.3.1 Goal Lifecycle

Our group defined a second model for Goal Reasoning, based on the concept of goal refinement, which we call the Goal Lifecycle [33]. Goal refinement, an extension of plan refinement [24], models the progressive refinement of goals through the addition of constraints. This is visualized in the Goal Lifecycle shown in Fig. 3.3 [33]. In this model, individual goals transition through stages of increasingly detailed *modes* by activating a series of refinement *strategies*. For example, goals and their initial constraints are introduced using the *formulate* strategy, while the *expand* strategy concerns the automated generation of plans for a given goal.

Briefly, the refinement strategies are:

1. *Formulate*: This creates a new goal and enters it into the Goal Lifecycle by defining its initial constraints, criteria, and prerequisites.
2. *Select*: This chooses which goal(s) to actively pursue; it ensures that the goals' prerequisites are met and that the agent has the needed resources to pursue them.
3. *Expand*: This generates one or more expansions (i.e., plans) to achieve a given goal, along with a set of expectations for each.
4. *Commit*: This picks a single expansion to pursue from the set of expansions created by the expand strategy.
5. *Dispatch*: This executes the committed expansion and defines the criteria by which a goal can be evaluated during execution.

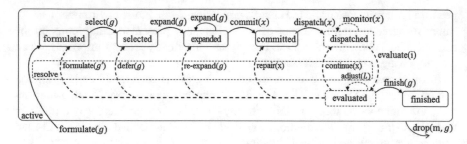

Fig. 3.3 The Goal Lifecycle [33] depicts the application of strategies to transition goals (and their associated constraints) through a sequence of modes. While the top-level strategies progress goals towards completion, the set of *resolve* strategies (e.g., *re-expand*) support goal adaptation, deferment, and reformulation

In addition to these strategies, which progressively add detail to the goal's definition, the Goal Lifecycle includes a set of strategies for detecting and reacting to events and changes during execution. After being dispatched, each goal can be actively *monitored* and, if problems are detected, or if an unexpected event occurs, the goal can be *evaluated*. As a result of this evaluation, the system may elect to *continue* the goal as is, *drop* the goal (as either completed or failed), or attempt to resolve the detected problems through one of several strategies (e.g., *repair*, *defer*). *Resolve* strategies transition a goal to an earlier mode before execution resumes.

In contrast to the GDA model, goal refinement provides a more explicit representation of the context in which a goal is pursued by a Goal Reasoning agent. This has benefits that relate to Trusted Autonomy. For example, contextual constraints can be used to guarantee that agents will behave according to a given specification, and can also be used to more clearly deliberate on and communicate details of their reasoning (i.e., for selecting the next strategy to apply). We discuss these topics in the following sections.

3.3.2   Guaranteeing the Execution of Specified Behaviors

Once a Goal Reasoning agent is provided with or self-selects a goal to pursue, it must use some combination of planning and control algorithms to undertake the actions necessary to achieve it. Careful design of these components can provide valuable capabilities for a Goal Reasoning agent. Here we describe the Situated Decision Process (SDP), which manages and executes goals for a team of autonomous vehicles. A more thorough description of the SDP and its components can be found in [33], and we describe an application of the SDP in Sect. 3.3.3.

The SDP (Fig. 3.4) takes as input goal updates (e.g., commands) from an operator and passes them to the Mission Manager, which performs Goal Reasoning operations using the Goal Lifecycle described in Sect. 3.3.1. Once the Mission Manager selects

Fig. 3.4 A conceptual design of the Situated Decision Process (SDP). The Mission Manager performs Goal Reasoning operations. It creates a schedule of actions for a team of vehicles, each of which operates a synthesized Finite State Automaton

a goal, it dispatches an expansion to the vehicles by creating a schedule of commands for them and passing that schedule to the Coordination Manager. The Coordination Manager interprets the schedule and passes the applicable commands to a Team Executive, which then assigns the commands to individual vehicles.

Each vehicle interprets their command as an input to a Finite State Automaton (FSA), which is automatically synthesized using a template. This template specifies the regions where the behaviors are to be executed, ensuring they are active in only appropriate areas, as well as any *mission sensors* that cause an automatic switching between behaviors when the vehicle observes a particular event. This yields a play-calling architecture, detailed in [5], which provides guarantees on the execution of the goals chosen by the Mission Manager. The execution of each command is predicated on the satisfaction of a set of pre-defined *health sensors*, which establish required conditions before the vehicle can pursue the commanded goal (e.g., the vehicle must have a sufficient amount of fuel to reach the goal location). If one or more of the health sensors are not satisfied, the FSA activates a *contingency behavior* that causes the vehicle to engage in a behavior aimed towards maintaining safety (e.g., landing an air vehicle) or fixing the health sensor (e.g., returning to a base station to refuel).

The FSAs used by the vehicles are synthesized from a temporal logic specification, and are guaranteed to satisfy this specification. Full details on the synthesis process we use in the SDP can be found in [26]. Briefly, the behavior of the vehicle is specified as a Linear Temporal Logic (LTL) formula:

$$\phi = \varphi_e \rightarrow \varphi_s, \tag{3.1}$$

where the behavior of system φ_s is specified in reaction to changes in environment φ_e. The specified behavior of the environment and system includes three components:

- The initial state: $\varphi_i^{\{e,s\}}$
- A set of safety constraints that restrict the transitions of the system: $\varphi_t^{\{e,s\}}$
- A set of goal conditions that must be satisfied infinitely often: $\varphi_g^{\{e,s\}}$

Formula ϕ is specified over a set of Boolean propositions that represent the state of the environment, as sensed by the vehicle, and the state of the system. The resulting FSA is synthesized automatically, in a manner that guarantees that the transitions given by the FSA will satisfy ϕ. Coupled with the play-calling templates that are used to generate the specification, the resulting FSA is guaranteed to activate behaviors that pursue the commanded goal, whenever its observed internal and external state allow it to do so.

This framework relates to Trusted Autonomy. While the SDP can adjust its goals autonomously, the pursuit of those goals is constrained to abide by specific guarantees. Such guarantees will affect how such an agent or group of agents is viewed and trusted within a larger system.

3.3.3   A Distributed Robotics Application

One interesting question concerning Goal Reasoning and Trusted Autonomy is how to design a Goal Reasoning system that clearly communicates how and why it chooses to change its goals. We have addressed this in Goal Reasoning with Information Measures (GRIM), a Goal Reasoning system that instantiates the Goal Lifecycle, described in Sect. 3.3.1. GRIM, which is a derivation of the SDP (Sect. 3.3.2), employs a single measure for assessing goal performance and communicates this to an operator. We briefly describe an application of GRIM here; a more complete description can be found in [23].

We applied GRIM to a simulated disaster relief scenario (Fig. 3.5) where a team of two autonomous vehicles must survey a set of regions to locate a local official and establish communications. Each of these three regions (labeled as an Airport and two Office Buildings) corresponds to an individual survey goal within the Goal Lifecycle, and is surveyed by following a series of waypoints. The Goal Reasoning process in GRIM is then framed with respect to the uncertainty left in the area survey, which is defined as the length of the search pattern that has yet to be traversed by the vehicles.

Figure 3.6 displays a graphical representation of four of the Goal Lifecycle strategies. In Fig. 3.6a, each of the three survey goals is *formulated* by generating constraints on the maximum allowable uncertainty over time. After each of the goals is *formulated*, GRIM *selects* a single goal (the Airport survey goal) to pursue, based on the constraints of each goal. The *selected* goal is then *expanded* by generating a set of plans to achieve it. The expectations of these plans (depicted as a change in the uncertainty over time) are shown in Fig. 3.6b. GRIM then *commits* to a single expansion, and *dispatches* that expansion to the vehicles. The expectations for the expansion, and a set of performance bounds that are generated as part of the *dispatch* strategy, are shown in Fig. 3.6c. Finally, Fig. 3.6d displays the execution performance over time, as obtained by the *monitor* strategy.

During execution, when the vehicle's performance is predicted to violate a goal constraint (as occurs in Fig. 3.6d when its ongoing execution reaches the worst-case

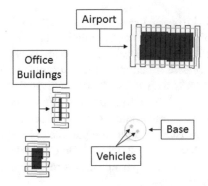

Fig. 3.5 Map of the survey regions for the disaster relief scenario used to demonstrate GRIM. Each of the three survey regions is covered by a waypoint pattern that the vehicles follow to search for a local official

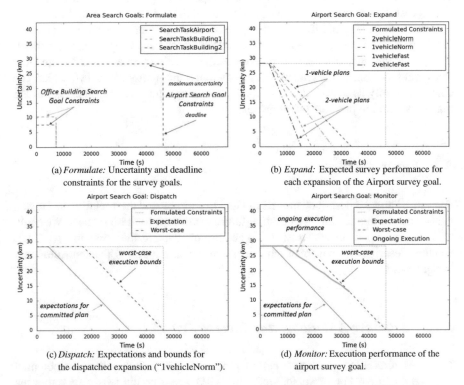

Fig. 3.6 Plots of selected strategies from the execution of GRIM for the disaster relief scenario

execution bound), GRIM triggers the *evaluate* strategy to determine what violation occurred. If the execution satisfies the completion criteria, the goal is marked as completed and *dropped*. If it instead violates the constraints on the goal, the goal is marked as failed and *dropped*. If neither of these has occurred, but the performance violates the execution bounds, a *resolve* strategy is activated in an attempt to adjust the goal (or its expansion) before continuing execution. The selected *resolve* strategy can transition the goal back to an earlier mode in the Goal Lifecycle. For example, it may *repair* the committed expansion by adjusting parameters that affect the expectations and bounds. Alternately, a *resolve* strategy may force GRIM to completely *re-expand* the goal to obtain a new set of expansions, before proceeding to *commit* to and *dispatch* one of the new expansions.

For further details on the operation of GRIM, please see [23], which describes an ablation study with the *resolve* strategies. We showed that they allow GRIM to perform Goal Reasoning during execution, improve its performance, and enable it to successfully complete more goals under uncertain and changing conditions.

Associating the Goal Lifecycle strategies with a single metric, as is done in GRIM, can be useful for multiple reasons. For example, it can be used to define clear decision points that increase the transparency of the decision process used by the Goal Reasoning system. For a system that has as much autonomy as GRIM, which can change not only its plans but also its goals, transparency in how those decisions are made may help to promote operator trust.

3.4 Future Topics

Section 3.2 and 3.3 described models of Goal Reasoning and their relation to Trusted Autonomy. In this section, we describe future extensions to these models. We begin with a discussion on inverse trust and its support of adaptive autonomy in Sect. 3.4.1, and then describe the concept of rebel agents and their relation to Trusted Autonomy in Sect. 3.4.2.

3.4.1 Adaptive Autonomy and Inverse Trust

The ASM and GRIM agents have certain properties that, arguably, have the potential to engender trust. For example, since the ASM agent continuously monitors the behavior of teammates, it can rapidly respond to any changes in their plans or goals. Similarly, since the GRIM agent monitors the progress of controlled vehicles with respect to goal constraints, it can automatically apply a *resolve* strategy when it recognizes that a constraint is projected to be violated. However, neither ASM nor GRIM agents use specific mechanisms to build or maintain operator trust.

Traditional computational trust metrics [35] are used to measure how much trust an agent has in another agent using information from past interactions or third-party feedback. These metrics allow an agent embedded in a human team to measure its trust in other teammates but do not allow it to measure how trustworthy it is from its teammates' perspective. For this reason, we developed an *inverse trust metric* [15] that allows an agent to estimate its own trustworthiness. While many factors that influence human-robot trust (discussed in more detail in Sect. 7.6) are not directly observable to an agent (e.g., a teammate's experience with other agents, or a teammate's internal evaluation of an agent), factors that are observable, such as the agent's performance, have been found to have the greatest influence on trust [19].

The inverse trust estimate allows an agent to evaluate its own performance and use that information to estimate the corresponding influence on its trustworthiness (i.e., increasing, decreasing, constant). Based on this estimate, an agent can reason about whether its current behavior is trustworthy or untrustworthy. In situations where the agent believes its behavior is untrustworthy, it can modify its behavior in an attempt to learn (and apply) a more trustworthy behavior, thus implementing a form of adaptive autonomy. Preliminary studies in limited simulations have shown that an agent using an inverse trust method can successfully adapt its behavior given implicit feedback [15], and can benefit further from explicit feedback [16] as well as the ability to generate explanations when it modifies its behaviors [14].

The primary benefit of our approach is that it gives the agent control over maintaining trust and does not require an exhaustive engineering effort to develop behaviors that will be trustworthy for all teammates, in all environments, and in all contexts. However, we have not yet integrated it with a Goal Reasoning agent, and to date it has only been examined in the context of an agent whose goals are static. Our plans for future work include testing variants of inverse trust in a Goal Reasoning agent in environments in which the operator can specify a variety of goals to achieve, and where unexpected situations can arise (thus motivating the need for self-selection of goals or recommendation of goal changes to the operator). We expect that our studies will demonstrate the utility of adaptive autonomy in interactive Goal Reasoning agents.

3.4.2 Rebel Agents

Rebel agents [10] represent a relatively novel research direction in the context of Goal Reasoning. Rebel agents can object to, or even completely reject, goals or associated courses of action that are assigned to them by external agents (human or artificial), and they challenge the general attitudes or behaviors of those other agents. For example, an operator may command a rebel agent to pursue a specified goal without knowledge of the agent's context or access to its information sources, in which case the agent may respond with a recommendation for an alternative goal (along with an explanation).

Several situations exist in which modeling a rebel agent that can adjust its goals (or plans) can be viewed as beneficial, including the following:

- *Divergent information sources*: The agent may have access to information currently not available to the operator, which requires immediate action that is incompatible with the assigned goal.
- *Moral conflict*: The agent may be endowed with a "moral conscience" model that conflicts with an assigned goal. This can be a factor for protest in human-robot interaction [8].
- *Diversity*: The agent may be intended to contribute to the diversity of its team so as to ensure that sufficiently varied points of view and alternative goals are considered. This direction is inspired by studies claiming that diverse teams tend to outperform non-diverse teams [21, 37] under certain circumstances.
- *Self-assessment*: The agent may assess its assigned task as not being a good match for its capabilities. This relates to studies in personality psychology where the strengths-based leadership approach [32] argues that every person (i.e., leaders or other team members) should be offered the opportunity to routinely conduct activities in line with their strengths.
- *Believability*: An agent playing a character in an interactive narrative or training simulation may be given (and refuse) a goal that undermines its believability [7].

Several of these situations assume that the agent has an internal motivation model that conflicts with an assigned goal (or plan). This model can be based on many factors, such as simulated memory, emotion, or social relationships. The agent's attitude towards a goal can change over time (due, for example, to changes in the environment or in the agent's knowledge of an operator's motivation), leading to incremental increase or decrease in the agent's inclination to rebel. The rebel agents may or may not be "aware" that they are rebelling; i.e., they may not be able to reason about the social implications and potential consequences of rebellion. For those agents that *are* rebellion-aware (e.g., social planning agents, as described in Chap. 4), an inner conflict may emerge between the drive to rebel based on the agent's own motivating factors and the anticipated consequence of rebellion.

In one definition of Trusted Autonomy, Abbass et al. [1] express trust in terms of vulnerability. A moment of rebellion is inherently one of vulnerability. By rebelling, an agent (1) makes itself vulnerable, and (2) creates vulnerability in the system it is (or was originally) part of. This anticipates some ways in which trust can be a factor for rebellion, such as the following:

- *Self-trust*: The agent's amount of trust in itself (e.g., its trust that it can accurately assess the current situation as warranting opposition).
- *Perceived trust*: The agent's model of how much other agents trust its judgment (e.g., based on its perceived expertise).
- *Risk*: The degree to which the agent trusts other agents to handle the vulnerabilities that rebellion creates for the rebel agent and the entire system.
- *Distrust*: The agent's distrust of other agents (i.e., its belief that it cannot entrust its vulnerabilities to them if the goal is to be achieved).

The relation between rebellion and trust is multifaceted. Some ways in which rebellion can impact trust include:

1. Rebellion can diminish the trust of other agents in the rebel agent (e.g., an operator may lose trust in an agent that refuses to pursue an assigned goal/objective).
2. Rebellion can increase the trust of other agents in the rebel agent (e.g., a rebel agent that displays expertise when rejecting a goal, by raising objections when appropriate, may be more trusted to act autonomously).
3. The way in which other agents behave following a situation of rebellion can impact the trust of the rebel agent in those other agents.

In our future work on Goal Reasoning agents that can rebel, we plan to first test them in scenarios in which agents may rebel because their operator does not have complete access to their current state (including sensor data). We wish to model methods through which the agent learns how to explain its rebellion and negotiate with its operator so as to maximize its confidence that it is pursuing a well-justified goal. For example, this may involve soliciting an explanation from the operator as to why a goal should be pursued, or providing an argument for rejecting it.

3.5 Conclusion

The topic of Goal Reasoning is an important one in Robotics, Intelligent Agents, and Artificial Intelligence. Creating autonomous systems that can deliberate on and change their own goals allows those systems more freedom to intelligently adapt their behaviors to unexpected events and changing conditions. This chapter presented two different models of Goal Reasoning. First, we presented the GDA model and a related method for goal selection using motivators, as well as an application of GDA in a human-robot teaming task. We also presented a model based on goal refinement, instantiated in the Goal Lifecycle, and discussed an architecture for placing guarantees on the behavior of the agents as well as an application in multi-agent robotics. Finally, we discussed two ongoing extensions to our Goal Reasoning work: adaptive autonomy using inverse trust and the study of rebel agents.

Goal Reasoning also has a number of close connections to the topic of Trusted Autonomy. Goal Reasoning agents can modify and change their goals autonomously, in addition to adjusting how they achieve those goals. As such, any trust of such systems must inherently extend to their capability to reason at the goal level. Goal Reasoning also provides opportunities to design systems that cultivate trust through transparency or through goals that actively account for the trust of other agents.

Acknowledgements We would like to thank Mark Roberts and Thomas Apker for their assistance with this chapter, as well as much of the underlying research. We would also like to thank NRL, ONR, and OSD ASD (R&E) for sponsoring our group's work on this topic.

References

1. H.A. Abbass, E. Petraki, K. Merrick, J. Harvey, M. Barlow, Trusted autonomy and cognitive cyber symbiosis: open challenges. Cogn. Comput. **8**(3), 385–408 (2016)
2. D.W. Aha (ed.), *Goal Reasoning: Papers from the ACS Workshop (Technical Report GT-IRIM-CR-2015-001)* (Georgia Institute of Technology, Institute for Robotics and Intelligent Machines, Atlanta, GA, USA, 2015)
3. D.W. Aha, M.T Cox, H. Muñoz-Avila (eds.), *Goal Reasoning: Papers from the ACS Workshop (Technical Report CS-TR-5029).* University of Maryland, Department of Computer Science, College Park, MD, USA, 2013
4. D.W. Aha, M. Klenk, H. Muñoz-Avila, A. Ram, D. Shapiro (eds.), *Goal-Directed Autonomy: Notes from the AAAI Workshop* (AAAI Press, Atlanta, GA, USA, 2010)
5. T.B. Apker, B. Johnson, L. Humphrey, LTL templates for play-calling supervisory control, in *Proceedings of AIAA Science and Technology Exposition* (2016)
6. G. Baldassare, M. Mirolli (eds.), *Intrinsically Motivated Learning in Natural and Artificial Systems* (Springer, Berlin, 2013)
7. J. Bates, The role of emotions in believable agents. Commun. ACM **37**(7), 122–125 (1994)
8. G. Briggs, M. Scheutz, "Sorry, I can't Do That": developing mechanisms to appropriately reject directives in human-robot interactions, in *2015 AAAI Fall Symposium Series* (2015)
9. J.Y.C. Chen, M.J. Barnes, A.R. Selkowitz, K. Stowers, S.G. Lakhmani, N. Kasdaglis, Human-autonomy teaming and agent transparency, in *Proceedings of the 21st International Conference on Intelligent User Interfaces* (ACM, Sonoma, USA, 2016), pp. 28–31
10. A. Coman, K. Gillespie, H. Muñoz-Avila, Case-based local and global percept processing for Rebel Agents, in *Proceedings of the 23rd Annual International Conference on Case-Based Reasoning, CEUR Workshop*, ed. by J. Kendall-Morwick (2015), pp. 23–32
11. M.T. Cox, Goal-driven autonomy and question-based problem recognition, in *Proceedings of 2nd Annual Conference on Advances in Cognitive Systems* (Baltimore, MD, USA, 2013), pp. 29–45
12. M.T. Cox, Z. Alavi, D. Dannenhauer, V. Eyorokon, H. Muñoz-Avila, D. Perlis, *MIDCA: A Metacognitive, Integrated Dual-Cycle Architecture for Self-Regulated Autonomy* (2016), pp. 3712–3718
13. D. Dannenhauer, H. Muñoz-Avila, Raising expectations in GDA agents acting in dynamic environments, in *Proceedings of the International Joint Conference on Artificial Intelligence* (AAAI Press, Buenos Aires, Argentina, 2015), pp. 2241–2247
14. M.W. Floyd, D.W. Aha, Incorporating transparency during trust-guided behavior adaptation, in *Proceedings of the 24th International Conference on Case-Based Reasoning* (Springer, Atlanta, GA, USA, 2016)
15. M.W. Floyd, M. Drinkwater, D.W. Aha, How much do you trust me? Learning a case-based model of inverse trust, in *Proceedings of the 22nd International Conference on Case-Based Reasoning* (Springer, Cham, 2014), pp. 125–139
16. M.W. Floyd, M. Drinkwater, D.W. Aha, Improving trust-guided behavior adaptation using operator feedback, in *Proceedings of the 23rd International Conference on Case-Based Reasoning* (Springer, Frankfurt, Germany, 2015), pp. 134–148
17. M.W. Floyd, M. Drinkwater, D.W. Aha, Trust-guided behavior adaptation using case-based reasoning, in *Proceedings of the Twenty-Fourth International Joint Conference on Artificial Intelligence* (AAAI Press, 2015)
18. K. Gillespie, M. Molineaux, M.W. Floyd, S.S. Vattam, D.W. Aha. Goal reasoning for an autonomous squad member, in *Proceedings of the Workshop on Goal Reasoning (held at the 3rd Conference on Advances in Cognitive Systems)*, ed. by D.W. Aha (Georgia Institute of Technology, Institute for Robotics and Intelligent Machines, 2015), pp. 52–67
19. P.A. Hancock, D.R. Billings, K.E. Schaefer, J.Y.C. Chen, E.J. De Visser, R. Parasuraman, A meta-analysis of factors affecting trust in human-robot interaction. Hum. Factors J. Hum. Factors Ergon. Soc. **53**(5), 517–527 (2011)

20. N. Hawes, A survey of motivation frameworks for intelligent systems. Artif. Intell. **175**(5), 1020–1036 (2011)
21. L. Hong, S.E. Page, Groups of diverse problem solvers can outperform groups of high-ability problem solvers. Proc. Natl. Acad. Sci. U.S.A. **101**(46), 16385–16389 (2004)
22. U. Jaidee, H. Muñoz-Avila, D.W. Aha, Case-based goal-driven coordination of multiple learning agents, in *Proceedings of the 21st International Conference on Case-Based Reasoning*, Saratoga Springs, NY, USA, 2013, pp. 164–178
23. B. Johnson, M. Roberts, T. Apker, D.W. Aha, Goal reasoning with information measures, in *Proceedings of the 4th Annual Conference on Advances in Cognitive Systems*, Evanston, IL, USA, 2016
24. S. Kambhampati, C.A. Knoblock, Q. Yang, Planning as refinement search: a unified framework for evaluating design tradeoffs in partial-order planning. Artif. Intell. **76**(1), 167–238 (1995)
25. M. Klenk, M. Molineaux, D.W. Aha, Goal-driven autonomy for responding to unexpected events in strategy simulations. Comput. Intell. **29**(2), 187–206 (2013)
26. H. Kress-Gazit, G.E. Fainekos, G.J. Pappas, Temporal-logic-based reactive mission and motion planning. IEEE Trans. Robot. **25**(6), 1370–1381 (2009)
27. K.E. Merrick, K. Shafi, Achievement, affiliation, and power: motive profiles for artificial agents. Adapt. Behav. **19**(1), 40–62 (2011)
28. M. Molineaux, M. Klenk, D.W. Aha, Goal-driven autonomy in a navy strategy simulation, in *Proceedings of the 24th AAAI Conference on Artificial Intelligence* (AAAI Press, Atlanta, GA, USA, 2010)
29. H. Muñoz-Avila, D.W. Aha, U. Jaidee, M. Klenk, M. Molineaux. Applying goal driven autonomy to a team shooter game, in *Proceedings of the 23rd International FLAIRS Conference*, Daytona Beach, FL, USA, 2010, pp. 465–470
30. D. Nau, Current trends in automated planning. AI Mag. **28**(4), 43–58 (2007)
31. J. Powell, M. Molineaux, D.W. Aha, Active and interactive learning of goal selection knowledge, in *Proceedings of the Twenty-Fourth Florida Artificial Intelligence Research Society Conference* (AAAI Press, 2011)
32. T. Rath, B. Conchie, *Strengths Based Leadership: Great Leaders, Teams, and Why People Follow* (Simon and Schuster, New York, 2008)
33. M. Roberts, T. Apker, B. Johnson, B. Auslander, B. Wellman, D.W. Aha, Coordinating robot teams for disaster relief, in *Proceedings of the 28th International FLAIRS Conference*, Hollywood, Florida, USA, 2015
34. M. Roberts, D. Borrajo, M.T. Cox, N. Yorke-Smith (eds.), *Goal Reasoning: Papers from the IJCAI Workshop*, New York, NY, USA, 2016
35. J. Sabater, C. Sierra, Review on computational trust and reputation models. Artif. Intell. Rev. **24**(1), 33–60 (2005)
36. R. Sun, Motivational representations within a computational cognitive architecture. Cogn. Comput. **1**(1), 91–103 (2009)
37. J. Surowiecki, *The Wisdom of Crowds* (Doubleday, New York, 2004)
38. Swaroop Vattam, Matthew Klenk, Matthew Molineaux, and David W Aha. Breadth of approaches to goal reasoning: a research survey, in *Goal Reasoning: Papers from the ACS Workshop (Techinical Report CS-TR-5029)*, ed. by D.W. Aha, M.T. Cox, H. Muñoz-Avila (University of Maryland, College Park, MD, 2013), p. 111
39. B.G. Weber, M. Mateas, A. Jhala, Applying goal-driven autonomy to starCraft, in *Proceedings of the 6th AAAI Conference on Artificial Intelligence and Interactive Digital Entertainment*, ed. by G. Michael Youngblood, V. Bulitko (AAAI Press, Stanford, CA, USA, 2010), pp. 101–106
40. B.G. Weber, M. Mateas, A. Jhala, Learning from demonstration for goal-driven autonomy, in *Proceedings of the 26th AAAI Conference on Artificial Intelligence*, Toronto, Canada, 2012, pp. 1176–1182
41. M. Wilson, J. McMahon, A. Wolek, D.W. Aha, B.H. Houston, Toward goal reasoning for autonomous underwater vehicles: responding to unexpected agents, in *Goal Reasoning: Papers from the IJCAI Workshop*, ed. by M. Roberts, D. Borrajo, M.T. Cox, N. Yorke-Smith (2016)

42. M. Wilson, M. Molineaux, D.W. Aha, Domain-independent heuristics for goal formulation, in *Proceedings of 26th International FLAIRS Conference* (AAAI Press, St. Pete Beach, USA, 2013), pp. 160–165

Chapter 4
Social Planning for Trusted Autonomy

Tim Miller, Adrian R. Pearce and Liz Sonenberg

4.1 Introduction

Early work on Trusted Autonomy (See Sect. 4.5) introduced the term *social autonomy* (See Chap. 1) to capture the idea that to be coordinated with other agents or keep its commitments, an agent must relinquish some of its autonomy, but that an agent that is sociable and responsible can still be autonomous: it would attempt to coordinate with others where appropriate and keep its commitments as much as possible, but it would exercise its autonomy in entering into those commitments in the first place [1].

It has been argued that human-machine trust can enhance performance in complex situations [2], and while we acknowledge there are many unanswered questions about the relationship between human-human trust, and human-machine trust, especially in the context of technology advances impacting machine capability for autonomy [3], we adopt the hopefully uncontroversial perspective that successful human-agent interaction demands that the agent behaves in an intuitive and explainable way from the perspective of the human.

So the work described here, on computational mechanisms for constructing and representing explainable plans in human-agent interactions, addresses one aspect of what it will take to meet the requirements of a trusted autonomous system. In turn, such properties are essential to enable the deployment of autonomous systems from the laboratory into production, such as in manufacturing assembly environments, assistive robotics, disaster management, defence applications, and self-driving cars.

Consider a simple example of a self-driving car that receives information that a road on its planned route is blocked. Re-planning the route to take a different road is straightforward, but the autonomy in the car should inform the passengers of this so

T. Miller (✉) · A. R. Pearce · L. Sonenberg
Department of Computing and Information Systems,
University of Melbourne, Melbourne, VIC 3010, Australia
e-mail: tmiller@unimelb.edu.au

A. R. Pearce
e-mail: adrianrp@unimelb.edu.au

L. Sonenberg
e-mail: l.sonenberg@unimelb.edu.au

© The Author(s) 2018
H. A. Abbass et al. (eds.), *Foundations of Trusted Autonomy*, Studies in Systems,
Decision and Control 117, https://doi.org/10.1007/978-3-319-64816-3_4

that they understand why an unusual route is taken. However, the autonomy should not inform the passengers if they are aware of this road closure already; for example, on the return trip.

We assert that scenarios such as this require *social planning*. Social planning is automated planning in which the planning agent maintains and reasons with an explicit model of the humans with which it interacts, including the human's goals, intentions(See Sect. 6.6), and belief, as well as their potential behaviours. Indeed, humans themselves use these concepts to make decisions that are intuitive, explainable, and acceptable to other people. This phenomenon is known as *Theory of Mind*(See Sect. 15.8), a term introduced by Premack and Woodruff in the context of the study of animal behaviour [4] and widely used since in philosophy, psychology amd cognitive science, e.g. [5, 6].

The state-of-the-art in artificial intelligence offers limited foundations on such constructs. Indeed, as articulated recently, challenges for artificial intelligence in the delivery of systems that can operate autonomously under some conditions, but cannot always complete an entire task on their own (so-called *semi-autononmous systems*), include the development of realtime activity and intent recognition techniques, the design of representations for human actions that are usable in the context of automated planning, integrated with interfaces that facilitate communication and transfer of control between the human and the machine, and supported by novel execution architectures [7]. The work described in this chapter addresses (in part) the second and fourth of these issues.

Specifically, we seek to build artificial agents that are able to fluidly operate in complex dynamic environments with humans, interacting in a 'human-intuitive' manner. We are developing building blocks towards the design of non-human agents whose actions can be trusted and understood by humans, and towards approaches that take these factors into account when designing the collaboration with humans.

The structure of this Chapter is as follows. So far we have offered an overview of the challenge of planning in human-agent teams, with a specific focus on social planning as one way to increase transparency and explainability, and hence a critical enabler of trust. Section 4.1 provides some high level background on (classical) planning and the motivation for social planning. Section 4.3 includes an introduction to a recent body technical work by the authors and collaborators in social planning - specifically in multi-agent epistemic planning [8–15]. In Sect. 4.4, we present two scenarios that illustrate the benefits of planning in the presence of nested belief reasoning and first-person multi-agent planning, hence indicating how social planning could be used as a means for planning human-agent interaction explicitly as part of the 'deliberation' cycle. Section 4.5 offers some brief summary remarks.

4.2 Motivation and Background

In this section, we outline some background material required to understand the chapter, as well as some motivation for our work.

Fig. 4.1 Conceptual model
of AI planning (from [16])

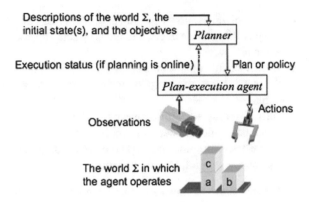

4.2.1   Automated Planning

Planning research in classical planning has yielded highly efficient mechanisms for
plan synthesis suiting single-agent scenarios. Figure 4.1 outlines a conceptual model
of AI planning. A planning problem is formulated as a tuple $\langle F, \mathcal{I}, \mathcal{G}, \mathcal{A} \rangle$, with the
following meanings:

1. F is a set of Boolean fluents describing the objects within the world of interest;
2. $\mathcal{I} \subseteq F$ is the initial state represented as the set of Boolean fluents that are true in
   the world before the plan-execution agent performs any actions;
3. $\mathcal{G} \subseteq F$ is a set of fluents describing the desired objectives, such as achieving a
   goal or performing a specific task; and
3. $\mathcal{A}$ is a set of actions, described as a pair containing a precondition specifying the
   fluents that must be true for that action to be executed, and the effects that action
   will have on the world, described as fluents that will become true or false.

The output consists of either a plan (a sequence of actions for the agent to perform)
or a policy (an action to perform for each reachable state).

A simple and commonly supported extension to classical planning is *conditional
effects*. A conditional effect of an action is of the form $(C \rightarrow l)$, in which C is a set
of fluents representing a condition, and l is a single fluent. The informal semantics
of such an effect is that *if* C held before the action was executed, then l holds after
the action is executed. A single action can have multiple such conditional effects.

Much research over the last three decades has focused on the problem of *offline
classical planning*, proposing compact state and transition encodings and effective
domain-independent heuristics. This has led to massive improvements in classical
planning tools, which can solve problems with hundreds of actions and large state
spaces ($\approx 2^{1000}$ states) from several milliseconds to just a few hours.

However, classical planning is the simplest of the domain-independent planning
problems, as it assumes the following:

- Deterministic events: The effects of all actions are deterministic — there is only one possible set of effects, and those effects happen each time the action is applied in the real world.
- Worlds change only as the result of an action: the only manner in which a world changes is when the planning agent executes an action — the world is otherwise static.
- Fully observable (omniscience): the state of the world is always fully observable — as such, when an action is applied, the agent can see the effects fully.
- Single actor (omnipotence): There are no other agents in the world, either cooperative, adversarial, and ambivalent.

Clearly, none of these assumption hold in the setting of real-world autonomy. However, more recently, research in the area of automated planning has focused on relaxing the problem description to enable a wider range of problems to be specified. In particular, planning in non-deterministic [17] and partially-observable [18] domains has matured to the point in which many problems in these domains can be solved efficiently offline, producing robust policies for execution. A key part of almost all solutions in this area is that a classical planning tool is used to solve part of the richer, underlying problem.

However, most planning research to date is still lacking in one key area: the consideration of other agents (human or otherwise) in the domain.

4.2.2 From Autistic Planning to Social Planning

To move into planning into multi-agent environments, agents must move out of the so-called *autistic* realm and into the *social* realm [19]. This means that a single agent reasoning in a multi-agent environment must have a Theory of Mind, considering the possible behaviours and mental states of others in the environment.

Building on recent analysis by Bolander and Herzig [20], we note that extending classical planning to the multi-agent case presents many new challenges:

1. Planners must track beliefs (or knowledge) of other agents, which are typically incomplete and only partially correct.
2. These beliefs include higher-order beliefs; that is, beliefs about other agents' beliefs about other agents' beliefs, etc. (as in Fig. 4.2).
3. Other agents have their own goals and intentions, which may be cooperative or competitive with our own, and these goals and intentions direct their actions, which influence our ability to achieve our own goals.
4. Chosen actions should be plausible or acceptable from the perspective of other agents; for example, in an adversarial setting in which an agent is attempting to conceal their real identity, their actions must conform to the identity attributed to them by their adversaries.

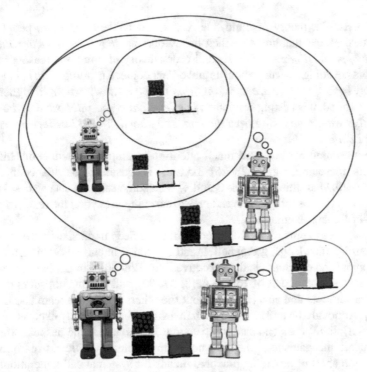

Fig. 4.2 Tracking others' beliefs about beliefs (taken from http://www2.compute.dtu.dk/~tobo/epistemic-planning-cph/index.php)

These present significant computational challenges: the actions of the other a-gents can induce a combinatorial explosion in the number of contingencies to be considered, making both the search space and the solution size exponentially larger, hence demanding novel methods [12, 21, 22].

The ability to hold a Theory of Mind to oneself and others, and to understand that others are doing the same, is important in many domains. Consider two fighter pilots seeking to disable an enemy radar defended by missiles. To do so, the pilots need to fool the enemy missile operators into believing that the two aircraft are attacking from the opposite direction to what they are truly attacking, in order to get close enough to the radar. Further, they need to attack simultaneously — one will destroy the radar while the other provides cover. However, they may be required to approach without communication, to reduce the chance of revealing their location. Thus, their agreed plan is to attack simultaneously only when they believe the enemy is deceived, *and* they believe that their team member believes that the enemy is deceived. To do this, they need to independently observe the same events as each other in the environment, and from these, update their theory of the others' mental state, as well as that of the enemy. Provided that both pilots are able to observe key events and understand that the observations of these events are common (known as *co-presence*), then they can coordinate their actions without communication.

In a first-order theory of mind, the reasoner considers that other people have beliefs, desires, etc. that influence their behaviour; e.g. *they believe we are attacking from the opposite direction*. In a second-order theory of mind, the reasoner allows that others are doing the same about us and other people; e.g. *my co-pilot believes that the enemy believe that we are attacking from the opposite direction*. In higher-order theories of mind, this nesting continues; *I believe that my co-pilot believes the enemy believe that we are attacking from the opposite direction, and I believe my co-pilot believes I believe this*.

Such reasoning has received much attention in empirical studies of children's and adults' reasoning, e.g. [6, 23–25] and there is considerable evidence that many adults have ToM abilities of levels 3 and 4, with some subjects succeeding in tasks requiring level 5 reasoning, yet even level 2 reasoning is beyond the reach of almost all state-of-the-art planning tools.

Multi-agent systems research has contributed a deep understanding of concepts such as group knowledge, group belief, and collective intention, often informed by philosophical and psychological perspectives, e.g. [26–30]. Studies have also examined computational models of ToM, e.g. [8, 31, 32], and also the impact of different levels of awareness that an agent has about the others acting in a team task context, e.g. [33]. Although the tools used in such investigations are highly expressive – typically description logics and rich multi-modal logics, and some bespoke algebraic belief update mechanisms – they are not accompanied by efficient reasoning engines, so fall short of providing practical means for systematically operationalising complex analyses.

Existing multi-agent planning tools that do take into account the beliefs, goals, intentions and capabilities of others, e.g. [34], consider a *third-person view*, in which a plan is constructed for a team, and each member is given their part to execute. When planning must be *distributed* amongst a team (including, when humans are to be in the loop), a semi-autonomous system must plan for its own actions while considering others explicitly - i.e. such reasoning demands a *first-person view*.

4.3 Social Planning

The authors, in conjunction with several collaborators, have made recent advances in this area; notably in the area of multi-agent epistemic planning. In this section, we overview two of the key advances made and provide a high-level technical overview of these. The two areas are:

1. *Efficient epistemic planning* — Bolander and Anderson [21] define the concept of *epistemic planning domains*, a generalisation of classical planning domains in which action models can have preconditions and effects on the (possibly nested) belief of others. They also show epistemic planning to be decidable in the single-agent case, but only semi-decidable in the multi-agent case.

In recent work, the authors, along with other collaborators, showed how restricted forms of epistemic knowledge bases can be used for efficient querying [10, 11, 15], and proposed a method that used these knowledge bases to take extend planning domains with higher-order belief operators, in a similar spirit to Bolander and Anderson's epistemic planning, and encode these as propositional planning problem [13]. The resulting encoding allows a large class of epistemic planning problems to be solved efficiently.

2. *First-person perspective multi-agent planning* — The authors and their collaborators propose a computational model for reasoning about and with others in multi-agent environments using heterogeneous *agent models* [8, 9], and subsequently instantiate this model as a non-deterministic planning problem [14]. The result is a planning tool that can produce policies for acting in a multi-agent environment, in which the policy has been compiled such that the agent considers the actions of others as it deliberates.

The latter item allows an agent to act in a multi-agent world considering the other agents' actions, while the former extends this with a Theory of Mind about the other agents' beliefs. Integrating these two pieces provides a tool for social planning: the ability to consider the possible behaviours and mental states (in this case, beliefs and goals) of others during the deliberation process.

4.3.1   A Formal Model for Multi-agent Epistemic Planning

In this section, we present a formal model for our multi-agent epistemic planning problem. This problem extends standard planning problems with the addition of epistemic fluents and multi-agent actions.

4.3.1.1   Epistemic Fluents

The notion of epistemic planning refers to the ability to reason about knowledge (or belief), rather than just about facts of the world. In the example of the two fighter pilots outlined in Sect. 4.2.2, these pilots are reasoning about the knowledge/beliefs of their partners as well as that of their adversary. Such reasoning is imperative for Theorem of Mind reasoning: to put oneself in the shoes of another, one must adopt their perspective of the world, including their understanding of the environment and others within it.

Epistemic logics extend standard propositional logics with *modal operators*, in which the mode of the formula represents the perspective of individual agents and groups of agents. First, we present some background material on epistemic and doxastic[1] logics that is required for this chapter. Throughout the remainder of this

---

[1] We use the term "epistemic" to refer to both knowledge and belief throughout the paper.

chapter, we will assume that the epistemic logic use is modal logic KD (see Fagin et al. [35] for a definition of this), and as such, is truly a belief operator, rather than a knowledge operator.

Due to the high computational complexity of epistemic logic, we adopt a simplified version of epistemic/doxastic logic by restricting modal formulae to *restricted modal literal* (RML) [36], proposed by Lakemeyer and Lespérance. An RML is defined using the following grammar:

$$\phi ::= p \mid [i]\phi \mid \neg\phi$$

where p is a propositional literal and i is an agent identified. Note that an RML cannot contain disjunctions, and is always in *negation normal form* (NNF). A set of RMLs, which is equivalent to their conjunction, is called a *proper epistemic knowledge base* (PEKB).

These RMLs are the fluents used in our epistemic planning problems: they offer an increase in expressiveness over propositional fluents, but as we will show later, they do not greatly increase the difficultly of solving the problem.

4.3.1.2   First-Person Multi-agent Planning

Similar to our earlier work [14], we define a first-person multi-agent planning problem as a tuple

$$\langle Ag, F, \mathcal{I}, \mathcal{G}_{i=0\cdots|Ag|-1}, \mathcal{A}_{i=0\cdots|Ag|-1}\rangle$$

where:

- Ag is the set of agents in the world, including the planning agent specially designated as 0;
- F is a set of epistemic fluents, in which each fluent is an RML;
- $\mathcal{I} \subseteq F$ is the initial state of the world;
- $\mathcal{G}_i \subseteq F$ is the goal for agent $i \in Ag$; and
- $\mathcal{A}_i$ is the finite set of actions agent i can execute.

Note the difference between this and the definition of classical planning outlined in Sect. 4.2: there is a set of agents associated with the problem definition, fluents can be epistemic, each agent has a goal, and actions are associated with particular agents.

Each action $a \in \mathcal{A}_i$ is a tuple of the form $\langle \text{Pre}_a, \text{Eff}_a \rangle$ where $\text{Pre}_a \subseteq F$ is the precondition that must hold for the action to be executed, and Eff_a is a *set* of one or more possible conditional effects, in which exactly one of the effects will hold after the execution of the action, but we do not know which until the action has been executed; that is, actions can be non-deterministic. We assume here that the non-deterministic effects are fully-observable; that is, the agents do not which outcome will occur, but they can observe the outcome immediately after the action is executed.

```
(:action share
      :derive-condition    (at $agent$ ?l)
      :parameters          (?a ?as - agent ?l - loc)
      :precondition        (and (at ?a ?l) [?a](secret ?as))
      :effect              (and
                                (forall ?a2 - agent
                                    (when       (at ?a2 ?l)
                                            [?a2](secret ?as))))
                           )

  )
```

Fig. 4.3 An epistemic PDDL description of sharing a secret

The set of all joint actions between agents is the cross product of all individual actions: $\mathcal{A} = \mathcal{A}_0 \times \cdots \times \mathcal{A}_{|Ag|-1}$. To model that it is possible for some agents to perform an action while others do not, (at least some) agents must be equipped with a "noop" (no operation) action, which has no effects.

Example 1 Consider the *Grapevine problem*, based on the well-known *gossip problem*, in which agents can move between rooms, share a secret piece of information in their room, but only those agents in the room will learn the secret when it is shared. The epistemic Planning Domain Description Language (PDDL) [37] extension of this action can be modelled as in Fig. 4.3.

In this example, ?l is a room, ?a is the agent sharing the secret, and ?as are the other agents in the room. The fluent [?a2](secret ?as) means that agent ?a2 believes fluent (secret ?as). Note that any agent can execute this action. Action preconditions can be used to restrict actions to only a subset of the agents in the domain.

The *derive condition* at the top of the action definition models the conditions of *mutual awareness*. Essentially, this says that for any agent in the room ?l, they will derive the effects of this action if the action is executed. In essence, they will be *aware* that the action has been executed and will see its effect. They will therefore know the secret, but also know that all other agents in the room know the secret.

The types of goals one could consider in this example are: to share one's secret with only a subset of the agents; to deceive a particular set of agents; or to have every agent share their secret with everyone else.

A solution to a first-person multi-agent planning problem is a *policy* $P : 2^F \to \mathcal{A}_0$, thus mapping a partial state (a set of fluents) to an action specifying which action the 0 agent should take in a state that satisfies the partial state.

4.3.2 Solving Multi-agent Epistemic Planning Problems

While multi-agent epistemic planning problems are significantly more expressive than standard classical or contingent planning problems, they often can be solved with some compilations to and modifications of existing — albeit advanced — planning technology.

As noted earlier, we solve this problem in two ways. First, we compile away the epistemic fluents in the planning problem into standard propositional fluents, such that any action defined using epistemic fluents can be compiled into an equivalent action and solved using an existing planner, such as a classical planner or non-deterministic planner. Second, by modifying an existing non-deterministic planning tool to consider multiple agents (without epistemic fluents), and then treating the effects of other agents actions as non-determinism in the environment. Thus, compiling a multi-agent epistemic planning problem into a multi-agent propositional planning problem and using this multi-agent planner, we can solve this rich class of problems.

4.3.2.1 Compiling Away Epistemic Fluents

There are several parts to the compilation – in this section we describe just the two most important: encoding consistent belief update; and encoding the perspective of other agents when the planning agent is unsure whether they witnessed an event. These both extend a *base encoding*, which strips away epistemic fluents are replaces them with propositional fluents suitable for our (non-epistemic) multi-agent planner. Technical details about this encoding can be found in Muise et al. [13]. In this section, we simply provide the intuition behind these via some examples.

Base Encoding. The base encoding describes a simple multi-agent planning problem that is not equivalent to the original problem. This encoding is then extended to deal with belief update and uncertain firing of events.

Put simply, the encoded problem takes the original problem and compiles it to an alternative problem such that each epistemic fluent in the action models, initial state, and goal is encoded into a proposition; that is, fluents of the form [?a]p are compiled to a_p. Thus, a_p represents the agent *a* believing *p* as a proposition. This replacement is nested for nested beliefs; for example, [?a][?b][?c]p is encoded as a_b_c_p. Negations of the form not([?a]p) are encoded as not_a_p.

Belief Update. In classical planning, belief update is straightforward: when a proposition becomes true, it is no longer false, and vice versa. However, in epistemic planning, the problem is not so simple. Consider the *Grapevine* example described in Example 1, in which agent 1 learns secret *s*, modelled as the epistemic fluent [?a]s. The propositional fluent a_s models this, however, we must also consider that if [?a]s is true, then so is not([?a] not(s)) — if agent *a* believes *s*, then is should not believe the negation of *s*. Thus, for every compiled action in which a_s becomes true, so too must not_a_not_s. This counters for epistemic actions in which not([?a] not(s)) is a precondition for example. If we add only a_s to the state, then not_a_not_s will not be true when that precondition is evaluated for another action. As such, the encoded model would not be equivalent without this modified belief update.

The reverse problem occurs if we want to no longer believe
not([?a] not(s)) – we must also remove [?a]s.

To solve these problems, one could modify our multi-agent underlying planner to
know that whenever a_s is true, then not_a_not_s must also be true. Instead, we
extend the base encoding by adding additional effects to actions that explicitly consider these situations, resulting in an encoding that faithfully encodes the dynamics
of the original problem.

Compiling this down not only allows us to keep the epistemic and multi-agent
parts of our solution loosely coupled – it means that our epistemic compilation tool
can be used for other problems and other planners that support PDDL,[2] such as other
classical planners, temporal planners, non-deterministic planners, etc.

Uncertain Firing. Consider again the *Grapevine* scenario, and an example in which
we model the trustworthiness of agents. We may have a model of the share action that only believes a secret an agent shares if we believe that agent is trustworthy. For this, we would use a *conditional effect* on the action of the form
[0]trustworthy(?a) --> [0]secret(?a), meaning that we only add
[0]secret(?a) to our state if [0]trustworthy(?a) was in our state before
the action was executed (recall that the planning agent is agent 0). This models what
is intended, but what if agent 0 is unsure whether agent *a* is trustworthy? That is,
neither [0]trustworthy(?a) nor [0]not(trustworthy(?a)) are in the
state. Should [0]secret(?a) be added to the state?

Our intuition is that the solution should be to *remove* [0]not (secret(?a))
(if it is in the state) if not([0]trustworthy(?a)) holds before the action executes. Note here that not([0]trustworthy(?a)) is not the same as [0]not
(trustworthy(?a)). In the latter, we model that agent 0 believes that *a* is not
trustworthy, while in the former we model that it is not the case that agent 0 believes
a is trustworthy – agent 0 may be unsure of agent *a*'s trustworthiness.

Thus, we model that if 0 is unsure whether *a* is trustworthy, then it should *not*
believe the secret, but it should at least no longer believe that the secret is false either:
it should be uncertain whether the secret is true or not.

4.3.2.2   Multi-agent Problems as Non-deterministic Problems

The difference between single-agent and multi-agent planning problems is clear: in
multi-agent planning problems, the agents must consider not only their own actions,
but actions of other agents as well. For example, consider the simple two-player game
Tic-Tac-Toe. When playing a move, we should not only consider whether we can get
three pieces in a row, which is trivial in a single-player version, but also whether our
opponent can block us or whether they can also get their own three pieces in a row.

One way to model other agents is to treat them as a dynamic environment. That
is, when we execute an action, and another agent can subsequently change the world,

---

[2]Note: the underlying planner must support *conditional effects* for our compilation to work.

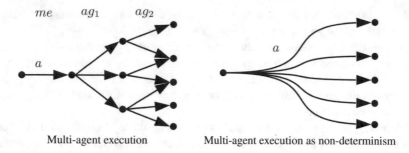

Multi-agent execution   Multi-agent execution as non-determinism

Fig. 4.4 Treating other agents' actions as non-determinism

we treat their action as arbitrary changes in a dynamic environment, such that it is as if the environment itself changed, rather than being explicitly changed by another agent.

Such an approach presents an opportunity to build on recent advances in *non-deterministic* planning [17] to extend planning technology to multi-agent environments. In non-deterministic planning, actions can have multiple possible effects, but the actual effect cannot be known until after the action is executed.

Using techniques in non-deterministic planning, we can cast the problem of planning in multi-agent environments as a non-deterministic planning task. Essentially, we can treat the actions of other agents in the environment as non-deterministic effects of our own actions.

Figure 4.4 outlines the intuition behind this idea. Figure 4.4: Left shows an agent *me* considering the execution of action a. If agents ag_1 and ag_2 will then subsequently perform actions, then from a deliberation perspective, the possible effects of executing action a should be consider as all possible effects of agents ag_1 and ag_2's actions. Figure 4.4: Right illustrates the non-deterministic treatment of this.

Modelling the problem like this results in a faithful encoding of the original problem; however, as noted earlier in Sect. 4.2, the actions of the other agents can induce a combinatorial explosion in the number of contingencies to be considered, making the search for solutions too high for all but the most trivial applications.

One way that we mitigate this problem is to consider the *intent* of the other agents in the scenario. That is, if we know/believe that the other agents have some particular intent, then we are able to reduce the branching factor by focusing the search only on those actions that are *plausible* given the other agents' intent.

For example, consider *Tic-Tac-Toe*. We know our opponent's goal: to win the game. Given this, if we are planning what to do in the state of the game shown in Fig. 4.5, where the opponent is **O** and we are player **X**. If it is our move, one possible move is to place **X** in the bottom right corner, which sets us up to win along the bottom row. We then need to consider player **Y**'s moves. A rational agent would consider that player **O**'s most plausible response is to play in the top-right corner, winning the game. Player **O**'s next most plausible move (if one can really consider

Fig. 4.5 A sample game
state of *Tic-Tac-Toe*

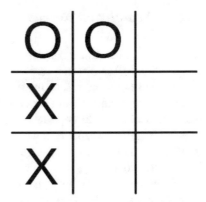

any such move as plausible!), would be to block the cell at the middle bottom, thus preventing us from winning.

Our search algorithm considers this by looking at the other agents' goals and using standard planning search heuristics to decide which actions are the best for the other agents when assessing what they may do. It uses these heuristics to rank the opponents move from most plausible to least plausible. Then, it considers the agent's most plausible action first at each stage of a scenario, until the search terminates. Then, it considers the next most plausible action, and so on.

Using the Tic-Tac-Toe example, our algorithm would consider player **Y** first playing top-right, then it would consider middle-bottom. Other moves are implausible and it is little value to explore them. The search continues until either: (a) the entire search terminates, in which case we have a complete solution and the plausibility ranking is meaningless; or more likely (b) a pre-defined time or memory budget is exhausted, at which point it has the best move considering the search space that it has explored.

This search strategy is highly effective in many domains, because it does not assume complete rationality of the other agents; nor that the model we have of the other agents is complete. That is, rather than determine exactly which action other agents will choose, it considers all, but only reasons about the effects of the most plausible ones. Given enough time and memory, this will result in a complete search, but for large problems, it focuses the search on those actions that are the most likely.

Technical details of the problem formulation, solution, and evaluation of this approach can be found in Muise et al. [14].

4.4   Social Planning for Human Robot Interaction

To demonstrate the benefits of social planning, we present two case studies involving semi-autonomous teams, which we have adapted to illustrate the benefits of planning in the presence of nested belief and first-person multi-agent planning.

4.4.1 Search and Rescue

Disaster response and management involves a number of important tasks, such as preparation before disaster, response and restoration. If we consider a scenario of a simplified search and rescue mission following a natural disaster, such as an earthquake, there are a series of tasks that must be undertaken to search for survivors and get them to the appropriate service, such as medical evacuation. As part of this, we can imagine a scenario in which two unmanned ground vehicles (UGVs) and a human operator are working as a three-member semi-autonomous team to locate and assess survivors.

The environment of this scenario consists of a set of buildings, organised according to a known map. However, the buildings may be damaged, leading to unexpected inaccessibility to search regions or locations. Buildings may contain survivors or could be empty, but this is initially unknown.

The human supervisor oversees the entire mission and coordinates the UGVs. They can interact with the agents controlling the vehicles by assigning goals, such as to search a particular building or to return back to particular base location. They can also query the agents on their current goals, intentions, and beliefs; including the nested beliefs of the other agent and the operator themselves.

The two UGVs have the same capabilities, which can be modelled as actions in epistemic PDDL, such as:

1. Moving to specific way-points, identified by coordinates on the map, including inside buildings.
2. Attempt to open doors to buildings/rooms (which may either succeed or fail).
3. Go into buildings and rooms, providing the doors are open.
4. Take a picture and upload it for assessment.
5. Drop a first-aid survival pack, provided that the agent believes there is a survivor in need of this.
6. Drop water to a survivor.
7. Lift a survivor onto one of the vehicles, which requires the assistance of the other vehicle.
8. Communicate with the other agent or the supervisor.

This final action — communication — is enabled by the epistemic actions. Communication can simply be modelled as an action with epistemic effects. For example, one agent can send a message to the others indicating that a particular door is blocked (e.g. by rubble).

Such an action can be modelled as follows:

Parameters   The parameters are the agents to which to send the message, and the location of the door that is blocked.

Precondition   The precondition is that the sending agent believes that the door is blocked, and believes the recipient does not believe that the door is blocked.

Effects   The effects are that the sending agent believes that the receiving agents believe that the door is blocked, and further, the sending agent believes that the receiving agent believes that the sending agent believes that the door is blocked. Similarly, an agent receiving such a message would believe that the door is blocked and that the sender believes that the door is locked.

Although this scenario is a simplification of a real search and rescue mission, it illustrates the implications that explicitly modelling these communication actions has on the scenario. We assert that these go some way towards improving the interactions between the semi-autonomous team, such as:

Efficiency   It may be that the other vehicle agent could want to use that door later as part of a plan, and now knows to plan a different route to get inside the building/room that does not used this blocked door.

Lower Communication Overhead   By explicitly representing the beliefs of others (and potentially their nested beliefs of the team), the amount of communication overhead can be reduced. For example, the precondition of the action above is that the receiving agents do not already believe this information. Thus, if the planning agent already has information noting that another agent beliefs the door is blocked, it will not send this information on. This is particularly important in semi-autonomous teams, in which human collaborators are more easily overloaded with information than their artificial team members.

Re-planning   Having an explicit model of the other agents' Theory of Mind, and being able to update this model, enables agents to identify that their expectation of their team members' behaviour is no longer valid, thus triggering them to re-assess or re-plan the intentions and plans of their team members.

Coordination   As outlined in the task of the fighter pilots in Sect. 4.2.2, providing updates of each others mental states allows agents to synchronise on joint actions that require e.g. simultaneous execution, such as lifting a survivor onto one of the vehicles to transport them back to a location with further medical assistance, which may require that each agent believes that the survivor is on the stretcher and believes that the other agent believes this as well.

Transparency   It provides some transparency to the human supervisor, informing them why the agent is not entering the room that it had originally planned, without (at least in some cases) the agent having to explicitly update the supervisor on its new plan.

Epistemic goals   Being able to model the epistemic effects allow us to pose epistemic goals, such as that agent A believes something is true while agent B believes the opposite — in other words, one of the agents is deceived.

While it is straightforward to model communication actions in other planning languages, the ability to model the *epistemic effects* of these actions, and have these effects represented as a Theory of Mind, enables additional possibilities over using propositional planning, particularly regarding coordination and transparency.

Fig. 4.6 Industrial painting
robots with optional human
inspection; presented in [39],
the video can be found at
http://tiny.cc/2aytrw

Tercio reschedules work in response to a Quality
Assurance request.

Further to this though, we assert that epistemic first-person multi-agent problem
is a more natural way to model these problems, compared to existing approaches,
such as keeping a separate model for each other agent [38].

In particular, the ability to model and reason about epistemic goals requires the
ability to model basic multi-agent epistemic effects — they cannot be captured with
separate models.

4.4.2 Collaborative Manufacturing

In Fig. 4.6 we consider a variant of a painting and assessment task presented in [39]
where robot actions are only *partially observable* to human operators. The task
involves the real-time scheduling of painting robots for the fuselage of an aeroplane.
Human operators optionally intervene in the painting process to assess the quality of
the painted surfaces.

The painting robots must adapt and re-schedule to the optional assessment of
panels by human operators, to allow the panels to have sufficient drying time and to
achieve the goal of painting the fuselage in the time allocated to the task. The temporal
constraints are captured in this task using a form of temporal constraint networks
termed simple temporal networks (STNs) [40].[3] The time required for the application
of each coat is captured using the STNs, along with the time before the (optional)
assessment of each coat; including the assessment time. From the perspective of
the painting robots, paths through STNs emerge according to the non-deterministic
choice of humans according to which panels they chose to assess. This forms a
branching tree, similar to Fig. 4.4: Left, as painting robots and human operators
interleave painting and assessment tasks. At any instant, there is a *minimal* STN that
achieves all of the tasks within a minimum time, which will be traversed according
to the optional assessments that are potentially performed by human operators in the
future.

---

[3] See [39] for the STN encoding details for this task and the video at http://tiny.cc/2aytrw.

Fig. 4.7 Industrial assembly
with human-robot
interaction; presented in [41],
the video can be found at
http://tinyurl.com/7n439cg

If the robots work too far down the fuselage from the human operators, humans cannot distinguish *which* panels the robots are painting. If nested belief is used during planning this allows robots to choose panels to paint which are observable to human assessors. The robots know that actions observable to the human operators allow the humans to infer the minimal STN. The robots therefore know that humans know the robots know the humans know the minimal STN. Humans can therefore understand the choice of panel robots make to paint; and can even take this choice into account in deciding which panel next to assess. Thus, theory of mind facilitates the robots to maintain human knowledge of which panel(s) are ready to assess. Social planning builds trust between robots and human operators—leading to goal achievement within shorter times.

In another industrial task, shown in Fig. 4.7, an assembly task is shown where a human and a robot share in a simple assembly task that involves placing fasteners then applying torque to each fastener. The robot shares the task by applying sealant to each hole ahead of placement of the fastener. The robot must be able to handle different preferences of human operators. For example, operators may choose to place all the fasteners first, then apply torque to each one. Alternatively operators may choose to place each fastener then apply torque to each one immediately following placement. The approach described in [41] shows an approach that can adapt to the preferences of humans using dynamic scheduling, the video can be found at http://tinyurl.com/7n439eg.

We adapt this task to utilise social planning. If we assume the goal is to minimise the overall time to complete the task, theory of mind facilitates robots to adapt to humans changing their preferred assembly behaviour part-way through the achievement of task, as they learn to perform the task within less time. Using theory of mind principles, the robot uses social planning in the knowledge that the human knows the robot knows the human has learned the shortest human-robot interleaving strategy. This enables the robot to perform other preparatory tasks, such as fetching and positioning the correct number of fasteners, further shortening the time to complete the task.

4.5 Discussion

We have presented an an outline of several principal elements of the emerging field of social planning. These include theory of mind, as we move to first-person perspective planning in a multi-agent setting, and we present a formal model for first-person multi-agent epistemic planning. We have covered two emerging solution techniques for solving multi-agent epistemic planning problems, including an approach for compiling away epistemic fluents, where multi-agent problems are posed as non-deterministic problems, for which solutions are quite well understood. Finally, we presented two case studies of semi-autonomous systems by adapting examples from the literature to utilise social planning and theory of mind principles to demonstrate the benefits for realising trusted autonomy. These examples demonstrate how social planning can used to improve the interaction between humans and robots in semi-autonomous teams.

The work forms an important step towards achieving trusted autonomy where the perspective of both humans and robots are explicitly modelled using a first-person theory of mind approach. There is excellent potential for the exploitation of recent developments in efficient epistemic and non-deterministic reasoning techniques. For example, recent techniques in proper epistemic databases such as 'knowing whether' [10] can be used to *establish the knowledge* of human operators during more complex tasks without knowing the knowledge itself, and the observability of *asynchronously occurring* actions can even be modelled [42]. Further work and experimentation is warranted to explore the application of these and other related techniques in social planning.

References

1. M.N. Huhns, D.A. Buell, Trusted autonomy. IEEE Internet Comput. **6**(3), 92–95 (2002)
2. H.A. Abbass, E. Petraki, K. Merrick, J. Harvey, M. Barlow, Trusted autonomy and cognitive cyber symbiosis: open challenges. Cogn. Comput. **8**(3), 385–408 (2016)
3. S. Wheeler, Trusted autonomy: conceptual developments in technology foresight. Technical Report DSTO-TR-3153, Defence Science and Technology Group (DSTG), 2015
4. D. Premack, G. Woodruff, Does the chimpanzee have a theory of mind? Behav. Brain Sci. **1**(4), 515–526 (1978)
5. J. Call, M. Tomasello, Does the chimpanzee have a theory of mind? 30 years later. Trends Cogn. Sci. **12**(5), 187–192 (2008)
6. A.I. Goldman, Theory of mind, in *Oxford Handbook of Philosophy and Cognitive Science*, Chap. 17, ed. by E. Margolis, R. Samuels, S. Stich (Oxford University Press, Oxford, 2012), pp. 201–213
7. S. Zilberstein, Building strong semi-autonomous systems, in *Proceedings of the Twenty-Ninth AAAI Conference on Artificial Intelligence, Austin, Texas, USA*, 25–30 Jan 2015, pp. 4088–4092
8. P. Felli, T. Miller, C.J. Muise, A.R. Pearce, L. Sonenberg, Artificial social reasoning: computational mechanisms for reasoning about others, in *Social Robotics—6th International Conference, ICSR 2014, Sydney, NSW, Australia, October 27–29, 2014. Proceedings* (2014), pp. 146–155

9. P. Felli, T. Miller, C.J. Muise, A.R. Pearce, L. Sonenberg, Computing social behaviours using agent models, in *Proceedings of the Twenty-Fourth International Joint Conference on Artificial Intelligence, IJCAI 2015, Buenos Aires, Argentina*, 25–31 July 2015, pp. 2978–2984

10. T. Miller, P. Felli, C.J. Muise, A.R. Pearce, L. Sonenberg, 'Knowing whether' in proper epistemic knowledge bases, in *Proceedings of the Thirtieth AAAI Conference on Artificial Intelligence, Phoenix, Arizona, USA*, 12–17 Feb 2016, pp. 1044–1050

11. T. Miller, C.J. Muise, Belief update for proper epistemic knowledge bases, in *Proceedings of the Twenty-Fifth International Joint Conference on Artificial Intelligence, IJCAI 2016, New York, NY, USA*, 9–15 July 2016, pp. 1209–1215

12. T. Miller, A. Pearce, L. Sonenberg, F. Dignum, P. Felli, C. Muise, Foundations of human-agent collaboration: situation-relevant information sharing, in *2014 AAAI Fall Symposium Series* (2014)

13. C.J. Muise, V. Belle, P. Felli, S.A. McIlraith, T. Miller, A.R. Pearce, L. Sonenberg, Planning over multi-agent epistemic states: a classical planning approach, in *Proceedings of the Twenty-Ninth AAAI Conference on Artificial Intelligence, Austin, Texas, USA*, 25–30 Jan 2015, pp. 3327–3334

14. C.J. Muise, P. Felli, T. Miller, A.R. Pearce, L. Sonenberg, Planning for a single agent in a multi-agent environment using FOND, in *Proceedings of the Twenty-Fifth International Joint Conference on Artificial Intelligence, IJCAI 2016, New York, NY, USA*, 9–15 July 2016, pp. 3206–3212

15. C.J. Muise, T. Miller, P. Felli, A.R. Pearce, L. Sonenberg, Efficient reasoning with consistent proper epistemic knowledge bases, in *Proceedings of the 2015 International Conference on Autonomous Agents and Multiagent Systems, AAMAS 2015, Istanbul, Turkey*, 4–8 May 2015, pp. 1461–1469

16. T.-C. Au, U. Kuter, D. Nau, Planning for interactions among autonomous agents, in *International Workshop on Programming Multi-Agent Systems* (Springer, Berlin, 2008), pp. 1–23

17. C.J. Muise, S.A. McIlraith, J. Christopher Beck, Improved non-deterministic planning by exploiting state relevance. In *Proceedings of the Twenty-Second International Conference on Automated Planning and Scheduling, ICAPS 2012, Atibaia, São Paulo, Brazil*, 25–19 June 2012

18. H. Palacios, H. Geffner, Compiling uncertainty away in conformant planning problems with bounded width. J. Artif. Intell. Res. **35**, 623–675 (2009)

19. F. Dignum, G.J. Hofstede, R. Prada. From autistic to social agents, in *Proceedings of the 12th International AAMAS Conference* (2014), pp. 1161–1164

20. T. Bolander, A. Herzig, Group attitudes and multi-agent planning: overview and perspectives (2014), Accessed at http://www.sintelnet.eu/wiki/reports/Thomas_Bolander_WG3_report.pdf

21. T. Bolander, M.B. Andersen, Epistemic planning for single- and multi-agent systems. J. Appl. Non Class. Logics **21**(1), 9–34 (2011)

22. A. Pearce, L. Sonenberg, P. Nixon, Toward resilient human-robot interaction through situation projection for effective joint action, in *Robot-Human Teamwork in Dynamic Adverse Environment: AAAI Fall Symposium* (2011), pp. 44–48

23. A. Brandenburger, X. Li, Thinking about thinking and its cognitive limits (2015), http://adambrandenburger.com/articles/papers/. Accessed Sept 2016

24. T. Kneeland, Identifying higher-order rationality. Econometrica **83**(5), 2065–2079 (2015)

25. H.D. Schlinger, Theory of mind: an overview and behavioral perspective. Psychol. Rec. **59**(3), 435–448 (2009)

26. M. Gilbert, Modelling collective belief. Synthese **73**(1), 185–204 (1987)

27. R. Hakli, Group beliefs and the distinction between belief and acceptance. Cogn. Syst. Res. **7**(2), 286–297 (2006)

28. L. Lismont, P. Mongin, On the logic of common belief and common knowledge. Theor. Decis. **37**(1), 75–106 (1994)

29. R. Tuomela, W. Balzer, Collective acceptance and collective social notions. Synthese **117**(2), 175–205 (1998)

30. H. van Ditmarsch, J. van Eijck, R. Verbrugge, Common knowledge and common belief, in *Discourses on Social Software*, ed. by J. van Eijck, R. Verbrugge, vol. 5 of *Texts in Logic and Games* (2009), pp. 99–122
31. L. Van Maanen, R. Verbrugge, A computational model of second-order social reasoning, in *Proceedings of the 10th International Conference on Cognitive Modeling* (2010), pp. 259–264
32. H. Weerd, R. Verbrugge, B. Verheij, Negotiating with other minds: the role of recursive theory of mind in negotiation with incomplete information, in *Autonomous Agents and Multi-Agent Systems* (2015), pp. 1–38
33. H. De Weerd, R. Verbrugge, B. Verheij, How much does it help to know what she knows you know? An agent-based simulation study. Artif. Intell. **199**, 67–92 (2013)
34. R.F. Kelly, A.R. Pearce, Asynchronous knowledge with hidden actions in the situation calculus. Artif. Intell. **221**, 1–35 (2015)
35. Y.M.R. Fagin, J.Y. Halpern, M.Y. Vardi, *Reasoning about Knowledge* (MIT Press, Cambridge, MA, 1995)
36. G. Lakemeyer, Y. Lespérance, Efficient reasoning in multiagent epistemic logics, in *European Conference on Artificial Intelligence* (2012), pp. 498–503
37. D. McDermott, M. Ghallab, A. Howe, C. Knoblock, A. Ram, M. Veloso, D. Weld, D. Wilkins, PDDL—The Planning Domain Definition Language (1998)
38. V.V. Unhelkar, J.A. Shah, Contact: Deciding to communicate during time-critical collaborative tasks in unknown, deterministic domains, in *Proceedings of the Thirtieth AAAI Conference on Artificial Intelligence, Phoenix, Arizona, USA*, 12–17 Feb 2016, pp. 2544–2550
39. M.C Gombolay, R. Wilcox, J.A Shah. Fast scheduling of multi-robot teams with temporospatial constraints, in *Robotics: Science and Systems IX*, Technische Universität Berlin, Germany, June 2013, pp. 49–56
40. R. Dechter, I. Meiri, J. Pearl, Temporal constraint networks. Artif. Intell. **49**(1–3), 61–95 (1991)
41. R. Wilcox, S. Nikolaidis, J. Shah, Optimization of temporal dynamics for adaptive human-robot interaction in assembly manufacturing, in *Robotics Science and Systems VIII* (2012), pp. 441–448
42. R.F. Kelly, *Asynchronous Multi-Agent Reasoning in the Situation Calculus*. Ph.D. University of Melbourne (2008)

Chapter 5
A Neuroevolutionary Approach to Adaptive Multi-agent Teams

Bobby D. Bryant and Risto Miikkulainen

5.1 Introduction

Multi-agent systems are a commonplace in social, political, and economic enterprises. Each of these domains consists of multiple autonomous parties cooperating or competing at some task. Multi-agent systems are often formalized for entertainment as well, with instances ranging from team sports to computer games. Games have previously been identified as a possible "killer application" for artificial intelligence [14], and a game involving multiple autonomous agents is a suitable platform for research into multi-agent systems as well.

The agents that comprise a multi-agent system can be either homogeneous or heterogeneous. Heterogeneous teams are often used for complex tasks because they allow agents to be specialized for sub-tasks (e.g. [2, 13, 29]). However, heterogeneous teams of sub-task specialists are brittle: if one specialist fails then the whole team may fail at its task. Moreover, when the agents in a team are programmed or trained to optimize a pre-specified division of labor, the team may perform inefficiently if the size of the team changes – for example, if more agents are added to speed up the task – or if the scope of the task changes.

For example, suppose you owned a team of ten reactor cleaning robots, and the optimal division of labor for the cleaning task required two sprayers, seven scrubbers, and one pumper (Fig. 5.1). If the individual robots were programmed or trained as sub-task specialists the team would be brittle and lacking in flexibility. Brittle, because the failure of a single spraying robot would reduce the entire team to half speed at the cleaning task, or the loss of the pumper robot would cause the team to fail entirely. Inflexible, because if a client requested a 20% speed-up for the task you would not be

B. D. Bryant (✉) · R. Miikkulainen
Department of Computer Sciences, University of Texas at Austin, Austin, USA
e-mail: bdbryant257@gmail.com

R. Miikkulainen
e-mail: risto@cs.utexas.edu

© The Author(s) 2018
H. A. Abbass et al. (eds.), *Foundations of Trusted Autonomy*, Studies in Systems, Decision and Control 117, https://doi.org/10.1007/978-3-319-64816-3_5

Fig. 5.1 *Left*: A heterogeneous team of cleaning robots is trained or programmed for sub-task specializations. Such a system is inflexible when the scope of the task changes, and brittle if key specialists are unavailable. *Right*: An Adaptive Team of Agents provides every agent with the capability of performing any of the necessary sub-tasks, and with a control policy that allows agents to switch between tasks at need. The resulting team is more flexible and less brittle than the heterogeneous team

able to simply send in 20% more robots; you would have to add ⌈20%⌉ more robots for each sub-task specialization, four more robots in all rather than two.

An alternative approach is to use a team of homogeneous agents, each capable of adopting any role required by the team's task, and capable of switching roles to optimize the team's performance in its current context. We call such a multi-agent architecture an *Adaptive Team of Agents* (ATA) [4]. An ATA is a homogeneous team that self-organizes a division of labor in situ so that it behaves as if it were a heterogeneous team. It changes the division dynamically as conditions change, and if composed of autonomous agents it must be able to organize the necessary divisions of labor without direction from a human operator.

Thus the ATA requires trusted autonomy. Within the team, individual agents must trust all the others to "do the right thing". Agents cannot select appropriate sub-tasks without some sort of assurance – possibly supported by observation – that the other members of the team are also selecting contextually appropriate sub-tasks. The owner of the team, whether in the context of robotics, simulation, or games, must also be able to trust the team as a whole to work out an effective division of labor in order to get the team's overall task done thoroughly and efficiently. That is, the team must pursue its owner's intent, and either the reality or appearance of intent may need to be instilled into the individual agents in order to achieve that. An ATA is robust because there are no critical task specialists that cannot be replaced by other members of the team; it is flexible because individual agents can switch roles whenever they observe that a sub-task is not receiving sufficient attention. If necessary, an agent can alternate between roles continuously in order to ensure that sufficient progress is made on all sub-tasks. Thus for many kinds of task an ATA could be successful even if there are fewer agents than the number of roles demanded by the task.

Such adaptivity is often critical for autonomous agents embedded in games or simulators. For example, games in the *Civilization*™ genre usually provide a "settler" unit type that is capable of founding new cities plus carrying out various construction tasks. Play of the game requires a division of labor among the settlers, and the details of the division vary with the number in play and the demands of the growing civilization – e.g. the choice between founding more cities versus constructing roads to connect the existing cities. If the settler units were heterogeneous, i.e. each recruited to perform only one specific task, there would be a great loss of flexibility and a risk of complete loss of support for a game strategy if all the settlers of a given type were eliminated. But in fact the game offers homogeneous settlers, and human players switch them between tasks as needed. For embedded autonomous settlers, that switching would have to be made by the settler units themselves: an Adaptive Team of Agents is desirable.

Here we explore the Adaptive Team of Agents experimentally, using genetic algorithms to train artificial neural networks (ANN) as the "brains" for the agents in a game, and find that it is possible to evolve an ATA with ANN controllers for a simple but non-trivial strategy game. The game, called *Legion II*, is described in Sect. 5.2, and the agents' control architectures are described in Sect. 5.3. The evolutionary mechanism used to train the game agents to behave as an adaptive team, called *Neuroevolution with Enforced Sub-Populations* (ESP), is described in Sect. 5.4. Methodological considerations are addressed in Sect. 5.5, and then experiments are reported in Sect. 5.6. Finally, discussion of the experimental results and an examination of future directions is given in Sect. 5.7.

5.2   The *Legion II* Game

Legion II is a discrete-state strategy game designed as a test bed for studying the division of labor in multi-agent systems. It requires a group of legions to defend a province of the Roman Empire against the pillage of a steady influx of barbarians. The legions are the agents under study; they are trained by the method described in Sect. 5.4, or by other methods reported elsewhere (e.g. [3, 5, 6]). The barbarians act according to a preprogrammed policy, to serve as a foil for the legions.

Legion II provides challenges similar to those provided by other games and simulators currently used for computational intelligence approaches to multi-agent learning research (e.g. [1, 15, 26, 27]), and is expandable to provide more complex learning challenges as research progresses. In its current incarnation it is incrementally more complex than a multi-predator/multi-prey game. It is conceptually similar to the pseudo-*Warcraft*™ simulator used in [1], differing primarily in its focus on the predators rather than the prey, and consequently in the details of scoring games.

The following subsections describe the components, features, and rules of the *Legion II* game, including the map, the game agents, and the method of calculating game scores.

5.2.1 The Map

Legion II is played on a planar map. The map is tiled with hexagonal cells in order to discretize the location and movement of objects in the game; in gaming jargon such cells are called *hexes* (singular *hex*). Several randomly selected map cells are distinguished as *cities*, and the remainder are considered to be *farmland* (Fig. 5.2).

The hex tiling imposes a six-fold radial symmetry on the map grid, defining the cardinal directions NE, E, SE, SW, W, and NW. This six-fold symmetry, along with the discretization of location imposed by the tiling, means that an agent's atomic moves are restricted to a discrete choice of jumps to one of the six cells adjacent

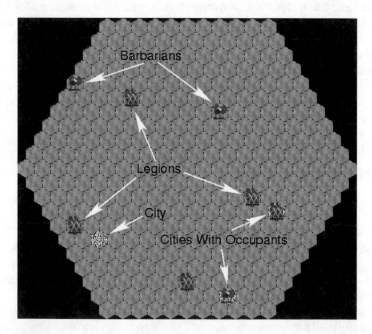

Fig. 5.2 A large hexagonal playing area is tiled with smaller hexagons in order to discretize the positions of the game objects. Legions are shown iconically as close pairs of men ranked behind large rectangular shields, and barbarians as individuals bearing an axe and a smaller round shield. Each icon represents a large body of men, i.e. a legion or a warband. Cities are shown in white, with any occupant superimposed. All non-city map cells are farmland, shown with a mottled pattern. The game is a test bed for multi-agent learning methods, whereby the legions must learn to contest possession of the playing area with the barbarians. (Animations of the *Legion II* game can be viewed at http://nn.cs.utexas.edu/keyword?ATA.)

to the agent's current location, and that the atomic move is always in one of the six cardinal directions. This map structure has important consequences for the design of the sensors and controllers for the agents in the game, which are described in detail in Sect. 5.3.

5.2.2 Units

There are two types of autonomous agents that can be placed on the map in a *Legion II* game: *legions* and *barbarians*. In accordance with gaming jargon these mobile agents are called *units* (singular *unit*) when no distinction needs to be made between the types.

Each unit is considered to be "in" some specific map cell at any time. A unit may move according to the rules described below, but its moves occur as an atomic jump from the cell it currently occupies to an adjacent one, not as continuous movement in Euclidean space.

The current position of each unit is shown by a sprite on the game map. In accordance with the jargon of the *Civilization* game genre, the sizes of the units are ambiguous. Thus the unit type called "a legion" represents a body of legionnaires, but is shown graphically as only a pair of men behind large rectangular shields. Similarly, the unit type called "a barbarian" represents a body of barbarians operating as a warband, but is shown graphically as only a single individual with axe and shield (Fig. 5.2).

The legions start the game already on the map, in randomly selected map cells. There are no barbarians in play at the start of the game. Instead, barbarians enter at the rate of one per turn, in a randomly selected unoccupied map cell.

5.2.3 Game Play

Legion II is played in turns. At the beginning of each turn a new barbarian is placed at a random location on the map. If the randomly generated location is already occupied by a unit, a new location is generated. This search continues until an unoccupied location for the new barbarian is found.

Thereafter, each legion is allowed to make a move, and then each barbarian in play makes a move. A unit's move can be a jump to one of the six adjacent map cells, or it can elect to remain stationary for the current turn. When all the units have made their moves the turn is complete, and a new turn begins with the placement of another new barbarian. Play continues for a pre-specified number of turns, 200 in all the experiments reported here.

All the units are autonomous; there is no virtual player that manipulates them as passive objects. Whenever it is a unit's turn to move, the game engine calculates the activation values for that unit's egocentric sensors, presents them to the unit's controller, and implements the choice of move signaled at the output of the unit's controller, as described in Sect. 5.3.

There are some restrictions on whether the move requested by a unit is actually allowed, and the restrictions vary slightly between the legions and the barbarians. The general restrictions are that only one unit may occupy any given map cell at a time, and no unit may ever move off the edge of the playing area defined by the tiling.

If a legion requests a move into an unoccupied map cell, or requests to remain stationary for the current turn, the request is immediately implemented by the game engine. If the legion requests moving into a map cell occupied by another legion, or requests a move off the edge of the map, the game engine leaves the legion stationary for that turn instead. If the legion requests moving into a map cell occupied by a barbarian, the game engine immediately removes the barbarian from play and then moves the legion as requested.

If a barbarian requests a move that is neither off-map nor into an occupied map cell, the request is immediately implemented by the game engine. If the barbarian requests a move into a map cell occupied by either a legion or another barbarian, the game engine leaves it stationary for the current turn. (Notice that this does not allow a barbarian to eliminate a legion from play the way a legion can eliminate a barbarian.) If the barbarian requests a move off the edge of the map, the game engine consults the barbarian's controller to see what its second choice would have been. If that second choice is also off-map then the game engine leaves the barbarian stationary for the current turn; otherwise, the secondary preference is implemented.

Barbarians are given their second choice when they request a move off the map, because their programming is very simple, and it is not desirable to leave them 'stuck' at the edge of the map during a game. Legions do not get the second-chance benefit; they are expected to learn to request useful moves.

5.2.4 Scoring the Game

The game score is computed as follows. The barbarians accumulate points for any pillaging they are able to do, and the legions excel by minimizing the amount of pillage points that the barbarians accumulate. At the end of every game turn, each barbarian in play receives 100 points for pillage if it is in a city, or only a single point otherwise. The points are totaled for all the barbarians each turn, and accumulated over the course of the game. When a barbarian is eliminated no points are forfeited, but that barbarian cannot contribute any further points to the total thereafter.

This scoring scheme was designed in order to force the legions to learn two distinct classes of behavior in order to minimize the barbarian's score. Due to the expensive point cost for the cities, the legions must keep the barbarians out of them, which they can easily do by garrisoning them. However, further optimization requires any legions beyond those needed for garrison duty to actively pursue and destroy the barbarians in the countryside. If they fail to do so, a large number of barbarians will accumulate in the countryside, and though each only scores one point of pillage per turn, their cumulative aggregate is very damaging to the legions' goal of minimizing the barbarian's score.

In principle the legions might be able to minimize the pillage by neglecting to garrison the cities and utilizing every legion to try to chase down the barbarians, but the random placement of the incoming barbarians means that they can appear behind the legions, near any ungarrisoned cities, and inflict several turns of the very expensive city pillaging before a legion arrives to clear them out. The barbarian arrival rate was by design set high enough to ensure that the legions cannot mop them up fast enough to risk leaving any cities ungarrisoned. Thus the legions *must* garrison the cities in order to score well, and any improvement beyond what can be obtained by garrisoning the cities can only come at the cost of learning a second mode of behavior, pursuing the barbarians.

For the purposes of reporting game scores the pillage points collected by the barbarians are normalized to a scale of [0, 100], by calculating the maximum possible points and scaling the actual points down according to the formula:

$$Score_{reported} = 100 \times Points_{actual}/Points_{possible} \qquad (5.1)$$

The result can be interpreted as a pillage rate, stated as a percentage of the expected amount of pillaging that would have been done in the absence of any legions to contest the barbarians' activities. Notice that from the legions' point of view, *lower* scores are better.

In practice the legions are never able to drive the score to zero. This fact is due in part to the vagaries of the starting conditions: if the random set-up places all the legions very distant from a city and a barbarian is placed very near that city on the first turn, there is nothing the legions can do to beat that barbarian into the city, no matter how well trained they are. However, a factor that weighs in more heavily than that is the rapid rate of appearance of the barbarians versus the finite speed of the legions. Since the legions and barbarians move at the same speed, it is difficult for the legions to chase down the barbarians that appear at arbitrary distances away. Moreover, as the legions thin out the barbarians on the map the average distance between the remaining barbarians increases, and it takes the legions longer to chase any additional barbarians down. Thus even for well-trained legions the game settles down into a dynamic equilibrium between the rate of new barbarian arrivals and the speed of the legions, yielding a steady-state density of barbarians on the map, and thus a steady-state accumulation of pillage counts after the equilibrium is achieved.

5.3   Agent Control Architectures

The legions and barbarians are controlled by policies that map egocentric sensory inputs onto a choice of the discrete actions allowed in the game. This section describes their sensors and controllers. The simpler sensors and controllers used by the barbarians are described first, then the more elaborate system used to control the legions.

5.3.1 Barbarian Sensors and Controllers

The legions are the only learning agents in the game, so the barbarians can use any simple pre-programmed logic that poses a suitable threat to the legions' interests. The barbarians' basic design calls for them to be attracted toward cities and repulsed from legions, with the attraction slightly stronger than the repulsion, so that the barbarians will take some risks when an opportunity for pillaging a city presents itself. This behavior is implemented by algebraically combining two "motivation" vectors, one for the attraction toward cities and one for the repulsion from legions:

$$\mathcal{M}_{\text{final}} = \mathcal{M}_{\text{cities}} + 0.9 \mathcal{M}_{\text{legions}} \tag{5.2}$$

Each vector consists of six floating point numbers, indicating the strength of the barbarian's "desire" to move in each of the six cardinal directions. The 0.9 factor is what makes the motivation to flee the legions slightly weaker than the motivation to approach the cities. After the combination, the peak value in the $\mathcal{M}_{\text{final}}$ vector indicates which direction the barbarian "most wants" to move. In situations where a second choice must be considered, the second-highest value in $\mathcal{M}_{\text{final}}$ is used to select the direction instead.

The values in the two arrays are derived, directly or indirectly, from the activation values in a simple sensor system. The barbarian's sensor system consists of two sensor arrays, one that detects cities and another that detects legions. Each array divides the world into six 60° non-overlapping egocentric fields of view. The value sensed for each field is:

$$s = \sum_i \frac{1}{d_i}, \tag{5.3}$$

where d_i is the distance to an object i of the correct type within that field of view. The distances are measured in the hex-tile equivalent of Manhattan distance, i.e. the length of the shortest path of map cells from the viewer to the object, not counting the cell that the viewer itself is in (Fig. 5.3).

For simplicity, if an object is exactly on the boundary between two fields of view, the sensors report it as being in the field to the clockwise of the boundary. Due to the relatively small map, no limit is placed on the range of the sensors.

Notice that this sensor architecture obscures a great deal of detail about the environment. It does not give specific object counts, distances, or directions, but rather only a general indication of how much opportunity or threat the relevant class of objects presents in each of the six fields of view.

Once these values have been calculated and loaded into the sensor arrays, the activations in the array that senses cities can be used directly for the $\mathcal{M}_{\text{cities}}$ vector in Eq. 5.2. $\mathcal{M}_{\text{legions}}$ can be derived from the values in the array that senses legions by permuting the values in the array to reverse their directional senses, i.e. the sensor activation for legions to the west can be used as the motivation value for a move to the east, and similarly for the other five cardinal directions. After the conversions

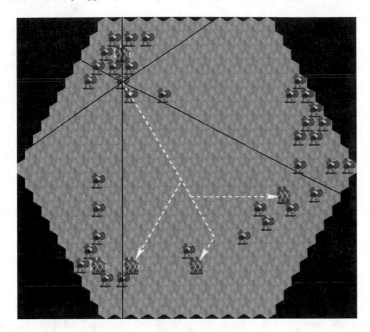

Fig. 5.3 The solid black lines show the boundaries of the six sensory fields of view for one barbarian near the northwest corner of the map. The boundaries emanate from the center of the map cell occupied by the barbarian and out through its six corners. The dashed white lines show the hexagonal Manhattan distances to the three legions in the SE field of view of the sensing barbarian. These lines are traced from the center to center along a path of map cells, and thus emanate through the sides of the hexagons rather than through the corners as the field boundaries do

from sensor activations to motivation vectors, $\mathcal{M}_{\text{final}}$ can be calculated and its peak value identified to determine the requested direction for the current move.

There is no explicit mechanism to allow a barbarian to request remaining stationary for the current turn. For simplicity the game engine examines a barbarian's location at the start of its move and leaves the barbarian stationary if it is already in a city. Otherwise the game engine calculates the values to be loaded into the barbarian's sensors, performs the numerical manipulations described above, and implements the resulting move request if it is not prohibited by the rules described in Sect. 5.2.3.

The resulting behavior, although simple, has the desired effect in the game. As suggested by Fig. 5.3, barbarians will stream toward the cities to occupy them, or congregate around them if the city is already occupied. Other barbarians will flee any roving legions, sometimes congregating in clusters on the periphery of the map. The barbarians are quick to exploit any city that the legions leave unguarded. They do, however, tend to get in each other's way when a legion approaches a crowd and they need to flee, resulting in many casualties, but that is perhaps an appropriate simulation of the behavior of undisciplined barbarians on a pillaging raid.

5.3.2 *Legion Sensors and Controllers*

Unlike the barbarians, the legions are required to learn appropriate behavior for their gameplay. They are therefore provided with a more sophisticated, trainable control system. The design includes a sensor system that provides more detail about the game state than the barbarians' sensors do, plus an artificial neural network "brain" to map the sensor inputs onto a choice of actions.

5.3.2.1 The Legions' Sensors

The legions are equipped with sensor systems that are conceptually similar to the barbarians', but enhanced in several ways. Unlike the barbarians, the legions have a sensor array for all three types of object in the game: cities, barbarians, and other legions. Also unlike the barbarians, each of those three sensor arrays are compound structures consisting of two six-element sub-arrays plus one additional element (Fig. 5.4), rather than the barbarian's simple six-element sensor arrays.

An array's two six-element sub-arrays are similar to the barbarians' sensor arrays, except that one only detects objects in adjacent map cells and the other only detects objects at greater distances. For the former, the game engine sets the array elements to 1.0 if there is an object of the appropriate type in the adjacent map cell in the appropriate direction, and to 0.0 otherwise. For the latter, the game engine assigns values to the elements slightly differently from the way it assigns values to the barbarian's sensors. First, it ignores objects at $d = 1$, since those are detected by the short-range array described above. Second, since the distances used in the calculations for this array are always greater than one, it deducts one from the distances used in the calculations, in order to increase the signal strength. That is, Eq. 5.3 used for the barbarians becomes:

$$s = \sum_i \frac{1}{d_i - 1}, \tag{5.4}$$

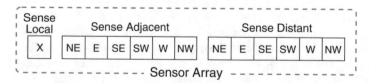

Fig. 5.4 The legions have three sensor arrays, one each for cities, barbarians, and other legions. Each of those three arrays consists of three sub-arrays as shown above. A single-element sub-array (left) detects objects co-located in the map cell that the legion occupies. Two six-element sub-arrays detect objects in the six radial fields of view; one only detects adjacent objects, and the other only detects objects farther away. These 13 elements of each of the three compound arrays are concatenated to serve as a 39-element input activation for an artificial neural network that controls the legion's behavior (Fig. 5.5)

for objects *i* of the correct type *and* at a distance greater than one, within that field of view.

A third difference is that sensed objects near a boundary between two sensor fields are not arbitrarily assigned to the field to the clockwise of the boundary. Instead, objects within 10° of the boundary (from the legion's perspective) have their signal split between the two fields. As a result each sensor array is effectively a set of 40° arcs of unsplit signals alternating with 20° arcs of split signals, though the aggregations result in an array of only six activation values. As with the barbarians' sensors, there is no range limit on this long-range sub-array.

The additional single sensor element in a compound array detects objects in the legion's own map cell: if an object of the appropriate type is present the game engine sets the value of this sensor to 1.0; otherwise it sets it to 0.0. However, a legion does not detect itself, and since the rules prevent multiple units from occupying the same map cell, the only time this local detection sensor is activated in practice is when the legion occupies a city. In principle the detect-local sensor could have been eliminated from the sensor arrays used to detect legions and barbarians, but identical arrays were used for all object types in order to simplify the implementation, and to make allowance for future game modifications that would allow "stacking" multiple units within a single map cell.

The full architecture of the compound sensors is shown in Fig. 5.4. The two sub-arrays contain six elements each, corresponding to the six cardinal directions. Thus together with the additional independent element, each array reports 13 floating point values ≥ 0.0 whenever a sense is collected from the environment. Since there is one compound sensor for each of the three types of game object, a legion's egocentric perception of the game state is represented by 39 floating point numbers.

5.3.2.2   The Legions' Controller Network

A legion's behavior is controlled by a feed-forward neural network. The network maps the legion's egocentric perception of the game state onto a choice of moves. Whenever it is a legion's turn to move, the game engine calculates the sensor values for the legion's view of the current game state and presents the resulting 39 floating point numbers to the input of the controller network. The values are propagated through the network, and the activation pattern at the network's output is decoded to determine the legion's choice of move for the current turn (Fig. 5.5).

The output layer of the networks consist of seven neurons, corresponding to the seven discrete actions available to the legions. When the input activations have been propagated through the network the activation pattern at the output layer is interpreted as an *action unit coding*, i.e. the action corresponding to the output neuron with the highest activation level is taken to be the network's choice of action for the current turn.

In addition to the sensory inputs, each neuron in the controller networks is fed by a bias unit with a fixed activation of +1.0 and a trainable weight to propagate the value into the neuron's accumulator. For the experiments reported below, the

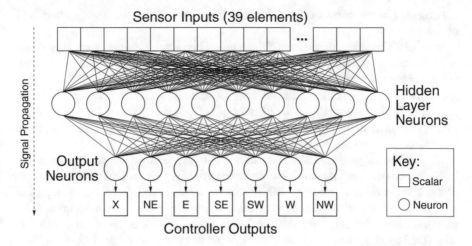

Fig. 5.5 During play the values obtained by a legion's sensors are propagated through an artificial neural network to create an activation pattern at the network's output. This pattern is then interpreted as a choice of one of the discrete actions available to the legion. When properly trained, the network serves as a "brain" for the legion as an autonomous agent in the game

controller network's hidden layer consisted of 10 neurons, which was found to be effective in preliminary survey experiments.

5.3.2.3 Properties of the Legions' Control Architecture

There are a number of important consequences of the adoption of this sensor/controller architecture for the legions, which the reader may wish to keep in mind while reading about the methodologies and experiments:

- *The sensor readings are egocentric.* For a given state of the game, each of the legions in play will perceive the map differently, depending on their individual locations on the map.
- *The sensors provide a lossy view of the map.* The legions have complete state information about their immediate neighborhood, but that is reduced to a fuzzy "feel" for the presence of more distant objects.
- *The legions must work with uninterpreted inputs.* There is a semantic structure to the sensor arrays, but that structure is not known to the legions: the sense values appear as a flat vector of floating point numbers in their controller networks' input layers. The significance of any individual input or set of inputs, or of any correlations between inputs, is something the legions must obtain via the learning process.
- *There is no explicit representation of goals.* None of the network inputs, nor any other part of the controller logic, provide a legion with any sort of objective.

Coherent higher-level behavior must be learned as a response to a sequence of inputs that vary over time.

- *The legions do not have memory.* The feed-forward controller networks do not allow for any saved state, so the legions are not able to learn an internal representation for goals to make up for the lack of externally specified goals. All actions are immediate reactive responses to the environment.

These various requirements and restrictions conspire to present the legions with a very difficult challenge if they are to learn to behave intelligently. The intent of the game designer, and any real or apparent intent on the part of the individual legions, must be instilled by means of the learning system. However, experience shows that properly trained artificial neural networks excel at producing the appearance of purposeful intelligent behavior (e.g. [1, 10, 17, 19, 20, 23]).

5.4   Neuroevolution With Enforced Sub-Populations (ESP)

For many agent control tasks the correct input-output mappings for the agents' controllers are not known, so it is not possible to program them or train them with supervised learning methods. However, controllers such as artificial neural networks can be evolved to perform a task in its actual context, discovering optimal mappings in the process. The use of a genetic algorithm to train an artificial neural network is called *neuroevolution*. Surveys of the field can be found in [24, 28]. An overview of the use of neuroevolution to learn egocentric input-output mappings for game agents' controllers can be found in [16].

One of the most empirically effective neuroevolutionary algorithms yet devised is *Neuroevolution with Enforced Sub-Populations* (NE-ESP, or usually just ESP) [9, 11]. The basic concept behind ESP is that each genetic representation specifies only a single neuron rather than an entire network, and a separate breeding population is maintained for each neuron in the network.

Evaluations cannot be made on a network's neurons in isolation, so the evaluations in ESP are done by drawing one neuron at random from each sub-population, assembling them into a complete network, evaluating the network as for any other neuroevolutionary algorithm, and ascribing that network's fitness score back to each of the individual neurons used to create it. When all the neurons in all the populations have been evaluated, selection and breeding is done independently within each sub-population. However, the fitness of an individual neuron depends not on its properties in isolation, but on how well it works together with neurons from the other populations. Thus the neurons in the sub-populations are subjected to cooperative coevolution [18, 21], and as evolution progresses they converge as symbiotic species into functional niches that work together in a network as a good solution to the target problem.

ESP was originally introduced for training fully recurrent networks as continuous-state controllers, e.g. for the inverted pendulum problem and the conceptually similar application to a finless rocket [12]. Both the application and the details of the ESP implementation used for *Legion II* are novel.

In *Legion II* ESP is used for learning to make a discrete choice among the legions' possible atomic actions. For the experiments reported here, the controller networks were non-recurrent feed-forward networks with a single hidden layer, as described in Sect. 5.3.2.2 (Fig. 5.5). A distinct sub-population was used for each position for a neuron in the network, regardless of which layer it is in; the representations in the populations held only the weights on the input side of the neurons (Fig. 5.6). In principle it is not necessary to provide separate neurons for the output layer; an architecture more similar to previous uses of ESP would have dispensed with those neurons and stored both the input and output weights in the representations of the hidden-layer neurons. That is in fact the mechanism used in the original *Legion I* experiments [4]. However, the introduction of new sub-populations for the output layer contributed to the improved scores in the experiments reported here.

Fitness evaluations were obtained during evolution by playing the current generation of controllers against randomly generated game setups; the set of possible game setups is so large that none ever have to be reused. A different sequence of training games was used for each independent run with a given parameterization in a given experiment. For fair evaluations within a single run, every neuron was evaluated against the same game before moving on to the next game. The methodology is described in more detail in Sect. 5.5.

When the ESP-mechanism is used, the actual fitness of a network is ascribed to each neuron used to construct it. As a result, the ascribed fitness is only an estimate of a neuron's "true" fitness; the "true" fitness is in fact ill-defined, since the neurons are only useful when associated with other neurons in the other populations. However, a reasonable estimate of the fitness of a neuron – *given that it will be used in a network with neurons from the other populations* – can be obtained by evaluating the neuron repeatedly, in networks comprised of independent random selections of neurons.

Thus for the experiments described here each neuron was evaluated on three different games per generation, and the three resulting fitness ratings were averaged to estimate the neuron's fitness. The associations of the neurons into networks were re-randomized before each of the three games so that the averaged fitness ratings would reflect the quality of a given neuron per se more than the quality of the other neurons it happened to be associated with in the network. Each of the three evaluations used a different game setup, and all of the neurons were evaluated on the same three games during the generation.

Since the training game setups differed continually from generation to generation, learning progressed somewhat noisily: a neuron that performed well on the training games in one generation might not perform well on the new training games of the following generation. However, neuroevolution with ESP is robust even when evaluations are somewhat noisy, and the use of three games per generation helped smooth the noise of the evaluations. The continually changing stream of training games from generation to generation required candidate solutions to generalize to novel game setups, or else risk having their constituent neurons be weeded out of the breeding population; if a network performed poorly on the game setup used during a given

Fig. 5.6 To apply the ESP method of neuroevolution for training the legions' controllers, a separate breeding population was maintained for each of the 17 neurons used in the controller network

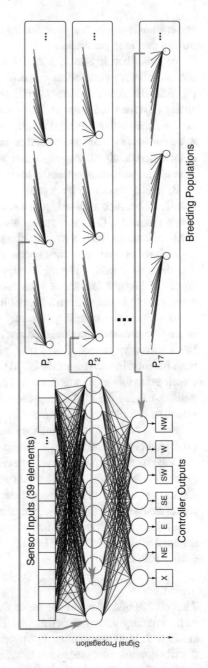

generation it received a poor fitness score, regardless of how well it had performed during previous generations.

As described in Sect. 5.5.2, an evaluation against the validation set was done at the end of every generation. For ordinary evolutionary mechanisms the network that performed best on the current generation's fitness evaluations would be chosen for the run against the validation set. However, the notion of "best network in the population" is ill-defined when the ESP mechanism is used, so for the *Legion II* experiments a *nominal best network* is defined as the network composed by selecting the most fit neuron from each sub-population. It was that nominal best network that was evaluated against the validation set at the end of each generation.

Breeding was done by a probabilistic method that strongly favored the most fit solutions, but also allowed less fit solutions to contribute to the next generation with low probability. The mechanism was as follows. When all the training evaluations for a generation were complete, the storage for the representations of the solutions in each sub-population was sorted from most fit to least fit, so that the most fit had the lowest index. Then each representation was replaced one at a time, starting from the highest index (i.e., the least fit neuron in the population). Two parents were selected with uniform probability over the indices less than or equal to the index of the representation currently being replaced. I.e., that representation or any more fit representation could be chosen as a parent. The two selections were made independently, so that it was possible for the same representation to be used for both parents; in such cases the child would differ from the parent only by mutations. (Notice that this mechanism *always* breeds the most fit neuron with itself at the final pairing.) Since the less fit representations were progressively eliminated from the effective breeding pool, the more fit solutions had more opportunities to contribute to the next population. Preliminary survey experiments showed that this mechanism produced better results than a simple elitist mechanism.

Once a pair of parents were selected they were bred with either 1-point or 2-point crossover, with a 50% chance for each. Only one child was produced from the crossover; the remaining genetic material was discarded. Each weight in the representation was then subjected to a mutation at a 10% probability, independently determined. Mutations were implemented as a delta to the current weight chosen from the exponential distribution (Eq. 5.5) with $\lambda = 5.0$, and inverted to be a negative delta with a 50% chance.

$$f(x, \lambda) = \lambda e^{-\lambda x}, x \geq 0 \qquad (5.5)$$

That choice of λ reduced the mean of the distribution, and was chosen on the basis of preliminary survey experiments. The deltas resulting from this distribution were small with high probability, but potentially very large with a low probability. That distribution allowed mutations to support both fine tuning of the weights and jumps to more distant regions of the solution space.

Training on the *Legion II* problem with neuroevolution makes progress asymptotically. For the experiments reported here, evolution was allowed to continue for 5000 generations, well out onto the flat of the learning curve, to ensure that comparisons and analyses were not made on undertrained solutions.

5.5   Experimental Methodology

The *Legion II* experiments followed the familiar methodology of using distinct training, validation, and test sets. However, procedural questions arise when applying that
methodology to a game such as *Legion II*. This section explains how those questions
were resolved for the experiments reported below.

5.5.1   Repeatable Gameplay

When training or testing by means of dynamic gameplay rather than static examples,
it is useful to have a definition for the concept of "the same game", e.g. to make
comparative evaluations of the performance of embedded game agents. However,
games that are genuinely identical with respect to the course of play are, in general,
impossible to generate, if they involve embedded game agents that learn: as the agents
learn, their behavior will change, and the changed behavior will cause the course of
the game to vary from earlier plays. For example, if the legions in *Legion II* fail to
garrison the cities during the early stages of training, the barbarians will occupy the
cities. But later during training, when the legions have learned to garrison the cities,
the details of the barbarians' behavior must also change in response – i.e., the city
will not be pillaged as before – even if there has been no change to the starting state
and the barbarians' control policy.

It is therefore useful to have a pragmatic definition of "the same game" for experimental work. Thus for *Legion II* two games are identified as "the same game" if
they use the same starting position for the cities and legions, and the same schedule
for barbarian arrivals. The schedule for arrivals includes both the time and the randomly selected position on the map. For all the games reported here the barbarian
arrivals were fixed at one per turn, so only their placement mattered for identifying
two games as being the same.

However, the randomized placement of the barbarians is not always repeatable:
as described in Sect. 5.2.3, if the position selected for placing a new barbarian on
the map is occupied, an alternative randomly selected position is used instead, and
re-tries continue until an empty map cell is found. But as described above, changes
to the legions' behavior will result in different game states at a given point in time
for various instances of "the same game", so a barbarian placement during one play
of the game may not be repeatable in another run using a different controller for the
legions. Therefore, for pragmatic reasons, "the same game" is defined for *Legion II*
to consider *only the first try* for the positioning of arriving barbarians; the additional
tries triggered by the unavoidable divergences of the game state are not considered
to make two games different.

This concept of "the same game" was used to create sets of games that were used
repeatedly during training and testing, as follows.

5.5.2 Training

Randomized learning algorithms such as neuroevolution do not always produce their best solution at the end of a fixed-length run; the random modifications to the representations in an evolutionary population can make the solutions worse as well as better. Therefore it is useful to have a mechanism for returning the best solution obtained at any time in the course of a run.

The commonly used mechanism is to evaluate candidate solutions periodically during training, and, if the top performer is better than any previously encountered during the run, to save that top performer as the potential output of the learning algorithm. At the end of the run, the most recently saved top performer is returned as the solution produced by the algorithm. The learning algorithm still *runs* for some fixed number of iterations that the experimenter deems sufficient for finding a good solution, but that solution may be discovered at any time during the run.

The periodic evaluation is performed against a *validation set*. When generalization is desired, the validation set must be independent of the training data; otherwise the algorithm will return a solution that is biased toward good performance on the training data at the expense of poorer performance on more general data of the same type. For supervised learning, the validation set normally takes the form of a reserved subset of the available training examples. However, when annotated examples are not available, such as when using evolutionary learning to learn a motor control task or a controller for an embedded game agent, the validation set can be a standardized set of example problems. The definition of "the same game" in *Legion II* allows construction of a distinctive set of games to serve as the validation set for *Legion II* learning tasks, and that is the mechanism used in the experiments reported here.

Therefore stopping was handled in the experiments by running the learning algorithms for a period deemed to be "long enough", and using the validation set mechanism to control which candidate was actually returned as the result of a run. The validation set for *Legion II* was a set of ten games. A set of ten games with independently generated starting positions and barbarian placement positions was judged to be a sufficient evaluation for generalization; larger sets adversely affect the run time of the evolutionary algorithm. The score for a controller's validation performance was defined as the average of the game scores obtained by play against the ten games of the validation set.

An evaluation was made against the validation set at the end of each generation, and the nominal best network saved if its validation score was better than at any previous generation. For a given run of the training program the same validation set was used at each evaluation period, to ensure consistent evaluations. However, the validation set was created independently for each run. The idea is that each run should represent an independent sample of the space of all possible runs, for a given parameterization of the learning algorithm. Since the random selection of a validation set is part of the the "possible world" of the run of a stochastic algorithm, its construction was allowed to vary from run to run, along with all the other stochastic decisions.

5.5.3 Testing

Each run of the learning algorithm returned a single neural network as its output. The networks were saved to files for later testing with a separate program; the test program spilled various run-time metrics to a file for analysis and plotting with the *R* statistical computing environment [22].

Tests of current performance were also conducted during the course of training, for the production of learning curves. Tests were only run on those generations where the evaluation on the validation set produced a new top performer, i.e. when measurable progress had been made. These were "side tests"; the learning algorithms ran the tests and spilled the results to a file for later analysis, but did not make any training decisions on the basis of the tests, to avoid biasing the training toward the test games.

Whether during the learning run or afterward, tests were run on a set of games constructed as the validation set was, but independent of both the validation set and the training games. As with the validation evaluations, the evaluation score for this composite test was defined as the average of the scores obtained on the individual games of the test set.

Unlike the validation set, the same test set was used for every independent run of every learning algorithm, to ensure that any differences in the test metrics were the result of differences in the solutions being examined rather than differences in the difficulty of independently generated test sets. The training-time evaluations on the test set are not as frequent as the evaluations on the validation set, so a larger test set could be used without unduly extending the run times of the training algorithms. Also, it is essential that the test set be an accurate model of the set of possible games; therefore a set of 31 games was used. (Statisticians deem a minimum of 30 samples necessary for characterizing a distribution when the measurements are not known *a priori* to fall into a normal distribution; it sometimes proves useful to use 31 rather than that minimum, so that there will be a clearly defined median for any measurement, to be used as a principled choice whenever it proves useful to plot or analyze a single "typical" example.)

5.6 Experiments

ANN controllers for the legions in *Legion II* were trained using ESP and the procedures described above. The homogeneity required by the ATA architecture was enforced by using the same controller to make all the legions' decisions during a game. The game parameters were set to require a division of labor to perform well: there were more legions than cities, the randomized placement of the barbarians and the 100:1 ratio of pillage between the cities and countryside made it essential to garrison the cities, and the large number of barbarians arriving over the course of the game made it essential to eliminate barbarians in the countryside as well, if pillage was to be minimized. With one barbarian arriving per turn, the count would ramp up from one to 200 over the course of a game in the absence of action by the legions, providing an average of ~100 pillage points per turn. With three cities each subject to an additional 100 pillage points per turn, pillaging the countryside can amount to ~1/4 of the worst possible score. Thus the legionary ATA must take actions beyond simply garrisoning the cities in order to minimize the pillage: a division of labor is required.

The following sections examine the results of the training experiment and the behavior produced in the legions.

5.6.1 Learning the Division of Labor

Hundreds of runs of neuroevolutionary learning on the *Legion II* problem, with a variety of learning parameters and a number of changes to the game rules and network architecture since the initial results reported in [4], have consistently performed well, where "well" is defined fuzzily as "learns to bring the pillage rate substantially below the 25% threshold" obtainable by a policy of static garrisons and no division of labor to support additional activity by the spare legions. For the experiments reported here, eleven independent runs of the base learning method with the parameters described in Sect. 5.4 (but independent streams of random numbers) produced a mean test performance score of 4.316%, with all falling in the range 3.5–6.0%. (Recall that there is no *a priori* expectation that a 0% pillage rate could be learned.) The scores on the games in the test set show that all eleven runs produced controllers that allowed the legions to reduce pillaging well below the 25% rate obtainable by garrisoning the cities and taking no further actions against the barbarians.

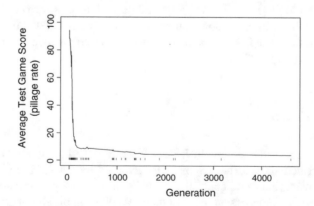

Fig. 5.7 The plot shows progress against the test set for the median performer of the eleven training runs. At each generation when progress was made on the validation set, the nominal best network was also evaluated against the test set. The hatch marks at the bottom of the plot identify those generations. Test scores for those generations (only) are connected with straight lines to improve visibility. The plot is not strictly monotonic because progress on the validation set does not strictly imply progress on the test set

As described in Sect. 5.5.3, performance against the test set was also checked during training so that learning progress can be examined. A typical learning curve is shown in Fig. 5.7. The learning curve is familiar from many machine learning applications (though inverted because lower scores are better), with fast initial learning tapering off into slower steady progress. Experience shows that learning by neuroevolution on the *Legion II* problem appears to progress asymptotically. There is no stair-step pattern to suggest that the legions' two modes of behavior were learned sequentially; observations confirm that the legions begin chasing barbarians even before they have learned to garrison all the cities rigorously.

The behavior of a trained controller network can be evaluated qualitatively by observing real-time animations of game play. In every case that has been observed, trained legions begin the game with a general rush toward the cities, but within a few turns negotiate a division of labor so that some of the legions enter the cities or remain near them as garrisons while the others begin to chase down barbarian warbands in the countryside. The only time the cities are not garrisoned promptly is when their positioning allows two of them mask the third from the legions' low-resolution sensors. However, even in those cases the third city is garrisoned as soon as one of the roaming legions pursues a barbarian far enough to one side to have a clear view of the third city so that it can "notice" that it is ungarrisoned. A feel for these qualitative context-aware behaviors can be obtained by comparing end-of-game screenshots taken early and late during a training run, as shown in Fig. 5.8. An animation of the trained legions' behavior can be found at http://nn.cs.utexas.edu/keyword?ATA.

The legions' division of labor can can also be examined by the use of run-time metrics. The test program was instrumented to record, after each legion's move, how

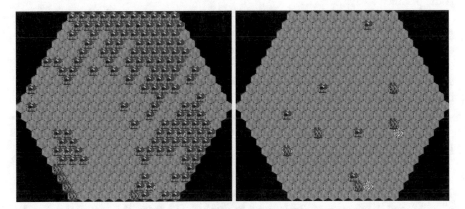

Fig. 5.8 Two end-of-game screenshots show the legions' performance before and after training. *Left*: Before training the legions move haphazardly, drift to an edge of the map, or sit idle throughout the game, thereby failing to garrison the cities and allowing large concentrations of barbarians to accumulate in the countryside. *Right*: After training the legions have learned to split their behavior so that three defend the three cities while the other two move to destroy most of the barbarians pillaging the countryside. The desired adaptive behavior has been induced in the team

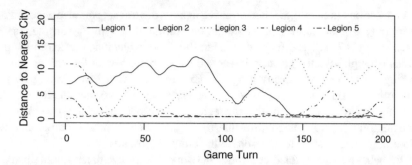

Fig. 5.9 The plot shows the distance from each legion to its nearest city over the course of a single game. The five legions start at random distances from the cities, but once some legion has had time to reach each of the three cities the team settles into an organizational split of three garrisons and two rovers. Note that the garrisons' average distance from their respective cities is not 0.0, because whenever there is only a single adjacent warband it is reasonably safe to exit the city long enough to eliminate it. The legions sometimes swap roles when a rover approaches a garrisoned city, e.g. Legions #3 and #4 just before turn 25. (The numbering of the legions is arbitrary; they are identical except for the happenstance of their starting positions. The lines have been smoothed by plotting the average of the values measured over the previous ten turns.)

far away it was from the nearest city. The result for allowing the median performer among the eleven trained networks to play the first game in the test set is shown in Fig. 5.9. The plot clearly shows that after a brief period of re-deploying from their random starting positions three of the legions remain very near the cities at all times while two others rove freely. The rovers do approach the cities occasionally, since that is where the barbarians primarily gather, but for most of the game they remain some distance away.

When a rover does approach a city there is sometimes a role swap with the current garrison, but the 3:2 split is maintained even after such swaps. However, the legions show surprisingly persistent long-term behavior for memoryless agents: the plot shows that Legion #1 acts as a rover for almost 3/4 of the game, and Legion #3, after starting as a garrison and then swapping roles with a rover, spends the final 7/8 of the game in that new role.

5.6.2   Run-Time Readaptation

The training games were parameterized to require the legions to organize a division of labor, and they successfully learned to do that. However, the motivation for the ATA multi-agent architecture in Sect. 5.1 calls for teams that can reorganize whenever a change in circumstances requires it. For example, if the pumper robot in the motivating example breaks down, one of the other robots should take over the task so that the team will not fail entirely. The legions in the *Legion II* game should also be able to reorganize at need.

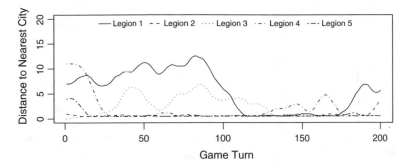

Fig. 5.10 In a second test, play follows the pattern of the previous test until turn 100, when a fourth city is added to the map. The team is forced to re-organize its division of labor so that there are now four garrisons and only a single rover. The role swaps continue, but they always leave four legions hovering near the cities. The typically lower average distance from the rover to its nearest city after the city is added is an artifact of the increased density of cities on the map. Legion #2 erroneously abandons its city near the end of the game; see the text for the explanation

That necessary ability was examined by modifying the test program to support the addition or removal of a city at the mid-point of the test games. When a city is added, the legions should reorganize into a team of four garrisons and one rover, or when a city is removed they should reorganize into a team of two garrisons and three rovers.

The result of adding a city is shown in Fig. 5.10. The plot again shows the median-performing controller's behavior on the first game in the test set. Since the legions' behavior is deterministic and the game's stochastic decisions are repeated, as described in Sect. 5.5.1, the game follows its original course exactly, up until the city is added at the mid-point of the game. Thereafter the team can no longer afford to have two rovers, and the plot shows the resulting reorganization. There are still role swaps in the second half of the game, but the swaps now always maintain four garrisons and a single rover.

The average distance from the rover to the cities is lower in the second half of the game. That is primarily an artifact of having more cities on the small map: regions that were once distant from all the cities no longer are, so even without any change in behavior the rover is expected to be nearer some city than before, on average. A second cause is an indirect result of the change in the team's organization. With only one rover in the field, the legions are not able to eliminate the barbarians as quickly as before, so during the second half of the game the concentration of barbarians on the map builds up to a higher level than previously. Since they tend to crowd around the cities and the roving legions tend to chase down the barbarians wherever they mass, the roving legion now has more reason to operate close to the cities.

The plot shows Legion #2 vacating the city it was garrisoning right at the end of the game. That is also an artifact of the increased density of the barbarians on the map. In ordinary play the trained legions are able to maintain a dynamic equilibrium between the rate of influx of the barbarians and the rate they are eliminated; the denser the barbarians are on the map, the easier it is for the rovers to catch some

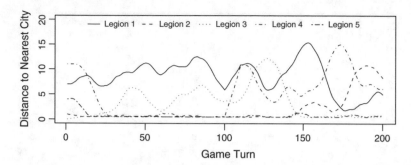

Fig. 5.11 In a third test, the mid-game reorganization experiment is repeated, except this time a city is *removed* from the map. That city had been garrisoned by Legion #4, which now finds itself far from the nearest city, and immediately adopts the role of a rover. (The steep ramp-up of its nearest-city distance on the plot is an artifact of the smoothing; the change is actually instantaneous when the city is suddenly removed from under the legion.) There is a role swap just before turn 150, but the team consistently keeps two legions very near the two remaining cities, with the other three roving at various distances

of them. However, when the add-city test causes one of the rovers to swap roles to garrison duty, that equilibrium can no longer be maintained by the single remaining rover, and the number of barbarians in play starts ramping up after the city has been added. Eventually their number oversaturates the legions' sensors – they have not seen such densities since early during training – and the legions begin to behave erratically. However, until their sensory input diverges quite far from what they were trained for, the legions are seen to exhibit the desired behavior.

The result of removing a city at the mid-point of a game is shown in Fig. 5.11. The play proceeds as before, until the city is removed at turn 100. At that point the legion formerly garrisoning that city finds itself far away from any, but it adopts the roving behavior rather than sitting idle or trying to crowd into one of the other cities, and it maintains that behavior for the remainder of the game. There is a role swap between two of the other legions later, but the team is always left with two legions hovering very near the cities on garrison duty, while the other three range over various distances in pursuit of the barbarians.

5.7 Discussion

The experiments show that the Adaptive Team of Agents is a feasible architecture for multi-agent systems, and that ATAs can be created by neuroevolutionary methods. The legions learned the desired variety of behavior, and the ability to organize a division of labor by individually adopting an appropriate choice of behaviors. They also learned to swap roles without disrupting the required organization of the team, both in the ordinary course of events and in response to a change in the scope of their task.

Such capabilities are essential for the agents embedded in many types of game or simulator. Games often provide multiple agents of some generic type – e.g. the settler type in the *Civilization* game – which must as individuals pursue differing activities that contribute to the success of the team rather than the individual. And those agents must be adaptive in their choice of activities, taking account of the game state, including the choices being made by their peers. Yet the scripted behavior of agents in commercial and open source games commonly fail at that requirement, making decisions that appear to take little account of context. For example, in strategy games, even when the agents are not autonomous and a higher-level AI is able to manipulate them according to some plan, individual agents are frequently observed to make piecemeal attacks that are suicidal due to a lack of supporting actions by their peers. To a human observer, such agents simply do not seem very intelligent. Appropriate computational intelligence methods should be able to take game intelligence beyond the brittleness, inflexibility, and narrowness of scripted activity, and for many games or simulators the context-awareness and adaptivity of the agents in an ATA will be a necessary part of any successful solution.

In the field of evolutionary robotics, Floreano et al. also examined homogeneous teams controlled by ANNs evolved by team selection, in a study of hypotheses for explaining biological altruism [8]. Altruism does not play an explicit role in *Legion II*, but their study found that a homogeneous team evolved by team selection performed better than three other architectures examined, producing robust altruistic behavior in the process. Altruism, when appropriate, is an important facet of trusted autonomy in multi-agent environments, and can contribute to the appearance of intelligent behavior as well.

It is interesting to note that the necessary adaptivity for our ATA was obtained using a simple feed-forward network for the legions' controllers. We know that artificial neural networks are powerful computing devices (see e.g. [7, 25]), and that genetic algorithms are able to train them to sophisticated behaviors (e.g. [1, 3, 15, 26, 27, 29]). To a first approximation it may be concluded that the *Legion II* controllers have been trained to partition the game's state space, as seen from an egocentric point of view, into two classes, and to choose a behavior on the basis of which class the current state observation falls in to. However, what they actually choose is one of seven atomic moves, none of which can be uniquely associated with either of the two behavior classes.

For an agent to pursue a coherent higher-level behavior across many game turns – i.e., to give an appearance of intent-driven behavior – would seem to require access to some internal state, i.e. an ability to "remember" what it is doing. Conjecturally, the *Legion II* agents have learned a workaround whereby they effectively store their internal state in the external environment. I.e., in addition to whatever else they learn during training, they learn a mapping from their egocentric view of the environment to a virtual representation of whatever internal state information is necessary for "remembering" what they are doing. The flow of information is in fact recurrent: the fact that the agents move within their environment causes a transformation of their next view of the environment. In an otherwise static environment those transformations would be deterministic; the presence of other agents in the *Legion II* environment makes them somewhat noisy. However, the potential noisiness in the state transitions may be greatly reduced if the agents also learn an implicit model of the other agents' behaviors. If they can predict how those other agents will act, they can learn to predict what effect their choice of actions will have on their next snapshot view of their world with high accuracy. Thus it is possible, in principle, for the pairing of a sufficiently powerful computational device with a sufficiently powerful learning mechanism to learn to use an egocentric view of the external state as if it were an internal state, for systems such as the *Legion II* game. Future work must pursue this concept to determine what the limits of such a mechanism are. Does a feed-forward network embedded in an environment that it can manipulate become as powerful as a Turing machine?

The *ad hoc* use of plots of an *ad hoc* metric for detecting the legions' division of labor in Sect. 5.6 also reveals a need for developing methods of behavior analysis. When studying agents in visible environments such as games and simulators, behavior is paramount [3]. Meaningful behavioral metrics are essential, and it would be useful to have methods that are abstract enough to be portable across application domains, and sensitive enough to detect similarities or differences in behavior when a domain involves more subtlety than a switch between two discrete behaviors. Work in this area is already underway, and will be a major component of the study of *visibly intelligent behavior* in the future.

The *Legion II* ATA experiments also revealed a special challenge for the application of computational intelligence methods to agent behavior problems. The goal of a simulation as understood by a machine learning algorithm – e.g. minimizing pillage in the *Legion II* game – may be satisfied by some abstract optimization, with little or no regard for the appearance of details of the learned behavior. For example, the legions in the *Legion II* ATA experiment learned to switch between appropriate roles on the basis of context, but some of the details of their behavior are not satisfactory to an observer. The garrisons' learned behavior often produced "mindless" oscillations in and out of their cities when there were no barbarians nearby to threaten pillage, and such behavior would likely be the subject of ridicule if seen in the behavior of the agents in a commercial game. In principle such details of behavior can be addressed by careful specification of the goals of the training regimen, such as an evolutionary reward function that penalizes undesirable behavior, but for applications as complex as a commercial game it may be as difficult to specify an appropriate reward function as it has proven to be to write a script that covers all situations adequately. Therefore work is underway on suppressing such oddities of behavior and inducing other desirable traits that will make agents *look* intelligent to observers, rather than merely acting out some abstractly optimal solution to the problem they have been trained for. (See [3] for an extensive preliminary treatment.)

5.8   Conclusions

The *Adaptive Team of Agents* is a viable architecture for systems based on the methods of computational intelligence, and has immediate applications in domains such as games and simulators, where autonomous agents must show flexible diversity of behavior at both the individual and team level. Such flexibility, supported by fulfilled trust among the members of a team, is a critical component of trusted autonomy in multi-agent systems, and is a key aspect of the sort of visibly intelligent behavior that viewers expect from agents that model real or imagined creatures or groups in some simulated world. Neuroevolution in particular can create such flexibility, along with other desired characteristics of visibly intelligent behavior. More powerful neuroevolutionary methods continue to be developed, and it can be expected that further work in applying them to the rich challenges of modern videogames will produce results of both practical and scientific merit.

Acknowledgements This work was supported in part by NSF grant IIS-0083776, Texas Higher Education Coordinating Board grant ARP-003658-476-2001, and a fellowship from the Digital Media Collaboratory at the IC^2 Institute at the University of Texas at Austin. Most of the CPU time for the experiments was made possible by NSF grant EIA-0303609, for the Mastodon cluster at UT-Austin.
Civilization is a registered trademark of Take-Two Interactive Software, Inc., and *Warcraft* is a trademark of Blizzard Entertainment, Inc. The images used in *Legion II*'s animated display are derived from graphics supplied with the FOSS game *Freeciv*, http://www.freeciv.org/.

References

1. A. Agogino, K. Stanley, R. Miikkulainen, Online interactive neuro-evolution. Neural Process. Lett. **11**, 29–38 (2000)
2. T. Balch, *Behavioral Diversity in Learning Robot Teams*. Ph.D. thesis, Georgia Institute of Technology, 1998. Technical Report GIT-CC-98-25
3. B.D. Bryant, *Evolving Visibly Intelligent Behavior for Embedded Game Agents*. Ph.D. thesis, Department of Computer Sciences, The University of Texas at Austin, Austin, TX, 2006
4. B.D. Bryant, R. Miikkulainen, Neuroevolution for adaptive teams, in *Proceeedings of the 2003 Congress on Evolutionary Computation (CEC 2003)*, vol. 3 (IEEE, Piscataway, NJ, 2003), pp. 2194–2201
5. B.D. Bryant, R. Miikkulainen, Evolving stochastic controller networks for intelligent game agents, in *Proceeedings of the 2006 Congress on Evolutionary Computation (CEC 2006)* (IEEE, Piscataway, NJ, 2006), pp. 3752–3759
6. B.D. Bryant, R. Miikkulainen, Exploiting sensor symmetries in example-based training for intelligent agents, in *Proceeedings of the 2006 IEEE Symposium on Computational Intelligence and Games (CIG'06)*, ed. by S.J. Louis, G. Kendall (IEEE, Piscataway, NJ, 2006), pp. 90–97
7. G. Cybenko, Approximation by superpositions of a sigmoidal function. Math. Control Signals Syst. **2**(4), 303–314 (1989)
8. D. Floreano, S. Mitri, A. Perez-Uribe, L. Keller, Evolution of altruistic robots, in *Computational Intelligence: Research Frontiers: IEEE World Congress on Computational Intelligence*, WCCI 2008, Hong Kong, China, June 1–6, 2008, Plenary/Invited Lectures, ed. by Jacek M. Zurada, Gary G. Yen, Jun Wang (Springer, Berlin, 2008), pp. 232–248

9. F. Gomez, *Robust Non-Linear Control Through Neuroevolution*. Ph.D. thesis, Department of Computer Sciences, The University of Texas at Austin, 2003
10. F. Gomez, R. Miikkulainen, 2-D pole-balancing with recurrent evolutionary networks, in *Proceedings of the International Conference on Artificial Neural Networks* (Springer, Berlin, 1998), pp. 425–430
11. F. Gomez, R. Miikkulainen, Solving non-Markovian control tasks with neuroevolution, in *Proceedings of the 16th International Joint Conference on Artificial Intelligence* (Morgan Kaufmann, San Francisco, CA, 1999), pp. 1356–1361
12. F. Gomez, R. Miikkulainen, Active guidance for a finless rocket using neuroevolution, in *Proceedings of the Genetic and Evolutionary Computation Conference* (Morgan Kaufmann, San Francisco, CA, 2003), pp. 2084–2095
13. T.D. Haynes, S. Sen, Co-adaptation in a team. Int. J. Comput. Intell. Organ. **1**, 231–233 (1997)
14. J.E. Laird, M. van Lent, Human-level AI's killer application: interactive computer games, in *Proceedings of the 17th National Conference on Artificial Intelligence* (AAAI Press, Menlo Park, CA, 2000)
15. S.M. Lucas, Cellz: a simple dynamic game for testing evolutionary algorithms, in *Proceeedings of the 2004 Congress on Evolutionary Computation (CEC 2004)* (IEEE, Piscataway, NJ, 2004), pp. 1007–1014
16. R. Miikkulainen, B.D. Bryant, R. Cornelius, I.V. Karpov, K.O. Stanley, C.H. Yong, Computational intelligence in games, in *Computational Intelligence: Principles and Practice*, Chap. 8, ed. by G.Y. Yen, D.B. Fogel (IEEE Computational Intelligence Society, Piscataway, NJ, 2006), pp. 155–191
17. D. Moriarty, R. Miikkulainen, *Learning sequential decision tasks*. Technical Report AI95-229, Department of Computer Sciences, The University of Texas at Austin, 1995
18. D.E. Moriarty, *Symbiotic Evolution of Neural Networks in Sequential Decision Tasks*. Ph.D. thesis, Department of Computer Sciences, The University of Texas at Austin, 1997. Technical Report UT-AI97-257
19. D.E. Moriarty, R. Miikkulainen, Discovering complex Othello strategies through evolutionary neural networks. Connection Sci. **7**(3), 195–209 (1995)
20. D.E. Moriarty, R. Miikkulainen, Evolving obstacle avoidance behavior in a robot arm, in *From Animals to Animats 4: Proceedings of the Fourth International Conference on Simulation of Adaptive Behavior*, ed. by P. Maes, M.J. Mataric, J.-A. Meyer, J. Pollack, S.W. Wilson (MIT Press, Cambridge, MA, 1996), pp. 468–475
21. M.A. Potter, K.A. De Jong, Evolving neural networks with collaborative species, in *Proceedings of the 1995 Summer Computer Simulation Conference* (1995)
22. R Development Core Team, *R: A language and environment for statistical computing*. R Foundation for Statistical Computing, Vienna, Austria, 2004
23. N. Richards, D. Moriarty, P. McQuesten, R. Miikkulainen, Evolving neural networks to play Go, in *Proceedings of the Seventh International Conference on Genetic Algorithms (ICGA-97, East Lansing, MI)*, ed. by T. Bäck (Morgan Kaufmann, San Francisco, CA, 1997), pp. 768–775
24. J.D. Schaffer, D. Whitley, L.J. Eshelman, Combinations of genetic algorithms and neural networks: a survey of the state of the art, in *Proceedings of the International Workshop on Combinations of Genetic Algorithms and Neural Networks*, ed. by D. Whitley, J. Schaffer (IEEE Computer Society Press, Los Alamitos, CA, 1992), pp. 1–37
25. H.T. Siegelmann, E.D. Sontag, Analog computation via neural networks. Theor. Comput. Sci. **131**(2), 331–360 (1994)
26. K.O. Stanley, B.D. Bryant, R. Miikkulainen, Evolving adaptive neural networks with and without adaptive synapses, in *Proceeedings of the 2003 Congress on Evolutionary Computation (CEC 2003)*, vol. 4 (IEEE, Piscataway, NJ, 2003), pp. 2557–2564
27. K.O. Stanley, B.D. Bryant, R. Miikkulainen, Real-time neuroevolution in the NERO video game. IEEE Trans. Evol. Comput. **9**(6), 653–668 (2005)
28. X. Yao, Evolving artificial neural networks. Proc. IEEE **87**(9), 1423–1447 (1999)
29. C.H. Yong, R. Miikkulainen, *Cooperative coevolution of multi-agent systems*. Technical Report AI01-287, Department of Computer Sciences, The University of Texas at Austin, 2001

Chapter 6
The Blessing and Curse of Emergence in Swarm Intelligence Systems

John Harvey

6.1 Introduction

We live in an increasingly complex and interconnected world, where there is an increasing need for autonomous systems that can control systems that are beyond the capabilities of human operators. To be useful, however, these autonomous systems must be able to be trusted, even in scenarios which cannot be predicted in advance. This is particularly important in safety critical systems where a mistake may lead to loss of life. At the same time, however, not taking advantage of the performance benefits of autonomous systems could also potentially lead to loss of life. One of the key issues to be addressed in developing trusted autonomous systems is dealing with the phenomenon of 'emergence', either by taking advantage of emergence or avoiding emergence.

In simple terms, emergence is behaviour at the global level that was not programmed in at the individual level and cannot be readily explained based on behaviour at the individual level. More formally, De Wolf identifies that "A system exhibits emergence when there are coherent emergents at the macro-level that dynamically arise from the interactions between the parts at the micro-level. Such emergents are novel w.r.t. the individual parts of the system" [1]. A well known example of emergence is the appearance of 'gliders' in Conway's *The Game of Life* [2]. The glider-like objects are an outcome of the code that controls the *The Game of Life* but the objects themselves were never explicitly 'designed in' as part of the code. In nature, the complex patterns displayed by flocks of birds and schools of fish are an emergent property of the interaction of many individual units without any centralised control.

Emergence is closely related to the concepts of 'complexity' and 'self-organisation'. Including both of these concepts, Goldstein defines emergence

J. Harvey (✉)
School of Engineering and Information Technology,
University of New South Wales, Canberra ACT 2913, Australia
e-mail: johnpaulharvey1@me.com

© The Author(s) 2018
H. A. Abbass et al. (eds.), *Foundations of Trusted Autonomy*, Studies in Systems, Decision and Control 117, https://doi.org/10.1007/978-3-319-64816-3_6

as "... the arising of novel and coherent structures, patterns and properties during the process of self-organization in complex systems" [3]. Complexity has been defined by Kennedy et al. as: "The interaction of many parts of a system, giving rise to behaviours and/or properties that are not found in the individual elements of the system" [4]. Or as Wolfram put it: "It is possible to make things of great complexity out of things that are very simple. There is no conservation of simplicity" [5]. Self organisation is defined by Camazine et al. as "... a process in which pattern at the global level of a system emerges solely from numerous interactions among the lower-level components of the system" [6]. Features of self organising systems that are essential to emergent behaviour are the existence of: positive feedback—that leads to amplification of fluctuations; negative feedback—to counterbalance amplification and provide stabilisation; multi stability—the coexistence of many stable states; and the existence of state transitions—leading to dramatic change of the system behaviour, i.e. 'bifurcations' in behaviour occur when some parameter/s are varied.

Goldstein [3] identifies five essential features of emergence:

- Radical novelty—novel behaviour occurs that cannot be predicted.
- Coherence or correlation—the novel behaviour has some level of coherence over time.
- Global or macro-level behaviour—coherence occurs at the macro level.
- Dynamical—the macro-level, while having some coherence in time, also evolves over time.
- Ostensive—emergent behaviours are recognised ostensively, i.e. by showing themselves.

While Goldstein identifies that emergence is inherently unpredictable, Fromm [7] proposes that there are four types of emergence, only two of which are unpredictable. The four types of emergence proposed by Fromm are shown in Table 6.1. Using Fromm's classification scheme, there is a clear gradation in the complexity of systems that display emergent behaviour, from the least complex in Type I, to the most complex in Type IV.

The following Sections will examine the implications of emergent behaviour in swarm intelligence systems, specifically in relation to their potential use in autonomous systems. As identified in Table 6.1, based on Fromm's classification scheme, swarm intelligence systems fall into Type II 'Weak and predictable' emergence.

Table 6.1 Fromm's classification of types of emergence

Type	Name	Predictability	Example
I	Nominal/ Intentional	Predictable	Ordinary machines such as clocks or steam engines
II	Weak	Predictable in principle	School of fish, flock of birds
III	Multiple	Not predictable	Stock markets, pattern formation in nature
IV	Strong	Not predictable in principle	Life and culture

6.2   Emergence in Swarm Intelligence

Swarm intelligence systems, based on the *local* interaction of a large number of relatively simple agents, display complex, goal-oriented behaviour at the *global* level. Swarm intelligence is defined by Kordon as "... coherence without choreography and is based on the emerging collective intelligence of simple artificial individuals" [8]. Swarm intelligence systems have proven useful in solving a wide range of complex, non-linear, real-world problems based on their ability to search complex problem spaces where other methods are unsuitable or ineffective.

Swarm intelligence systems, commonly comprising large numbers of relatively simple, homogenous agents, are one form of multi-agent systems. Alternate implementations exist. For example in Chap. 5, Bryant and Miikulainen examine the advantages and disadvantages of homogenous versus heterogenous agents and the benefits of adaptability of the agents.

Examples of swarm intelligence systems relevant to trusted autonomy include swarm robotics [9–12], control of groups of unmanned aerial vehicles [13–16], control of autonomous land and underwater vehicles [17, 18], network switching [19], economic load dispatch [20], the control of switching networks [21], and the control of chaotic non-linear networks [22].

The 'intelligence' displayed by swarm intelligence systems is an emergent property of the system, without any form of external control, synchronous clock or shared memory and in the absence of any system-wide communication mechanism [23].

While the emergent behaviour of swarm intelligence systems has proven useful in solving complex real-world problems, as Parunak notes: "Neither self-organization nor emergence is necessarily good" [15]. Emergence, therefore, can be both a blessing and a curse in the application of swarm intelligence techniques to develop trusted autonomous systems.

6.3   The 'Blessing' of Emergence

The emergent behaviour of swarm intelligence systems can be a 'blessing' in some complex problem solving situations, based on a number of advantages that emergent behaviour offers. The first of these advantages is simplicity: individual agents tend to be quite simple, yet together they can produce very complicated behaviour. This means that programming is easy as the complexity of individual agents is low [24–26]. And because agents are relatively simple, programming errors are less likely and debugging and validation of performance of the individual agents is relatively simple.

The second is robustness: swarming systems are able to continue to operate, albeit at a lower performance, even though there are failures in some individuals or disturbances in the environment [12, 27]. Robustness also comes from the lack of centralised control, which means there is no single point of failure.

The third is flexibility: the system is self-adjusting, able to adapt quickly to changing circumstances without changing individual agents' behaviour [12, 26, 27]. Closely related to flexibility is the concept of environment integration: environmental dynamics are directly integrated into swarm's behaviour, and can enhance system performance [25].

The fourth is scalability: the swarm can operate using different swarm sizes with little if any change to coordination mechanisms. Processing requirements, therefore, tend to increase linearly as the swarm size increases [12, 26].

The fifth is autonomy: swarm intelligence systems operate without external control or supervision, providing the capacity to control systems that are too complex or a require a response beyond the capacity of human involvement [17–19].

The sixth is parallelism: swarm intelligence systems inherently use parallel computation for problem solving [19].

Together, these factors make the emergent nature of swarm intelligence systems attractive for solving complex problems that cannot be broken down into simple parts. They can therefore be attractive for use in autonomous systems and the advantages they offer potentially contribute towards trust of the system.

6.4   The 'Curse' of Emergence

The emergent behaviour of swarm intelligence systems can also be a 'curse' in some complex problem solving situations, based on a number of inherent limitations of swarm intelligence systems. These limitations can lead to lack of trust in autonomous systems that rely on swarm intelligence, which, in turn, rely on emergence.

The first of these limitations is the challenge of predicting the behaviour of swarm intelligence systems. Fromm categorises swarming systems as Type II emergent behaviour and predictable in principle, but in practice predictability is difficult to achieve [7]. A simple swarming/not-swarming prediction may be possible, predicting the detailed characteristics of swarming behaviour, however, is more challenging. Predictability is particularly important in relation to phase boundaries where fundamental changes in behaviour occur [28]. As Wright et al. note, in real-world systems ".. the presence of undesirable behaviours that are a result of unforeseen non-linear interactions with the different components of these systems ... can have catastrophic consequences ..." [29]. If predictability cannot be guaranteed, at least within acceptable bounds, swarm intelligence systems will not be used for safety critical applications *a priori*. In one approach to improve the predictability of swarming systems, Harvey et al. have used measures typically associated with chaotic dynamics to quantify and predict swarming behaviour [30, 31].

The second limitation, and closely related to that of unpredictability, is the inability to control the behaviour of swarm intelligence systems [28]. As Everitt and Hutter note in Chap. 3, "... with increasing autonomy and responsibility, and with increasing intelligence and capability, there inevitably comes a risk of systems causing substantial harm." Control of swarming systems is inherently difficult due to the emergent, non-linear nature of their dynamics. Lack of control may be unacceptable in some problem solving areas where safety is critical. The inherent absence of centralised/higher-level control of swarming systems means that control of behaviour must be achieved indirectly, through the rules that control individual agent behaviour or the parameters that 'tune' the rules. Developing appropriate rules at the individual level can be a complex task. As Chevrier notes, the complexity is "...proportional to the distance between the simplicity of individuals and the complexity of the collective property" [32]. Choosing parameters to achieve a particular behaviour outcome is also a difficult task and in many cases may not be possible [12, 33, 34]. An alternate approach is to adjust parameters until a particular behaviour is achieved based on an objective measure of behaviour using an optimisation routine [35]. Another possible approach is to incorporate dynamic tuning—effectively a form of adaptation—in the model but this considerably increases the complexity of the agents and potentially the processing overhead and unpredictability of the behaviour of the system [36].

The third major limitation of swarm intelligence systems relates to the time required to reach a solution, which may limit the usefulness of swarm intelligence systems for on-line control tasks and time-critical tasks [33]. Options to improve the time to obtain a solution include increasing the number of swarm members and increasing the complexity of the members, for example, by incorporation of adaptation of members. These same changes, however, can also increase the processing time to converge to a solution. Balancing these competing factors is itself a complex optimising task which still may not lead to an acceptable outcome in the time required.

6.5   Taking Advantage of the Good While Avoiding the Bad

As systems become too complex and/or too dynamic for human control, some form of trusted autonomy will be required. In attempting to control such complex systems, emergent behaviour is likely, and probably necessary. Paranuk observes, therefore, that what is required are principles for designing and developing systems whose emergent behaviour is beneficial, or at least benign [15].

Swarm intelligence systems have shown they can be beneficial in solving complex real-world problems. This beneficial behaviour is dependent on emergence but currently processes are not available to guarantee behaviour will be benign in all possible circumstances. In an effort to take advantage of the benefits of swarm intelligence systems, while avoiding the limitations, Winfield et al. [37] have introduced the concept of "swarm engineering" which they see as a fusion of dependable systems engineering and swarm intelligence. They acknowledge the need to validate the behaviour of such systems but argue there is no reason that validating swarm intelligence systems should be any more complex than validating other complex systems. Winfield et al. discuss two key features of a system in relation to dependability: 'liveness', which relates to the swarm doing the right things; and 'safety' which relates to the swarm not doing the wrong thing. The two concepts are related but not the same thing. As Winfield et al. note: "A system that is provably safe could, for example, do the wrong thing safely" [37].

Promising mathematical modelling approaches have been developed to validate the 'liveness' aspect of swarm intelligence systems. In the context of swarm robotics examples include: Lancaster who uses networks of simple probabilistic graphs to predict swarm behaviour [38]; Dixon et al. who have investigated the verification of swarms using temporal logic and model checking [39]; and Brambilla et al. who have introduced an approach to the top-down design and verification of swarms via formal specification and model checking [40]. Less progress has been achieved on validation of safety aspects but Harper has shown the potential for using Lyapunov stability techniques [41].

But even if 'liveness' and 'safety' aspects can be unambiguously determined, there is still a body of work to be conducted to determine what 'trusted' means in the real world. As Devitt notes in Chap. 10, "We have different thresholds for trust depending on the risk of the decisions that have to be made and this in turn depends on impact of decisions." Consider a scenario where a swarm of robots is tasked to find all the survivors after a disaster. If the robots find 90% of the survivors but can be guaranteed not to injure anyone in the search process—can that system be considered 'trusted'. What if the swarm of robots can find 99% of survivors but there is a 10% chance of injuring a survivor during the search—would that system be trusted? Which would be the most trusted?

6.6   Conclusion

There is an increasing need for autonomous systems to control an increasingly complex world. To solve real world problems, however, autonomous systems must be able to be trusted. Swarm intelligence systems are one form of autonomous systems that have proven useful in controlling complex real-world systems. The intelligence displayed by these systems is an emergent property of swarming systems. The emergent behaviour of these systems is both a blessing and a curse. The emergent behaviour provides the potential to solve problems that may not be able to be solved by other means. But without the ability to verify and trust the emergent behaviour of swarm intelligence systems in the full range of situations in which they will be applied, there will be strict limits to their applicability in real-world systems. This is particularly important in safety critical systems.

References

1. T. De Wolf, T. Holvoet, *Emergence versus self-Organisation: Different Concepts but Promising when Combined* (Springer, Berlin, 2005)
2. J. Conway, The game of life. Sci. Am. **223**(4), 4 (1970)
3. J. Goldstein, Emergence as a construct: history and issues. Emergence **1**(1), 49–72 (1999)
4. J. Kennedy, J.F. Kennedy, R.C. Eberhart, *Y* (Shi, Morgan Kaufmann, Swarm intelligence, 2001)
5. S. Wolfram, Cellular automata as models of complexity. Nature **311**(5985), 419–424 (1984)
6. S. Camazine, *Self-organization in biological systems* (Princeton University Press, Princeton, 2003)
7. J. Fromm. Types and forms of emergence. *arXiv preprint nlin/0506028*, 2005
8. A.K. Kordon, *Swarm Intelligence: The Benefits of Swarms*, (Springer, 2010), pp. 145–174 3540699104
9. C. Blum, R. Groß, *Swarm Intelligence in Optimization and Robotics* (Springer, Berlin, 2015), pp. 1291–1309
10. M. Brambilla, E. Ferrante, M. Birattari, M. Dorigo, Swarm robotics: a review from the swarm engineering perspective. Swarm Intell. **7**(1), 1–41 (2013)
11. A. Oliveira, N. de Sá, L. Nedjah, L. de Macedo, Mourelle, Distributed efficient localization in swarm robotic systems using swarm intelligence algorithms. Neurocomputing **172**, 322–336 (2016)
12. W. Liu, A.F.T. Winfield, Modeling and optimization of adaptive foraging in swarm robotic systems. Int. J. Robotics. Res. **29**(14), 1743–1760 (2010)
13. B. Crowther, X. Riviere, Flocking of autonomous unmanned air vehicles. Aeronaut. J. **107**(1068), 99–109 (2003)
14. P. Gaudiano, B. Shargel, E. Bonabeau, B.T. Clough, Swarm intelligence: a new c2 paradigm with an application to control swarms of uavs. Technical report, 2003
15. H.V.D. Parunak, L.M. Purcell, F.C.S. SIX, and M.R. O'Connell, Digital pheromones for autonomous coordination of swarming uav's, in *American Institute of Aeronautics and Astronautics (AIAA) First Technical Conference and Workshop on Unmanned Aerospace Vehicles, Systems, and Operations*, 2002
16. J. Wu, J. Wang, Y. Cao, Y. Cao, X. Shi, *Research of Multi-UAVs Communication Range Optimization Based on Improved Artificial Fish-Swarm Algorithm* (World Scientific, Singapore, 2015)
17. J.-J. Li, R.-B. Zhang, Y. Yang, *Multi AUV Intelligent Autonomous Learning Mechanism Based on QPSO Algorithm* (Springer, Berlin, 2015), pp. 60–67
18. F. Ulbrich, S.S. Rotter, R. Rojas, Adapting to the traffic swarm: Swarm behaviour for autonomous cars. in*Handbook of Research on Design, Control, and Modeling of Swarm Robotics*, 2015, p. 263
19. I. Kassabalidis, M.A. El-Sharkawi, R.J. Marks, P. Arabshahi, A.A. Gray, Swarm Intelligence for Routing in Communication Networks, Vol. 6. IEEE, 2001
20. H.M. Dubey, M. Pandit, B.K. Panigrahi, M. Udgir, Economic load dispatch by hybrid swarm intelligence based gravitational search algorithm. Int. J. Intell. Syst. Appl. **5**(8), 21–32 (2013)
21. G. Herbert, A. Tanner, G.J. Pappas, Jadbabaie, Flocking in fixed and switching networks. IEEE Trans. Autom. Control **52**(5), 863–868 (2007)
22. O. Chua, Boids control of chaos. Int. J. Bifurcation Chaos **17**(02), 427–444 (2007)
23. G. Beni, J. Wang, *Swarm Intelligence in Cellular Robotic Systems.* (Springer, 1993), pp. 703–712
24. E. Bonabeau, M. Dorigo, G. Theraulaz, *Swarm Intelligence: From Natural to Artificial Systems* (Oxford University Press, Oxford, 1999)
25. D. Qu, H. Qu, Y. Liu. *Emergence in Swarming Pervasive Computing and Chaos Analysis.* IEEE, 2006
26. E. Şahin, *Swarm Robotics: From Sources of Inspiration to Domains of Application* (Springer, Berlin, 2005), pp. 10–20

27. A. Kundu, C. Ji. *Swarm Intelligence in Cloud Environment*, (Springer, 2012), pp. 37–44 3642309755

28. Y. Wu, J. Su, H. Tang, H. Tianfield. Analysis of the emergence in swarm model based on largest lyapunov exponent. *Mathematical Problems in Engineering*, 2011

29. W.A. Wright, R.E. Smith, M. Danek, Pillip Greenway, A generalisable measure of self-organisation and emergence. In *Artificial Neural Networks—ICANN 2001*, (Springer, Berlin, 2001.), pp. 857–864

30. J. Harvey, K. Merrick, H.A. Abbass, Application of chaos measures to a simplified boids flocking model. Swarm Intelligence **9**(1), 23–41 (2015)

31. J. Harvey, K. Merrick, H.A Abbass. Quantifying swarming behaviour, in *International Conference on Swarm Intelligence, Bali*, 2016

32. C. Vincent, *From Self-Organized Systems to Collective Problem Solving* (Springer, Berlin, 2004)

33. H. Ahmed, J. Glasgow, Swarm intelligence: concepts, models and applications. Technical Report 2012-585, School Of Computing, Queens University, Kingston, Ontario, Canada K7L3N6, 2012

34. Z. Yuan, M.A.M. De Oca, M. Birattari, T. Stützle, Continuous optimization algorithms for tuning real and integer parameters of swarm intelligence algorithms. Swarm Intell. **6**(1), 49–75 (2012)

35. N. Zaera, D. Cliff. *Not Evolving Collective Behaviours in Synthetic Fish*. Citeseer, 1996

36. P. Vannucci, Ale-pso: an adaptive swarm algorithm to solve design problems of laminates. Algorithms **2**(2), 710–734 (2009)

37. A.F.T. Winfield, C.J. Harper, J. Nembrini, *Towards Dependable Swarms and a New Discipline of Swarm Engineering* (Springer, Berlin, 2005)

38. J.P. Lancaster Jr., *Predicting the Behavior of Robotic Swarms in Discrete Simulation* (Kansas State University, Manhattan, 2015)

39. C. Dixon, A. Winfield, M. Fisher, *Towards Temporal Verification of Emergent Behaviours in Swarm Robotic Systems* (Springer, Berlin, 2011)

40. M. Brambilla, A. Brutschy, M. Dorigo, M. Birattari, Property-driven design for robot swarms: a design method based on prescriptive modeling and model checking. ACM Trans. Autonomous Adaptive Syst. (TAAS) **9**(4), 1556–4665 (2015)

41. C. Harper, A. Winfield, Designing behaviour based systems using the space-time distance principle. *Towards Intelligent Mobile Robots (Timr)*, (2001)

Chapter 7
Trusted Autonomous Game Play

Michael Barlow

7.1 Introduction

Just as play is the engine that gives a game life, so autonomy and trust have always been fundamental requirements for and enablers of all play.

Let us deal with autonomy first. Santayana [1] defines play as "..whatever is done spontaneously and for its own sake". Salen Tekinbas & Zimmerman [2] state "Play is free movement within a more rigid structure."- almost, in fact, a definition for autonomy itself. Gilmore [3] defines the term play as: "Play refers to those activities which are accompanied by a state of comparative pleasure, exhilaration, power and the feeling of self-initiative." Speaking of games Schell [4] states that "Games are entered willfully", while Avedon & Sutton-Smith [5] state "Games are an exercise of voluntary control systems...".

What then of trust? Huizinga [6] in his seminal study of play across cultures and periods of history defined a core feature which he coined as the Magic Circle. The magic circle delineates the mental space or universe created by players of a game. It defines the boundary between the real world and the game world. Critically, 'inside' the circle is safe - trusting and trusted - play of the game. This powerful concept has come to underpin much of the modern theory around game design.

So, trust and autonomy as individual concepts underpin all game play. Autonomy, because players of their own free will and volition choose to play the game. As Sid Meiers famously said [7] "[a game is] a series of interesting choices." - the operative word being choice. Trust, because players trust they are entering a shared virtual space defined by the rules and objectives of the game and that the other players - whether opponents or teammates - will share the 'purity' of that purpose, the willingness to abide by the rules and play the game 'for its own sake'. This psychological state or attitude is known as a lusory attitude [8].

M. Barlow (✉)
School of Engineering and IT, UNSW, Canberra, Australia
e-mail: m.barlow@adfa.edu.au

© The Author(s) 2018
H. A. Abbass et al. (eds.), *Foundations of Trusted Autonomy*, Studies in Systems,
Decision and Control 117, https://doi.org/10.1007/978-3-319-64816-3_7

The concept of a game is very broad; encompassing as it does sport in its multifarious forms (from individual to team, motor through to track-and-field, on or under water or in the air or on a field, ancient versus modern, extreme, etc.); board (e.g., Monopoly or Ticket to Ride), table-top (e.g., Warhammer 40 K or Dungeons & Dragons), and card (e.g., Bridge or Poker); gambling in its various forms; social (e.g., drinking games or storytelling); games of childhood and the playground (e.g., 'tip' and chasings, hide and seek, brandings, etc.); as well as the primary focus of this chapter, digital games (i.e., including the categories or labels of computer game, video game, console, mobile or smartphone and web). Numerous people ranging from academics working in the field of game studies, through practicing game designers have provided definitions for what is a game. While definitions such as Costikyan's [9] "An interactive structure of endogenous meaning that requires players to struggle toward a goal" or Tracy Fullerton's [10] "A game is a closed, formal system, that engages players in a structured conflict, and resolves its uncertainty in an unequal outcome." are succinct and subtly nuanced; Schell's 1st definition from his Game Design book [4] as a set of traits or attributes seems one of the most complete and least controversial or open to argumentation. Paraphrasing "Games: Are entered willfully; have goals; have conflict; have rules; can be won and lost; are interactive; have challenge; can create their own internal meaning; engage players; and are closed, formal system." McGonigal also shares a similar approach; though her list is even shorter [11]: "When you strip away the genre differences and the technological complexities, all games share four defining traits: a goal, rules, a feedback system, and voluntary participation."

But what is the significance of Trusted Autonomy to digital game design & play? What challenges and opportunities exist (for Trusted Autonomy) in the new types of technologies and game types & game play that are emerging?

7.2 TA Game AI

One long-standing, and apparently obvious area where Trusted Autonomy could make a true impact in game execution and game environments is AI - Artificial Intelligence - opponents, team-mates and characters to play along-side-of, against, and to inhabit the imaginary game worlds. A superficial glance at the intersections of computational intelligence and renowned cultural games of the intellect such as Chess or Go would seem to indicate that the 'AI challenge is solved'. In particular IBM's Deep Blue triumph over chess grandmaster Gary Kasparov in 1997 [12], and most recently Google's AlphaGo triumph over Go grandmaster Lee Sedol in early 2016 [13] mark turning points for computational intelligence; showing its ability to exceed the highest levels of human performance in abstract games of reasoning.

While there is not yet consensus within the academic community about the scope, range, type and enablers of human intelligence; there does seem to be broad agreement that human intelligence is much more than simply logical and mathematical - with visual/spatial, inter-personal (emotional), linguistic (language), and kinesthetics (bodily) being widely recognised (e.g., the theory of multiple intelligence [14]; or the Cattell-Horn-Carroll theory of intelligence [15]).

Further, and clearly, many of the games humans play require and utilise intelligence other than logical & mathematical. Indeed most, but not all, games are social activities and often at multiple levels (e.g., consider a team-sport where there are social elements at the intra-team and inter-team levels, and between players and officials, and perhaps between players and spectators). To express it another way, the greater share of games require a focus upon and awareness of the other participants of the play activity. In a multi-intelligence view this entails a minimum level of social and emotional intelligence to be an effective participant in the play (game) activity. In particular this requires active sensing and situational awareness of the other participants and modelling of their motivations, objectives & goals, and future intent and actions (e.g., an application of theory of the mind [16]). These are exactly the same attributes and requirements for effective Trusted Autonomy teaming of humans and computational intelligence actors.

As such advances in the technological underpinnings of Trusted Autonomy; and in particular those centring upon machine modelling and understanding of human intent, behaviour, trust, and emotional state; will find ready application as richer, more responsive, AI in digital games. That is game AI displaying a broader range of human behaviours and intelligence, with those behaviours more responsive and appropriate to the actions and choices of the human players. Computational (AI) team-mates and opponents that are indistinguishable from humans (Viz. Turing Test [17]), and richer NPCs (Non-Player-Characters - AI protagonists and entities in the game world) who truly interact with and respond to a player's in-game actions.

7.3 TA Game

Beyond in-game AI, the game 'itself' (i.e., as a system) could and should display the same level of awareness of and adaptability to the player and his or her state. The implications of such an approach for the way games are designed and developed are profound.

Jess Schell [4], in his book on game design, makes clear that "The [game] designer creates an experience … The game is not the experience … The experience rises out of a game." In other words the game serves as a vehicle - constructed by the game designer and their team - to transmit an experience to the player of that game. In Schell's words [4]: "And it is this that makes game design so very hard … we are far removed from what we are actually trying to create. We create an artefact that a player interacts with, and cross our fingers that the experience that takes place during the interactions is something they will enjoy." Further, that 'experience' is by its nature subjective and unique to each player, as it must of necessity be filtered and interpreted through the lens of each individual player's personality, tastes, intellect and current state (at the time of play), and motivation.

Game design and game development as practiced today is intensely a priori in nature. The designer conceives the core play mechanics (the way the world will work - its 'physics' - and its challenges), the setting, the story, the interface, the characters that inhabit, and the locations of the world, the look and feel. Through the development process these are further elaborated - 3D models are created for terrain, vegetation, buildings and characters; dialog is written for characters and recorded by voice actors; music is composed and recorded; logic is devised for game AI and coded by programmers, etc. etc. All this is decided, codified, and locked-in via a resource intensive development process (typically for AAA games across a period of 12–24 months and through the skills of 50–150 specialists) before the standard player ever experiences the game. Further, patches & DLC (Down Loadable Content) aside, that content is immutable and the same for every player who experiences that game.

But what if the game was aware enough of the player and their play goals, & self-aware enough to work in partnership with the player to create the best possible experience for that player at that time? What if the game software/system and the player were in a Trusted, Autonomous configuration with the goal of creating that optimum experience?

This would require a significant re-engineering of a game. Significant resources would need to be dedicated to embedding computational intelligence within the game - monitoring & modelling the player from moment-to-moment, prediction of desirable future game choices and environments to maintain engagement by the player, and JIT (Just In Time) creation of game content (stories, challenges, characters, dialogs, locations). This is extremely challenging. On the other hand there would be significant savings in the reduction or elimination of all the resources dedicated to development of game assets (the levels, dialog, music etc.) prior to release. Further, each game would ideally be a unique, tailored, and bespoke experience prepared (potentially, spun moment to moment; a kind of spontaneous but synchronized and continuing 'jam session' or conversation between the player and the game) with the individual player's motivations for play and individual tastes in mind.

Certain enabling technologies exist as building blocks for this vision. The concept of Flow [18] offers a framework in which to evaluate player engagement and interest as the game and player undergo changes. Procedural content generation [19] is a computational approach for the generation of simulation or game content - including terrain, vegetation, architecture, and even story (e.g., [20]). Bringing these foundational technologies together with low-cost and minimally intrusive sensing of the player's state (ideally physiological and EEG coupled with the already available in-game choices/actions) is challenging but not insurmountable.

7.4   TA Game Communities

One of the more popular forms of online game play today are what are known as MMO (Massively Multi-player Online) and simply Multiplayer (typically 1st person shooter) games. As an example the game League of Legends boast 27 million players each day (67 million players each month and a daily maximum simultaneous players of around 7.5 million - Riot Games [21]), while steamcharts.com [22],- a website that tracks player numbers in all games managed by the Steam application[1] - shows two other games with simultaneous player numbers in the hundreds of thousands (Counter Strike: Global Offensive - peak number of simultaneous players in July 2016 of 636,056 and over 255 million hours played in that month; DOTA 2 - peak number of simultaneous players of over a million for July 2016; and number of hours for July over 443 million) and dozens more with over 10 million hours of play for July 2016 (e.g., Rust, Team Fortress 2, Rocket League, Grand Theft Auto V, Arma 3, ARK: Survival Evolved). This is but a subset of popular online multiplayer games (e.g., War Thunder, World of Tanks, World of Warcraft also have very large player bases).

Arguably the most serious challenge for these online communities of players (and the companies that provide and profit from the games) is what is known as toxic behaviour [23]. Toxic behaviour includes harassment of fellow players (verbal abuse that includes racist, sexist, and sexually offensive language) and deliberate 'griefing' - the sabotage or corruption of a game being played by the perpetrator and others. Certain games and their communities (including League of Legends) earn a reputation as particularly toxic - deterring new players (newbies) from joining, and leading experienced players to quit the community; though this is a problem shared by all such games (and indeed online communities) to greater and lesser extents. Through the lens of Trusted Autonomy this challenge of toxic online behaviour and communities is one of creating a Trusted Autonomous environment for those players and the community. One in which a player can choose to join a game at random (the dominant form of play of these form of games, and known often as 'solo queueing'; as opposed to joining a game as a pre-configured team) and trust that the Magic Circle is being maintained by the game and that the other players are there to play with a lusory attitude.

The very scale of these games (e.g., for League of Legends several hundred thousand simultaneous games - each of 20–45 min - at any instant) and their player bases require an automated, computational-based solution. One which monitors and models each player's behaviour and motivation in the short-term (e.g., within each game) and on a longer-term basis. Most critically a game capable of supporting a range of different motivations-for-play and able to offer roles and opportunities for players with different goals and motivations for play - while still maintaining a fair, enjoyable and safe gaming experience for all participants.

---

[1] Steam is a desktop PC (primarily) based application for purchasing and maintaining a library 'in the cloud' of digital games. It is the single most popular and highest volume tool for this purpose in the Windows PC environment and manages some of the most successful/popular games - but not all such games.

7.5 TA Mixed Reality Games

At the time of writing Pokemon Go from Niantic/Nintendo (The Pokemon Company [24]) has been released for less than one month (in most of the world less than that period). A mobile (Android or iOS) based game, it has proven a social phenomenon receiving massive amounts of media coverage due to its impact upon social behaviour and use of public spaces, and proving to be immensely popular (e.g., App Annie reports over 100 million downloads of the application in the first 3 weeks of release [25]).

Pokemon Go is labelled an Augmented Reality game; but is more accurately a mixed reality, or indeed Reality Augmented Game (RAG; i.e., the virtual environment of the game is the primary focus and stimulus, while the real-world serves to augment that imaginary space). Abstracting out particulars, players move through the physical environment but observe a virtualised version of the physical world upon their mobile device. It is in this virtual space that players interact with monsters (wild Pokemon) and locations of interest (gyms and PokeStops - these later corresponding to actual locations in the physical world such as a museum or other major building). Hence the primary cognitive focus of players is upon the virtual space displayed upon their device - not the real-world around them.

Already there can be found multiple news stories concerning misadventure suffered by players of Pokemon Go - deaths (struck by a vehicle, wandered into a dangerous area of the city and murdered), injury (falling off a cliff or being struck by falling debris), robbery (criminals waiting at out-of-the-way locations frequented by players and robbing them of their cash and personal effects) - as well as less traumatic but also non-trivial social impacts (stresses on public services and facilities, disturbances to residents and special locations in the real-world such as cemeteries or places of particular reverence).

As with the previous examples, there is a clear need to recast such mixed reality games - where so much of a player's finite cognitive capacity is dedicated to the game or imaginary space - as a trusted autonomous relationship between player(s) and the game. Future versions of RAGs should work to provide a safe - trusted - experience for the players. This in turn will require a computationally intelligent game; one not just aware of the virtual world but the physical world through which the player moves; and which utilises that information to keep its player safe (and the world safe from the actions of its player).

7.6 Discussion: TA Games

Several examples of significant changes to the way games are designed, developed and played have been proposed based upon adopting a Trusted Autonomy framework or approach. In particular Trusted Autonomous AI to play with, against and as occupants of the virtual works; Trusted Autonomous games that self-modify to present the play opportunities of most interest to the players; Trusted Autonomous online game communities that offer fulfilment for different player types while maintaining a safe and trustworthy environment; and Trusted Autonomous Reality Augmented Games (RAGs) that keep their players safe in the real-world while much of their players' attention is focused upon the virtual space.

Common to all approaches is the need for games to become 'smart' and become 'aware' - to be cognisant (sense or monitor) the player, their community, and environment; and to use computational techniques to dynamically alter their own behaviour and interaction with the player so as to satisfy those goals.

The challenges are large; but with continuing advances in sensor technologies, modelling techniques & algorithms, computational power, big data and cloud computing, and HCI this vision is realisable and promises an exciting new era for digital gaming.

References

1. G. Santayana, *The Sense of Beauty Being the Outlines of Æthetic Theory* (Charles Scribner's sons, New York, 1896)
2. K. Salen, E. Zimmerman, *Rules of Play: Game Design Fundamentals* (MIT Press, Cambridge USA, 2004)
3. J.B. Gilmore. *Child's Play*, chapter Play: A special behavior. (Wiley, New York, 1971), pp. 311–325
4. J. Schell, *The Art of Game Design: A Book of Lenses* (CRC Press, USA, 2008)
5. E.M. Avedon, B. Sutton-Smith, *The Study of Games* (Wiley, Hoboken, 1971)
6. J. Huizinga, *Homo Ludens: A Study of the Play Element in Culture* (Beacon Press Books, Boston USA, 1950)
7. A. Rollings, D. Morris, *Game Architecture and Design with Cdrom* (Coriolis, Scottsdale, Arizona, 2000)
8. B. Suits. *2005: The Grasshopper: Games, Life and Utopia.* Toronto: University of Toronto Press. Repr. Peterborough, Broadview Press, ON 2005
9. G. Costikyan. I have no words & i must design, http://www.costik.com/nowords2002.pdf, 2002. Accessed 2/8/2016
10. T. Fullerton. *Game Design Workshop: A Playcentric Approach to Creating Innovative Games.* Morgan Kaufmann, 2008
11. J. McGonigal, *Reality is Broken: Why Games Make us Better and how they can Change the World* (Penguin Books, London, 2011)
12. E. Gibney, Google ai algorithm masters ancient game of go. Nature **529**(7587), 445–446 (2016)
13. T. Chouard. The go files: ai computer clinches victory against go champion. Nature, News & Comments, http://www.nature.com/news/the-go-files-ai-computer-clinches-victory-against-go-champion-1, March 2016. Accessed 1/8/2016
14. H. Gardner, *Frames of Mind: The Theory of Multiple Intelligences* (Basic Books, New York, 1993)
15. K.S. McGrew. *Contemporary Intellectual Assessment: Theories, Tests, and Issues*, chapter The Cattell-Horn-Carroll Theory of Cognitive Abilities: Past, Present, and Future, (Guilford Press, New York, 2005) pp. 151–179
16. D. Premack, G. Woodruff, Does the chimpanzee have a theory of mind? Behavioral Brain Sci. **1**(04), 515–526 (1978)
17. A.M. Turing, Computing machinery and intelligence. Mind **59**(236), 433–460 (1950)

18. M. Czsentmihalyi, *Flow: The Psychology of Optimal Experience* (Harper and Row, New York, 1990)
19. M. Hendrikx, S. Meijer, J. Van Der Velden, A. Iosup. Procedural content generation for games: a survey. *ACM Trans Multimedia Computing, Communications Appl (TOMM)*, 9(1):1, (2013)
20. P. Gervás, Computational drafting of plot structures for russian folk tales. Cognitive Comput. **8**(2), 187–203 (2016)
21. R. Games. Our games, http://www.riotgames.com/our-games, 2016. Accessed 4/8/2016
22. Steamcharts. Top games by current players, http://steamcharts.com/top. Accessed 4/8/2016
23. B. Maher, Can a video game company tame toxic behaviour? Nature **531**(7596), 568 (2016)
24. The Pokemon Company. Pokemon go, http://www.pokemongo.com/, 2016. Accessed 4/8/2016
25. S. Singh. Pokemon go: An opportunity, not a threat. App Annie, July 29 2016, https://www.appannie.com/insights/mobile-strategy/pokemon-go-an-opportunity-not-a-threat/, 2016. Accessed 4/8/2016

Part II
Trust

Chapter 8
The Role of Trust in Human-Robot Interaction

Michael Lewis, Katia Sycara and Phillip Walker

8.1 Introduction

Robots and other complex autonomous systems offer potential benefits through assisting humans in accomplishing their tasks. These beneficial effects, however, may not be realized due to maladaptive forms of interaction. While robots are only now being fielded in appreciable numbers, a substantial body of experience and research already exists characterizing human interactions with more conventional forms of automation in aviation and process industries.

In human interaction with automation, it has been observed that the human may fail to use the system when it would be advantageous to do so. This has been called *disuse (underutilization or under-reliance)* of the automation [97]. People also have been observed to fail to monitor automation properly (e.g. turning off alarms) when automation is in use, or they accept the automation's recommendations and actions when inappropriate [71, 97]. This has been called *misuse, complacency, or over-reliance*. Disuse can decrease automation benefits and lead to accidents if, for instance, safety systems and alarms are not consulted when needed. Another maladaptive attitude is automation bias [33, 55, 77, 88, 112], a user tendency to ascribe greater power and authority to automated decision aids than to other sources of advice (e.g. humans). When the decision aid's recommendations are incorrect, automation bias may have dire consequences [2, 78, 87, 89] (e.g. errors of omission, where the

M. Lewis (✉) · P. Walker
Department of Information Sciences, University of Pittsburgh,
Pittsburgh, PA, USA
e-mail: ml@sis.pitt.edu

P. Walker
e-mail: pmw19@pitt.edu

K. Sycara
Robotics Institute School of Computer Science,
Carnegie Mellon University, Pittsburgh, PA, USA
e-mail: katia@cs.cmu.edu

© The Author(s) 2018
H. A. Abbass et al. (eds.), *Foundations of Trusted Autonomy*, Studies in Systems,
Decision and Control 117, https://doi.org/10.1007/978-3-319-64816-3_8

user does not respond to a critical situation, or errors of commission, where the user does not analyze all available information but follows the advice of the automation).

Both naïve and expert users show these tendencies. In [128], it was found that skilled subject matter experts had misplaced trust in the accuracy of diagnostic expert systems. (see also [127]). Additionally the Aviation Safety Reporting System contains many reports from pilots that link their failure to monitor to excessive trust in automated systems such as autopilots or FMS [90, 119]. On the other hand, when corporate policy or federal regulations mandate the use of automation that is not trusted, operators may "creatively disable" the device [113]. In other words: disuse the automation.

Studies have shown [64, 92] that trust towards automation affects reliance (i.e. people tend to rely on automation they trust and not use automation they do not trust). For example, trust has frequently been cited [56, 93] as a contributor to human decisions about monitoring and using automation. Indeed, within the literature on trust in automation, complacency is conceptualized interchangeably as the overuse of automation, the failure to monitor automation, and lack of vigilance [6, 67, 96]. For optimal performance of a human-automation system, *human trust in automation should be well-calibrated*. Both disuse and misuse of the automation has resulted from improper calibration of trust, which has also led to accidents [51, 97].

In [58], trust is conceived to be an "attitude that an agent (automation or another person) will help achieve an individual's goals in a situation characterized by uncertainty and vulnerability." A majority of research in trust in automation has focused on the relation between automation reliability and operator usage, often without measuring the intervening variable, trust. The utility of introducing an intervening variable between automation performance and operator usage, however, lies in the ability to make more precise or accurate predictions with the intervening variable than without it. This requires that trust in automation be influenced by factors in addition to automation reliability/performance. The three dimensional (Purpose, Process, and Performance) model proposed by Lee and See [58], for example, presumes that trust (and indirectly, propensity to use) is influenced by a person's knowledge of what the automation is supposed to do (purpose), how it functions (process), and its actual performance. While such models seem plausible, support for the contribution of factors other than performance has typically been limited to correlation between questionnaire responses and automation use. Despite multiple studies of trust in automation, the conceptualization of trust and how it can be reliably modeled and measured is still a challenging problem.

In contrast to automation where system behavior has been pre-programmed and the system performance is limited to the specific actions it has been designed to perform, autonomous systems/robots have been defined as having intelligence-based capabilities that would allow them to have a degree of self governance, which enables them to respond to situations that were not pre-programmed or anticipated in the design. Therefore, the role of trust in interactions between humans and robots is more complex and difficult to understand.

In this chapter, we present the conceptual underpinnings of trust in Sect. 8.2, and then discuss models of, and the factors that affect, trust in automation in Sects. 8.3 and

8.4, respectively. Next, we will discuss instruments for measuring trust in Sect. 8.5, before moving on to trust in the context of human-robot interaction (HRI) in Sect. 8.6 both in how humans influence robots, and vice versa. We conclude in Sect. 8.7 with open questions and areas of future work.

8.2 Conceptualization of Trust

Trust has been studied in a variety of disciplines (including social psychology, human factors, and industrial organization) for understanding relationships between humans or between human and machine. The wide variety of contexts within which trust has been studied leads to various definitions and theories of trust. The different context within which trust has been studied has led to definitions of trust as an attitude, an intention, or a behavior [72, 76, 86]. Both within the inter-personal literature and human-automation trust literature, a widely accepted definition of trust is lacking [1]. However, it is generally agreed that trust is best conceptualized as a *multidimensional psychological attitude* involving beliefs and expectations about the trustee's trustworthiness derived from experience and interactions with the trustee in situations involving uncertainty and risk [47]. Trust has also been said to have both *cognitive and affective features*. In the interpersonal literature, trust is also seen involving affective processes, since trust development requires seeing others as personally motivated by care and concern to protect the trustor's interests [65]. In the automation literature, cognitive (rather than affective) processes may play a dominant role in the determination of trustworthiness, i.e., the extent to which automation is expected to do the task that it was designed to do [91]. In the trust in automation literature, it has been argued that trust is best conceptualized as an attitude [58] and a relatively well accepted definition of trust is: "...an attitude which includes the belief that the collaborator will perform as expected, and can, within the limits of the designer's intentions, be relied on to achieve the design goals" [85].

8.3 Modeling Trust

The basis of trust can be considered as a set of attributional abstractions (trust dimensions) that range from the trustee's competence to its intentions. Muir [91] combined the dimensions of trust from two works ([4] and [100]). Barber's model [4] is in terms of human expectations that form the basis of trust between human and machine. These expectations are persistence, technical competency, and fiduciary responsibility. Although in the subsequent literature, the number and concepts in the trust dimensions vary [58], there seems to be a convergence on the three dimensions—*Purpose, Process, and Performance* [58]—mentioned earlier, along with correspondences of those to earlier concepts, such as the dimensions in [4], and those of *Ability, Integrity, and Benevolence* [76]. *Ability* is the trustee competence in performing expected actions, *benevolence* is the trustee intrinsic and positive intentions towards the trustor, and *integrity* is trustee's adherence to a set of principles that are acceptable to the trustor [76].

Both trust in automation [92] and interpersonal relations literature [37, 53, 84, 107] agree that trust relations are *dynamic* and varying over time. There are three phases that characterize trust over time: *trust formation*, where trustors choose to trust trustees and potentially increase their trust over time, *trust dissolution*, where trustors decide to lower their trust in trustees after a trust violation has occurred, and *trust restoration* where trust stops decreasing after a trust violation and gets restored (although potentially not to the same level as before the trust violation). Early in the relationship, the trust in the system is based on the predictability of the system's behavior. Work in the literature has shown shifts in trust in response to changes in properties and performance of the automation [56, 91]. When the automation was reliable, operator trust increased over time and vice versa. Varying levels of trust were also positively correlated with the varying levels of automation use. As trust decreased, for instance, manual control became more frequent. As the operator interacts with the system, he/she attributes dependability to the automation. Prolonged interaction with the automation leads the operator to make generalizations about the automation and broader attributions about his belief in the future behavior of the system (faith). There is some difference in the literature as to when exactly faith develops in the dynamic process of trust development. Whereas [100] argue that interpersonal trust progresses from predictability to dependability to faith, [92] suggest that for trust in automation, faith is a better predictor of trust early rather than late in the relationship.

Some previous work has explored trust with respect to *automation versus human trustee* [64]. Their results indicate (a) the dynamics of trust are similar, in that faults diminish trust both towards automation or another human, (b) the sole predictor of reliance on automation was the difference between trust and self-confidence, and (c) participants, in human-human experiments, were more likely to delegate a task to a human when the human was thought to have a low opinion of their own trustworthiness. In other words, when participants thought their own trustworthiness in the eyes of others was high, they were more likely to retain control over a task. However, trustworthiness played no role when the collaborative partner was an automated controller, i.e. only participants' own confidence in their performance determined their decision to retain/obtain control. Other work on trust in humans versus trust in automation [61] explored the extent to which participants trusted identical advice given by an expert system under the belief that it was given by a human or a computer. The results of these studies were somewhat contradictory however. In one study, participants were more confident in the advice of the human (though their agreement with the human advice did not vary versus their agreement on the expert system's advice), while in the second study, *participants agreed more with the advice of the expert system, but had less confidence in the expert system.* Similar contradictory results have been shown in HRI studies, where work indicated that errors by a robot did not affect participants' decisions of whether or not to follow the advice of a robot [111], yet did affect their subjective reports of the robot's reliability and trustworthiness [104]. Study results by [71], however, indicated that reliance on a *human* aid was reduced in situations of higher risk.

8.4 Factors Affecting Trust

The factors that are likely to affect Trust in automation have generally been categorized as those pertaining to *automation*, the *operator*, and the *environment*. Most work on factors that have been empirically researched pertains to characteristics of the automation. Here we briefly present relevant work on the most important of these factors.

8.4.1 System Properties

The most important correlates of use of automation have been system reliability and effects of system faults. Reliability typically refers to automation that has some error rate—for example, misclassifying targets. Typically this rate is constant and data is analyzed using session means. Faults are typically more drastic, such as controller that fails making the whole system behave erratically. Faults are typically single events and studied as time series.

System reliability: Prior literature has provided empirical evidence that there is a relationship between trust in automation and the automation's reliability [85, 96–98, 102]. Research shows [86] that declining system reliability can lead to systematic decline in trust and trust expectations, and most crucially, these changes can be measured over time. There is also some evidence that only the most recent experiences with the automation affect trust judgments [51, 56].

System faults: System faults are a form of system reliability, but are treated separately because they concern discrete system events and involve different experimental designs. Different aspects of faults influence the relation between trust and automation. Lee and Moray [56] showed that in the presence of continual system faults, trust in the automation reached its lowest point only after six trials, but trust did recover gradually even as faults continued. The magnitude of system faults has differential effects on trust (smaller faults had minimal effect on trust while large faults negatively affected trust and were slower to recover the trust). Another finding [92] showed that faults of varying magnitude diminished trust more than large constant faults. Additionally, it was found that when faults occurred in a particular subsystem, the corresponding distrust did spread to other functions controlled by the same subsystem. The distrust did not, however, spread to independent or similar subsystems.

System predictability: Although system faults affect the trust in the automation, this happens when the human has little *a priori* knowledge about the faults. Research has shown that when people have prior knowledge of faults, these faults do not necessarily diminish trust in the system [64, 102]. A plausible explanation is that knowing that the automation may fail reduces the uncertainty and consequent risk associated with use of the automation. In other words, predictability may be as (or more) important as reliability.

System intelligibility and transparency: Systems that can explain their reasoning will be more likely to be trusted, since they would be more easily understood by their users [66, 117, 121, 122]. Such explanatory facility may also allow the operator to query the system in periods of low system operation in order to incrementally acquire and increase trust.

Level of Automation: Another factor that may affect trust in the system is its level of automation (i.e. the level of functional allocation between the human and the system). It has been suggested [91, 93] that system understandability is an important factor for trust development. In their seminal work on the subject [116], Sheridan and Verplank propose a scale for assessing the level of automation in a system from 0 to 10, with 0 being no autonomy and 10 being fully autonomous. Since higher levels of automation are more complex, thus potentially more opaque to the operator, higher levels of automation may engender less trust. Some limited empirical work suggests that different levels of automation may have different implications for trust [86]. Their work based on Level 3 [116] automation did not show same results when conducted with Level 7 (higher) automation.

8.4.2 Properties of the Operator

Propensity to trust: In the sociology literature [105] it has been suggested that people have different propensity to trust others and it has been hypothesized that this is a stable personality trait. In the trust in automation literature, there is very limited empirical work on the propensity to trust. Some evidence is provided in [97] suggests that operator's overall propensity to trust is distinct from trust towards a specific automated system. In other words, it may be the case that an operator has high propensity to trust in automation in general, but faced with a specific automated system, their trust may be very low.

Self Confidence: Self-confidence is a factor of individual difference and one of the few operator characteristics that has been studied in the trust in automation literature. Work in [57] suggested that when trust was higher than self-confidence, automation, rather than manual control would be used and vice versa when trust was lower than self-confidence. However, later work [86], which was conducted with a higher level of automation than [57], did not obtain similar results. It was instead found that trust was influenced by properties of the system (e.g., real or apparent false diagnoses) while self-confidence was influenced by operator traits and experiences (e.g. whether they had been responsible for accidents). Furthermore, it was also found that self-confidence was not affected by system reliability. This last finding was also suggested in the work of [64] which found that self-confidence was not lowered by shifts in automation reliability.

Individual Differences and Culture: It has been hypothesized, and supported by various studies, that individual differences [57, 74, 80, 119] and culture [50] affect the trust behavior of people. The interpersonal relations literature has identified many different personal characteristics of a trustor, such as self-esteem [105, 106], secure attachment [17], and motivational factors [54] that contribute to the different stages in the dynamics of trust. Besides individual characteristics, socio-cultural factors that contribute to differences in trust decisions in these different trust phases have also been identified [8, 10, 32, 37]. For example, combinations of socio-cultural factors that may result in quick trust formation (also called "swift trust" formation in temporary teams [83]) are time pressure [25] and high power distance with authority [16]. People in high power distance (PD) societies expect authority figures to be benign, competent and of high integrity. Thus people in high power distance societies will engage in less vigilance and monitoring for possible violations by authority figures. To the extent then that people of high PD cultures perceive the automation as authoritative, they should be quick to form trust. On the other hand, when violations occur, people in high PD cultures should be slow to restore trust once violations have occurred [11]. Additionally, it has been shown [79] via replication of Hofstede's [45] cultural dimensions for a very large-scale sample of pilots, that even in such a highly specialized and regulated profession, national culture still exerts a meaningful influence on attitude and behavior over and above the occupational context.

To date, only a handful of studies consider cultural factors and potential differences in the context of trust in automation, with [99, 125] and [22] being exceptions. As the use of automation gets increasingly globalized, it is imperative that we gain an understanding on how trust in automation is conceptualized across cultures and how it influences operator reliance and use of automation, and overall human-system performance.

8.4.3 Environmental Factors

In terms of environmental factors that influence trust in automation, risk seems most important. Research in trust in automation suggests that reliance on automation is modulated by the risk present in the decision to use the automation [101]. People are more averse to using the automation if negative consequences are more probable and, once trust has been lowered, it takes people longer to re-engage the automation in high-risk versus low risk situations [102]. However, knowing the failure behavior of the automation in advance may modify the perception of risk, in that people's trust in the system does not decrease [101].

8.5   Instruments for Measuring Trust

While a large body of work on trust in automation and robots has developed over the past two decades, standardized measures have remained elusive with many researchers continuing to rely on short idiosyncratically worded questionnaires. Trust (in automation) refers to a cognitive state or attitude, yet it has most often been studied *indirectly through its purported influence on behavior often without any direct cognitive measure*. The nature and complexity of the tasks and failures studied has varied greatly ranging from simple automatic target recognition (ATR) classification [33], to erratic responses of a controller embedded within a complex automated system [57] to robots misreading QR codes [30]. The variety of reported effects (automation bias, complacency, reliance, compliance, etc.) mirror these differences in tasks and scenarios [27] and [28] have criticized the very construct of trust in automation on the basis of this diversity as an unfalsifiable "folk model" without clear empirical grounding. Although the work cited in the reply to these criticism in [98] as well as the large body of work cited in the review by [96] have begun to examine the interrelations and commonalities of concepts involving trust in automation, empirical research is needed to integrate divergent manifestations of trust within a single task/test population so that common and comparable measures can be developed.

Most "measures" of trust in automation since the original study [92] have been created for individual studies based on face validity and have not in general benefited from the same rigor in development and validation that has characterized measures of interpersonal trust. "Trust in automation" has been primarily understood through its analogy to interpersonal trust and more sophisticated measures of trust in automation have largely depended on rationales and dimensions developed for interpersonal relations, such as ability, benevolence, and integrity.

Three measures of trust in automation, Empirically Derived (ED), Human-Computer Trust (HTC), and SHAPE Automation Trust Index (SATI) have benefited from systematic development and validation. The Empirically Derived 12 item scale developed by [46] was systematically developed, subjected to a validation study [120] and used in other studies [75]. In [46], they developed their scale in three phases beginning with a word elicitation task. They extracted a 12-factor structure used to develop a 12-item scale based on examination of clusters of words. The twelve items roughly correspond to the classic three dimensions: benevolence (purpose), integrity (process), and ability (performance).

The Human-Computer Trust (HTC) instrument developed in [72] demonstrated construct validity and high reliability within their validation sample and has subsequently been used to assess automation in air traffic control (ATC) simulations, most recently in [68]. Subjects initially identified constructs that they believed would affect their level of trust in a decision aid. Following refinement and modification of the constructs and potential items, the instrument was reduced to five constructs (reliability, technical competence, understandability, faith, and personal attachment). A subsequent principal components analysis limited to five factors found most scale items related to these factors.

The SHAPE Automation Trust Index, SATI, [41] developed by the European Organization for the Safety of Air Navigation is the most pragmatically oriented of the three measures. Preliminary measures of trust in ATC systems were constructed based on literature review and a model of the task. This resulted in a seven dimensional scale (reliability, accuracy, understanding, faith, liking, familiarity, and robustness). The measure was then refined in focus groups with air traffic controllers from different cultures rating two ATC simulations. Scale usability evaluations, and construct validity judgments were also collected. The instrument/items have reported reliabilities in the high 80s but its constructs have not been empirically validated.

All three scales have benefited from empirical study and systematic development yet each has its flaws. The ED instrument in [46], for instance, addresses trust in automation in the abstract without reference to an actual system and as a consequence appears to be more a measure of propensity to trust than trust in a specific system. A recent study [115] found scores on the ED instrument to be unaffected by reliability manipulations that produced significant changes in ratings of trust on other instruments. The HTC was developed from a model of trust and demonstrated agreement between items and target dimensions but stopped short of confirmatory factor analysis. Development of the SATI involved the most extensive pragmatic effort to adapt items so they made sense to users and captured aspects of what users believed contributed to trust. However, SATI development neglected psychometric tests of construct validity.

A recent effort [21, 23] has led to a general measure of trust in automation validated across large populations in three diverse cultures, US, Taiwan and Turkey, as representative of Dignity, Face, and Honor cultures [63]. The Cross-cultural measure of trust is consistent with the three (performance, purpose, process) dimensions of [58, 81] and contains two 9 item scales, one measuring the propensity to trust as in [46] and the other measuring trust in a specific system. The second scale is designed to be administered repeatedly to measure the effects of manipulations expected to affect trust while the propensity scale is administered once at the start of an experiment. The scales have been developed and validated for US, Taiwanese, and Turkish samples and are based on 773 responses (propensity scale) and 1673 responses (specific scale).

The Trust Perception Scale-HRI [114, 115] is a psychometrically-developed 40 item instrument intended to measure human trust in robots. Items are based on data collected identifying robot features from pictures and their perceived functional characteristics. While development was guided by the triadic (human, robot, environment) model of trust inspired by the meta-analysis in [43], a factor analysis of the resulting scale found four components corresponding roughly to capability, behavior, task, and appearance. Capability and behavior correspond to two of the dimensions commonly found in interpersonal trust [81] and trust in automation [58], while appearance may have a special significance for trust in robots. The instrument was validated in same-trait and multi-trait analyses producing changes in rated trust associated with manipulation of robot reliability. The scale was developed based on 580 responses and 21 validation participants.

The HRI Trust Scale [131] was developed from items based on five dimensions (team configuration, team process, context, task, and system) identified by 11 subject matter experts (SMEs) as likely to affect trust. A 100 participant Mechanical Turk sample was used to select 37 items representing these dimensions. The HRI Trust Scale is incomplete as a sole measure of trust and is intended to be paired with Rotter's [105] interpersonal trust inventory when administered. While Lee and See's dimensions [58] other than "process" are missing from the HRI scale, they are represented in Rotter's instrument.

Because trust in automation or robots is an attitude, self-report through psychometric instruments such as these provides the most direct measurement. Questionnaires, however, suffer from a number of weaknesses. Because they are intrusive, measurements cannot be conveniently taken during the course of a task but only after the task is completed. This may suffice for automation such as ATR where targets are missed at a fixed rate and the experimenter is investigating the effect of that rate on trust [33], but it does not work in measuring moment to moment trust in a robot reading QR codes to get its directions [30].

8.6   Trust in Human Robot Interaction

Robots are envisioned to be able to process many complex inputs from the environment and be active participants in many aspects of life, including work environments, home assistance, battlefield and crisis response, and others. Therefore, robots are envisioned to transition from tool to teammate as humans transition from operator to teammate in an interaction more akin to human-human teamwork. These envisioned transitions raise a number of general questions: How would human interaction with the robot be affected? How would performance of the human-robot team be affected? How would human performance or behavior be affected? Although there are numerous tasks, environments, and situations of human-robot collaboration, in order to best clarify the role of trust we distinguish two general types of interactions of humans and robots: *performance-based interactions*, where the focus is on the *human influencing/controlling the robot* so it can perform useful tasks for the human, and *social-based interactions*, where the focus is on how the *robot's behavior influences the human's beliefs and behavior*. In both these cases, the human is the trustor and the robot the trustee. In particular, in performance based interactions there is a particular task with a clear performance goal. An example of performance-based interactions is where human and robot collaborate in manufacturing assembly, or a UAV performing surveillance and recognition of victims in a search and rescue mission. Here measures of performance could be accuracy and timing to complete the task. On the other hand, in social interactions, the performance goal is not as crisply defined. An example of such a task is the ability of a robot to influence a human to reveal private knowledge, or how a robot can influence a human to take medicine or do useful exercises.

8.6.1   Performance-Based Interaction: Humans Influencing Robots

A large body of HRI research investigating factors thought to affect behavior via trust, such as reliability, rely strictly on behavioral measures without reference to trust. Meyer's [82] expected value (EV) theory of alarms provides one alternative by describing the human's choice as one between compliance (responding to an alarm) and reliance (not responding in the absence of an alarm). The expected values of these decisions are determined by the utilities associated with an uncorrected fault, the cost of intervention and the probabilities of misses (affecting reliance) and false alarms (affecting compliance). Research in [31], for example, investigated the effects of unmanned aerial vehicle (UAV) false alarms and misses on operator reliance inferred from longer reaction times for misses and compliance inferred from shorter reaction times to alarms. While reliance/compliance effects were not found, higher false alarm rates correlated with poorer performance on a monitoring task, while misses correlated with poorer performance on a parallel inspection task. A similar study by [20] of unmanned ground vehicle (UGV) control found participants with higher perceived attentional control were more adversely affected by false alarms (under-compliance) while those with low perceived attentional control were more strongly affected by misses (over-reliance). Reliance and compliance can be measured in much the same way for homogeneous teams of robots as illustrated by a follow up study of teams of UGVs [19] of similar design and results. A similar study [26] involved multiple UAVs manipulating ATR reliability and administering a trust questionnaire, again finding that ratings of trust increased with reliability.

Transparency, common ground, or shared mental models involve a second construct ("process" [58] or "integrity" [76]) believed to affect trust. According to these models, the extent to which a human can understand the way in which an autonomous system works and predict its behavior will influence trust in the system. There is far less research on effects of transparency, with most involving level of automation manipulations. An early study [60] in which all conditions received full information found best performance for an intermediate level of automation that facilitated checks of accuracy (was transparent). Participants, however, made substantially greater use of a higher level of automation that provided an opaque recommendation. In this study, ratings of trust were affected by reliability but not transparency. More recent studies have equated transparency with additional information providing insight into robot behavior. Researchers in [9] compared conditions in which participants observed a simulated robot represented on a map by a status icon (level of transparency 1), overlaid with environmental information such as terrain (level 2), or with additional uncertainty and projection information (level 3). Note that these levels are distinct from Sheridan's Levels of Automation mentioned previously. What might appear as erratic behavior in level 1, for example, might be "explained"' by the terrain being navigated in level 2. Participant's ratings of trust were higher for levels 2 and 3. A second study manipulated transparency by comparing minimal (such as static image) contextual (such as video clip) and constant (such as video) information for a simulated robot team mate with which participants had intermittent interactions but found no significant differences in trust. In [126], researchers took a different approach to transparency by having a simulated robot provide "explanations" of its actions. The robot guided by a POMDP model can make different aspects of its decision making such as beliefs (probability of dangerous chemicals in building) or capabilities (ATR has 70% reliability) available to its human partner. Robot reliability affected both performance and trust. Explanations did not improve performance but did increase trust among those in the high reliability condition. As these studies suggest, reliability appears to have a large effect on trust, reliance/compliance, and performance, while transparency about function has a relatively minor one, primarily influencing trust. The third component of trust in robot's "purpose" [58] or "benevolence" [76] has been attributed [69, 70, 95] to "transparency" as conveyed by appearance discussed in Sect. 8.6.2. By this interpretation, matching human expectations aroused by a robot's appearance to its purpose and capabilities can make interactions more transparent by providing a more accurate model to the human.

Studies discussed to this point have treated trust as a dependent variable to be measured at the end of a trial and have investigated whether or not it had been affected by characteristics of the robot or situation. If trust of a robot is modified through a process of interaction, however, it must be continuously varying as evidence accumulates of its trustworthiness or untrustworthiness. This was precisely the conception of trust investigated by Lee and Moray [56] in their seminal study but has been infrequently employed since. An recent example of such a study is reported in [29] where a series of experiments addressing temporal aspects of trust involving levels of automation and robot reliability have been conducted using a robot navigation and barriers task. In that task, a robot navigates through a course of boxes with labels that the operator can read through the robot's camera and QR codes presumed readable by the robot. The labels contain directions such as "turn right" or "U turn". In automation modes, robots follow a predetermined course with "failures" appearing to be misread QR codes. Operators can choose either the automation mode or a manual mode in which they determine the direction the robot takes. An initial experiment [29] investigated the effects of reliability drops at different intervals across a trial, finding that decline in trust as measured by post trial survey was greatest if the reliability decline occurred in the middle or final segments. In subsequent experiments, trust ratings were collected continuously by periodic button presses indicating increase or decrease in trust. These studies [30, 49] confirmed the primacy-recency bias in episodes of unreliability and the contribution of transparency in the form of confidence feedback from the robot.

Work in [24] collected similar periodic measures of trust using brief periodically presented questionnaires to participants performing a multi-UAV supervision task to test effects of priming on trust. These same data were used to fit a model similar to that formalized by [39] using decision field theory to address the decision to rely on the automation/robot's capabilities or to manually intervene based on the balance between the operator's self-confidence and her trust in the automation/robot. The model contains parameters characterizing information conveyed to operator, inertia in changing beliefs, noise, uncertainty, growth-decay rates for trust and self-confidence, and an inhibitory threshold for shifting between responses. By fitting these parameters to human subject data, the time course of trust (as defined by the model) can be inferred. An additional study of UAV control [38] has also demonstrated good fits for dynamic trust models with matches within 2.3% for control over teams of UGVs. By predicting effects of reliability and initial trust on system performance, such models might be used to select appropriate levels of automation or provide feedback to human operators. In another study involving assisted driving [123], the researchers use both objective (car position, velocity, acceleration, and lane marking scanners) and subjective (gaze detection and foot location) to train a mathematical model to recognize and diagnose over-reliance on the automation. The authors show that their models can be applied to other domains outside automation-assisted driving as well.

Willingness to rely on the automation has been found in the automation literature to correlate with user's self-confidence in their ability to perform the task [57]. It has been found that if a user is more confident in their own ability to perform the task, they will take control of the automation more frequently if they perceive that the automation does not perform well. However, as robots are envisioned to be deployed in increasingly risky situations, it may be the case that a user (e.g. a soldier) may elect to use a robot for bomb disposal irrespective of his confidence in performing the task. Another factor that has considerably influenced use of automation is user workload. It has been found in the literature that users exhibit over-reliance [7, 40] on the automation in high workload conditions.

Experiments in [104] show that people over-trusted a robot in fire emergency evacuation scenarios conducted with a real robot in a campus building, although the robot was shown to be defective in various ways (e.g. taking a circuitous route rather then the efficient route in guiding the participant in a waiting room before the emergency started). It was hypothesized by the experimenters that the participants, having experienced an interaction with a defective robot, would decrease their trust (as opposed to a non-defective robot), and also that participants' self-reported trust would correlate with their behavior (i.e their decision to follow the robot or not). The results showed that, in general, participants did not rate the non-efficient robot as a bad guide, and even the ones that rated it poorly still followed it during the emergency. In other words, trust rating and trust behavior were not correlated. Interestingly enough, participants in a previous study with similar scenarios of emergency evacuation *in simulation* by the same researchers [103] behaved differently, namely participants rated less reliant simulated robots as less trustworthy and were less prone to follow them in the evacuation. The results from the simulation studies of emergency evacuation, namely positive correlation between participants' trust assessment and behavior, are similar to results in low risk studies [30]. These contradictory results point strongly that more research needs to be done to refine robot, operator and task-context variables and relations that would lead to correct trust calibration, and better understanding of the relationship between trust and performance in human robot interaction.

One important issue is how an agent forms trust in agents it has not encountered before. One approach from the literature in multiagent systems (MAS) investigates how trust forms in ad hoc groups, where agents that had not interacted before come together for short periods of time to interact and achieve a goal, after which they disband. In such scenarios, a decision tree model based on both trust and other factors (such as incentives and reputation) can be used [13]. A significant problem in such systems, known as the *cold start problem*, is that when such groups form there is little to no prior information on which to base trust assessments. In other words, how does an agent choose who to trust and interact with when they have no information on any agent? Recent work has focused on bootstrapping such trust assessments by using stereotypes [12]. Similar to stereotypes used in interpersonal interactions among humans, stereotypes in MAS are quick judgements based on easily observable features of the other agent. However, whereby human judgements are often clouded by cultural or societal biases, stereotypes in MAS can be constructed in a way that maximizes the accuracy. Further work by the researchers in [14] shows how stereotypes in MAS can be spread throughout the group to improve others' trust assessments, and can be used by agents to detect unwanted biases received from others in the group. In [15], the authors show how this work can be used by organizations to create decision models based on trust assessments from stereotypes and other historical information about the other agents.

8.6.1.1   Towards Co-adaptive Trust

In other studies [129, 130], Xu and Dudek create an online trust model to allow a robot or other automation to assess the operator's trust in the system while a mission is ongoing, using the results of the model to adjust the automation behavior on the fly to adapt to the estimated trust level. Their end goal is *trust-seeking adaptive robots*, which seek to actively monitor and adapt to the estimated trust of the user to allow for greater efficiency in human-robot interactions. Importantly, the authors combined common objective, yet indirect, measures of trust (such as quantity and type of user interaction), with a subjective measure in the form of periodical queries to the operator about their current degree of trust.

In an attempt to develop an objective and direct measure of trust the human has in the system, the authors of [36] use a mathematical decision model to estimate trust by determining the expected value of decisions a trusting operator would make, and then evaluate the user's decisions in relation to this model. In other words, if the operator deviates largely from the expected value of their decisions, they are said to be less trusting, and vice versa. In another study [108], the authors use two-way trust to adjust the relative contribution of the human input to that of the autonomous controller, as well as the haptic feedback provided to the human operator. They model both robot-to-human and human-to-robot trust, with lower values of the former triggering higher levels of force feedback, and lower values of the latter triggering a higher degree of human control over that of the autonomous robot controller. The authors demonstrate their model can significantly improve performance and lower the workload of operators when compared to previous models and manual control only.

These studies help introduce the idea of "inverse trust". The inverse trust problem is defined in [34] as determining how "an autonomous agent can modify it's behavior in an attempt to increase the trust a human operator will have in it". In this paper, the authors base this measure largely on the number of times the automation is interrupted by a human operator, and uses this to evaluate the autonomous agent's assessment of change in the operator's trust level. Instead of determining an absolute numerical value of trust, the authors choose to have the automation estimate *changes* in the human's trust level. This is followed in [35] by studies in simulation validating their inverse trust model.

8.6.2   Social-Based Interactions: Robots Influencing Humans

Social robotics deals with humans and robots interacting in ways humans typically interact with each other. In most of these studies, the robot—either by its appearance or its behavior—influences the human's beliefs about trustworthiness, feelings of companionship, comfort, feelings of connectedness with the robot, or behavior (such as whether the human discloses secrets to the robot or follows the robot's recommendations). This is distinct from the prior work discussed, such as ATR, where a robot's actions are not typically meant to influence the feelings or behaviors of its operator. These social human-robot interactions contain affective elements that are closer to human-human interactions. There is a body of literature that looked at how robot characteristics affected ratings of animacy and other human-like characteristics, as well as trust in the robot, without explicitly naming a performance or social goal that the robot would perform. It has been consistently found in the social robotics literature that people tend to judge robot characteristics, such as reliability and intelligence, based on robot appearance. For example, people ascribe human qualities to robots that look more anthropomorphic. Another result of people's tendency to anthropomorphize robots is that they tend to ascribe animacy and intent to robots. This finding has not been reported just for robots [109] but even for simple moving shapes [44, 48]. Kiesler and Goetz [52] found that people rated more anthropomorphic looking robots as more reliable. Castro-Gonzalez et al. [18] investigated how the combination of movement characteristics with body appearance can influence people's attributions of animacy, liekeability, trustworthiness, and unpleasantness. They found that naturalistic motion was judged to be more animate, but only if the robot had a human appearance. Moreover, naturalistic motion improved ratings of likeability irrespective of the robot's appearance. More interestingly, a robot with human-like appearance was rated as more disturbing when its movements were more naturalistic. Participants also ascribe personality traits to robots based on appearance. For instance, in [118], robots with spider legs were rated as more aggressive whereas robots with arms rated as more intelligent than those without arms. Physical appearance is not the only attribute that influences human judgment about robot intelligence and knowledge. For example, [59] found that robots that spoke a particular language (e.g. Chinese) were rated higher in their purported knowledge of Chinese landmarks than robots that spoke English.

Robot appearance, physical presence [3], and matched speech [94] are likely to engender trust in the robot [124] found that empathetic language and physical expression elicits higher trust [62] found that highly expressive pedagogical interfaces engender more trust. A recent meta-analysis by Hancock et al. [43] found that robot characteristics such as reliability, behaviors and transparency influenced people's rating of trust in a robot. Besides these characteristics, the researchers in [43] also found that anthropomorphic qualities also had a strong influence on ratings of trust, and that trust in robots is influenced by experience with the robot.

Martelato et al. [73] found that if the robot is more expressive, this encourages participants to disclose information about themselves. However, counter to their hypotheses, disclosure of private information by the robot, a behavior that the authors labelled as making the robot more vulnerable, did not engender increased willingness to disclose on the part of the participants. In a study on willingness of children to disclose secrets, Bethel et al. [5] found in a qualitative study that preschool children were found to be as likely to share a secret with an adult as with a humanoid robot.

An interesting study is reported in [111], where the authors studied how errors performed by the robot affect human trustworthiness and willingness of the human to subsequently comply with the robot's (somewhat unusual) requests. Participants interacted with a home companion robot, in the experimental room that was the pretend home of the robot's human owner in two conditions, (a) where the robot did not make mistakes and (b) where the robot made mistakes. The study found that the participants' assessment of robot reliability and trustworthiness was decreased significantly in the faulty robot condition; nevertheless, the participants were not substantially influence in their decisions to comply with the robot's unusual requests. It was further found that the nature of the request (revocable versus irrevocable) influenced the participants' decisions on compliance. Interestingly, the results in this study also show that participants attributed less anthropomorphism when the robot made errors, which contradict those found by an earlier study the same authors had performed [110].

8.7 Conclusions and Recommendations

In this chapter we briefly reviewed the role of trust in human-robot interaction. We draw several conclusions, the first of which is that there is no accepted definition of what "trust" is in the context of trust in automation. Furthermore, when participants are asked to answer questions as to their level of trust in a robot or software automation, they are almost never given a definition of trust, leaving open the possibility that different participants are viewing the question of trust differently. From a review of the literature, it is apparent that robots still have not achieved full autonomy, and still lack the attributes that would allow them to be considered true teammates by their human counterparts. This is especially true because the literature is largely limited to simulation, or to specific, scripted interactions in the real world. Indeed, in [42], the authors argue that without human-like mental models and a sense of agency, robots will never be considered equal teammates within a mixed human-robot team. They argue that the reason researchers include robots in common HRI tasks is due to their ability to complement the skills of humans. Yet, because of the tendency of humans to anthropomorphize things they interact with, the controlled interactions researchers develop for HRI studies are more characteristic of human-human interactions. While this tendency to anthropomorphize can be helpful in some cases, it poses a serious risk if this naturally gives humans a higher degree of trust in robots than is warranted. The question of how a robot's performance influences anthropomorphization is also unclear—with recent studies finding conflicting results [110, 111].

There is a general agreement that the notion of trust involves vulnerability of the trustor to the trustee in circumstances of risk and uncertainty. In the performance-based literature, where the human is relying on the robot to do the whole task or part of the task, it is clear that the participant is vulnerable to the robot with respect to the participant's performance in the experimental task. In most of the studies in social robotics, however, where the robot is trying to get the participant to do something (e.g. comply with instructions to throw away someone else's mail, or disclose a secret) it is not clear that the participant is truly vulnerable to the robot (unless we regard breaking a social convention as making oneself vulnerable), merely enjoying the novelty of robots, or feeling pressure to follow experimental procedure. Therefore, the notion that was measured in those studies may not have been trust in the sense that the term is defined in the trust literature. For example in [104], where participants showed compliance with a robot guide even when reliability was ranked lower after an error, the researchers admit several confounding factors (e.g., participants did not have enough time to deliberate). The findings on human tendencies to ascribe reliability, trustworthiness, intelligence and other positive characteristics to robots may prohibit correct estimation of robot's abilities and prevent correct trust calibration. This is dangerous especially since the use of robots is envisioned to increase, especially in high risk situations such as emergency response and the military.

This overview enables us to provide several recommendations for how future work investigating trust in human-autonomy and human-robot interaction would proceed. First, it would be useful for the community to have a clear definition in each study as to what autonomy and what teammate characteristics the robot in the study possesses. Second, it would be useful for each study to define the notion of trust the author's espouse, as well as which dimensions of the notion of trust they believe are relevant to the task being investigated. The experimenters should also try to understand, via surveys or other means, what definition of trust the participants have in their heads. A possible idea is that experimenters could even give their definition of trust to the participants and see how this may affect the participants' answers.

Another recommendation is that, given the novelty of robots for the majority of the population, along with the well-known fact from in-group/out-group studies that people seem to be influenced very easily and for trivial reasons, it would be useful to perform longer duration studies to investigate the transient nature of trust assessments. In other words, how does trust in automation change as a function of how familiar users are with the automation and how much they interact with it over time? One could imaging someone unfamiliar with automation or robots placing a high degree of trust in them due to prior beliefs (which may be incorrect). Over time, this implicit trust may fade as they work more with automation and realize that it is not perfect.

Furthermore, we believe in a need to increase research in the multi-robot systems area, as well as the area of robots helping human teams. As the number of robots increase and hardware and operation costs decrease, it is inevitable that humans will be interacting with larger numbers of robots to perform increasingly complex tasks. Furthermore, trust in larger groups and collectives of robots is no doubt influenced by different factors—specifically those regarding the robots' behaviors—in addition to single robot control. Similarly, there is little work investigating how multiple humans working together with robots affect each others' trust levels, which needs to be addressed.

Finally, it would be helpful for the community to define a set of task categories of human-robot interaction with characteristics that involve specific differing dimensions of trust. Such characteristics could be the degree of risk to the trustor, the degree of uncertainty, the degree of potential gain, whether the trustor's vulnerability is to the reliability of the robot, or the robot's integrity or benevolence. Other studies should expand on the notion of co-adaptive trust to improve how robots assess their own behavior and how it affects the trust in them by their operator. As communication is key to any collaborative interaction, research should not focus merely on how the human sees the robot, but also how the robot sees the human.

Acknowledgements This work is supported by awards FA9550-13-1-0129 and FA9550-15-1-0442.

References

1. B. D. Adams, D. J. Bryant, and R.D. Webb. Trust in teams: Literature review. Technical Report Technical Report CR-2001-042, Report to Defense and Civil Insitute of Environmental Medicine. Humansystems Inc., 2001
2. Eugenio Alberdi, Andrey Povyakalo, Lorenzo Strigini, Peter Ayton, Effects of incorrect computer-aided detection (cad) output on human decision-making in mammography. Academic radiology **11**(8), 909–918 (2004)
3. Wilma A Bainbridge, Justin Hart, Elizabeth S Kim, and Brian Scassellati. The effect of presence on human-robot interaction. In *RO-MAN 2008-The 17th IEEE International Symposium on Robot and Human Interactive Communication*, pages 701–706. IEEE, 2008
4. Bernard Barber. *The logic and limits of trust*. Rutgers University Press, 1983
5. Cindy L Bethel, Matthew R Stevenson, and Brian Scassellati. Secret-sharing: Interactions between a child, robot, and adult. In *Systems, man, and cybernetics (SMC), 2011 IEEE International Conference on*, pages 2489–2494. IEEE, 2011
6. CE Billings, JK Lauber, H Funkhouser, EG Lyman, and EM Huff. Nasa aviation safety reporting system. Technical Report Technical Report TM-X-3445, NASA Ames Research Center, 1976

7. P. David, Biros, Mark Daly, and Gregg Gunsch. The influence of task load and automation trust on deception detection. Group Decision and Negotiation **13**(2), 173–189 (2004)
8. Iris Bohnet, Benedikt Hermann, Richard Zeckhauser, The requirements for trust in gulf and western countries. Quarterly Journal of Economics **125**, 811–828 (2010)
9. Michael W. Boyce, Jessie Y.C. Chen, Anthony R. Selkowitz, and Shan G. Lakhmani. Effects of agent transparency on operator trust. In *Proceedings of the Tenth Annual ACM/IEEE International Conference on Human-Robot Interaction Extended Abstracts*, HRI'15 Extended Abstracts, pages 179–180, New York, NY, USA, 2015. ACM
10. B. Marilynn, Brewer and Roderick M Kramer. The psychology of intergroup attitudes and behavior. Annual review of psychology **36**(1), 219–243 (1985)
11. Joel Brockner, Tom R Tyler, and Rochelle Cooper-Schneider. The influence of prior commitment to an institution on reactions to perceived unfairness: The higher they are, the harder they fall. *Administrative Science Quarterly*, pages 241–261, 1992
12. Chris Burnett, Timothy J Norman, and Katia Sycara. Bootstrapping trust evaluations through stereotypes. In *Proceedings of the 9th International Conference on Autonomous Agents and Multiagent Systems: volume 1-Volume 1*, pages 241–248. International Foundation for Autonomous Agents and Multiagent Systems, 2010
13. Chris Burnett, Timothy J Norman, and Katia Sycara. Decision-making with trust and control in multi-agent systems. *Twenty Second International Joint Conference on Artificial Intelligence*, **10**:241–248, 2011
14. Chris Burnett, Timothy J Norman, and Katia Sycara. Stereotypical trust and bias in dynamic multiagent systems. *ACM Transactions on Intelligent Systems and Technology (TIST)*, **4**(2):26, 2013
15. Chris Burnett, Timothy J Norman, Katia Sycara, and Nir Oren. Supporting trust assessment and decision making in coalitions. IEEE Intelligent Systems **29**(4), 18–24 (2014)
16. Dale Carl, V Gupta, and Mansour Javidan. Culture, leadership, and organizations: The globe study of 62 societies, 2004
17. Jude Cassidy, Child-mother attachment and the self in six-year-olds. Child development **59**(1), 121–134 (1988)
18. Álvaro Castro-González, Henny Admoni, Brian Scassellati, Effects of form and motion on judgments of social robots' animacy, likability, trustworthiness and unpleasantness. International Journal of Human-Computer Studies **90**, 27–38 (2016)
19. Jessie YC Chen, Michael J Barnes, and Michelle Harper-Sciarini. Supervisory control of multiple robots: Human-performance issues and user-interface design. *IEEE Transactions on Systems, Man, and Cybernetics, Part C (Applications and Reviews)*, **41**(4):435–454, 2011
20. J.Y.C. Chen, P.I. Terrence, Effects of imperfect automation and individual differences on concurrent performance of military and robotics tasks in a simulated multitasking environment. Ergonomics **52**(8), 907–920 (2009)
21. Shih-Yi Chien, Michael Lewis, Sebastian Hergeth, Zhaleh Semnani-Azad, and Katia Sycara. Cross-country validation of a cultural scale in measuring trust in automation. In *Proceedings of the Human Factors and Ergonomics Society Annual Meeting*, volume 59, pages 686–690. SAGE Publications, 2015
22. Shih-Yi Chien, Michael Lewis, K. Sycara, J-S. Liu, and A. Kumru. Influence of cultural factors in dynamic trust in automation. In *Proceedings of the Systems, Man, and Cybernetics Society*, 2016
23. Shih-Yi Chien, Zhaleh Semnani-Azad, Michael Lewis, and Katia Sycara. Towards the development of an inter-cultural scale to measure trust in automation. In *International Conference on Cross-Cultural Design*, pages 35–46. Springer, 2014
24. Andrew S Clare, Mary L Cummings, and Nelson P Repenning. Influencing trust for human–automation collaborative scheduling of multiple unmanned vehicles. *Human Factors: The Journal of the Human Factors and Ergonomics Society*, **57**(7):1208–1218, 2015
25. Carsten K.W. De Dreu, Peter, J Carnevale. Motivational bases of information processing and strategy in conflict and negotiation. Advances in experimental social psychology **35**, 235–291 (2003)

26. Ewart de Visser, Raja Parasuraman, Adaptive aiding of human-robot teaming effects of imperfect automation on performance, trust, and workload. Journal of Cognitive Engineering and Decision Making **5**(2), 209–231 (2011)
27. Sidney Dekker, Erik Hollnagel, Human factors and folk models. Cognition, Technology & Work **6**(2), 79–86 (2004)
28. Sidney W.A. Dekker, David, D Woods. Maba-maba or abracadabra? progress on human-automation co-ordination. Cognition, Technology & Work **4**(4), 240–244 (2002)
29. Munjal Desai. *Modeling trust to improve human-robot interaction.* PhD thesis, University of Massachusetts Lowell, 2012
30. Munjal Desai, Poornima Kaniarasu, Mikhail Medvedev, Aaron Steinfeld, and Holly Yanco. Impact of robot failures and feedback on real-time trust. In *Proceedings of the 8th ACM/IEEE international conference on Human-robot interaction*, pages 251–258. IEEE Press, 2013
31. Stephen R Dixon and Christopher D Wickens. Automation reliability in unmanned aerial vehicle control: A reliance-compliance model of automation dependence in high workload. *Human Factors: The Journal of the Human Factors and Ergonomics Society*, **48**(3):474–486, 2006
32. The identification of culturally endorsed leadership profiles, Peter W Dorfman, Paul J Hanges, and Felix C Brodbeck. Leadership and cultural variation. Culture, leadership, and organizations: The GLOBE study of **62**, 669–719 (2004)
33. Mary T Dzindolet, Linda G Pierce, Hall P Beck, and Lloyd A Dawe. The perceived utility of human and automated aids in a visual detection task. *Human Factors: The Journal of the Human Factors and Ergonomics Society*, **44**(1):79–94, 2002
34. Michael W Floyd, Michael Drinkwater, and David W Aha. Adapting autonomous behavior using an inverse trust estimation. In *International Conference on Computational Science and Its Applications*, pages 728–742. Springer, 2014
35. Michael W Floyd, Michael Drinkwater, and David W Aha. Learning trustworthy behaviors using an inverse trust metric. In *Robust Intelligence and Trust in Autonomous Systems*, pages 33–53. Springer, 2016
36. Amos Freedy, Ewart DeVisser, Gershon Weltman, and Nicole Coeyman. Measurement of trust in human-robot collaboration. In *Collaborative Technologies and Systems, 2007. CTS 2007. International Symposium on*, pages 106–114. IEEE, 2007
37. A. Fulmer and Gelfand M. *Models for intercultural collaboration and negotiation*, chapter Dynamic trust processes: trust dissolution, recovery and stabilization. Springer, 2012
38. Fei Gao, Andrew S Clare, Jamie C Macbeth, and ML Cummings. Modeling the impact of operator trust on performance in multiple robot control. In *Spring Symposium AAAI*, 2013
39. Ji Gao and John D Lee. Extending the decision field theory to model operators' reliance on automation in supervisory control situations. *IEEE Transactions on Systems, Man, and Cybernetics-Part A: Systems and Humans*, **36**(5):943–959, 2006
40. Kate Goddard, Abdul Roudsari, and Jeremy C Wyatt. Automation bias: a systematic review of frequency, effect mediators, and mitigators. *Journal of the American Medical Informatics Association*, **19**(1):121–127, 2012
41. P Goillau, C Kelly, M Boardman, and E Jeannot. Guidelines for trust in future atm systems-measures. *EUROCONTROL, the European Organization for the Safety of Air Navigation*, 2003
42. Victoria Groom, Clifford Nass, Can robots be teammates?: Benchmarks in human-robot teams. Interaction Studies **8**(3), 483–500 (2007)
43. Peter A Hancock, Deborah R Billings, Kristin E Schaefer, Jessie YC Chen, Ewart J De Visser, and Raja Parasuraman. A meta-analysis of factors affecting trust in human-robot interaction. *Human Factors: The Journal of the Human Factors and Ergonomics Society*, **53**(5):517–527, 2011
44. Fritz Heider, Marianne Simmel, An experimental study of apparent behavior. The American Journal of Psychology **57**(2), 243–259 (1944)
45. Geert Hofstede, Gert Jan Hofstede, and Michael Minkov. *Cultures and organizations: Software of the mind*, volume 2. Citeseer, 1991

46. Jiun-Yin Jian, Ann M Bisantz, and Colin G Drury. Foundations for an empirically determined scale of trust in automated systems. International Journal of Cognitive Ergonomics **4**(1), 53–71 (2000)
47. R. Gareth, Jones and Jennifer M George. The experience and evolution of trust: Implications for cooperation and teamwork. Academy of management review **23**(3), 531–546 (1998)
48. Wendy Ju and Leila Takayama. Approachability: How people interpret automatic door movement as gesture. *International Journal of Design*, **3**(2), 2009
49. Poornima Kaniarasu, Aaron Steinfeld, Munjal Desai, and Holly Yanco. Robot confidence and trust alignment. In *Proceedings of the 8th ACM/IEEE international conference on Human-robot interaction*, pages 155–156. IEEE Press, 2013
50. Kristiina Karvonen, Lucas Cardholm, and Stefan Karlsson. Designing trust for a universal audience: a multicultural study on the formation of trust in the internet in the nordic countries. In *HCI*, pages 1078–1082, 2001
51. C. Kelly, M. Boardman, P. Goillau, *and E Jeannot* (A literature review. European Organization for the Safety of Air Navigation, Guidelines for trust in future atm systems, 2003)
52. Sara Kiesler and Jennifer Goetz. Mental models of robotic assistants. In *CHI'02 extended abstracts on Human Factors in Computing Systems*, pages 576–577. ACM, 2002
53. H. Peter, Kim, Kurt T Dirks, and Cecily D Cooper. The repair of trust: A dynamic bilateral perspective and multilevel conceptualization. Academy of Management Review **34**(3), 401–422 (2009)
54. W. Arie, Kruglanski, Erik P Thompson, E Tory Higgins, M Atash, Antonio Pierro, James Y Shah, and Scott Spiegel. To" do the right thing" or to" just do it": locomotion and assessment as distinct self-regulatory imperatives. Journal of personality and social psychology **79**(5), 793 (2000)
55. Charles Layton, Philip J Smith, and C Elaine McCoy. Design of a cooperative problem-solving system for en-route flight planning: An empirical evaluation. *Human Factors: The Journal of the Human Factors and Ergonomics Society*, **36**(1):94–119, 1994
56. John Lee, Neville Moray, Trust, control strategies and allocation of function in human-machine systems. Ergonomics **35**(10), 1243–1270 (1992)
57. D. John, Lee and Neville Moray. Trust, self-confidence, and operators' adaptation to automation. International journal of human-computer studies **40**(1), 153–184 (1994)
58. John D Lee and Katrina A See. Trust in automation: Designing for appropriate reliance. *Human Factors: The Journal of the Human Factors and Ergonomics Society*, **46**(1):50–80, 2004
59. Sau-lai Lee, Ivy Yee-man Lau, S Kiesler, and Chi-Yue Chiu. Human mental models of humanoid robots. In *Proceedings of the 2005 IEEE international conference on robotics and automation*, pages 2767–2772. IEEE, 2005
60. Terri Lenox, Michael Lewis, Emilie Roth, Rande Shern, Linda Roberts, Tom Rafalski, and Jeff Jacobson. Support of teamwork in human-agent teams. In *Systems, Man, and Cybernetics, 1998. 1998 IEEE International Conference on*, volume 2, pages 1341–1346. IEEE, 1998
61. F Javier Lerch, Michael J Prietula, and Carol T Kulik. The turing effect: The nature of trust in expert systems advice. In *Expertise in context*, pages 417–448. MIT Press, 1997
62. James C Lester, Sharolyn A Converse, Susan E Kahler, S Todd Barlow, Brian A Stone, and Ravinder S Bhogal. The persona effect: affective impact of animated pedagogical agents. In *Proceedings of the ACM SIGCHI Conference on Human factors in computing systems*, pages 359–366. ACM, 1997
63. Angela K.-Y. Leung, Dov Cohen, Within-and between-culture variation: individual differences and the cultural logics of honor, face, and dignity cultures. Journal of personality and social psychology **100**(3), 507 (2011)
64. Stephan Lewandowsky, Michael Mundy, Gerard Tan, The dynamics of trust: comparing humans to automation. Journal of Experimental Psychology: Applied **6**(2), 104 (2000)
65. J. David, Lewis and Andrew Weigert. Trust as a social reality. Social forces **63**(4), 967–985 (1985)
66. Michael Lewis, Designing for human-agent interaction. AI Magazine **19**(2), 67 (1998)

67. James Llinas, Ann Bisantz, Colin Drury, Younho Seong, and Jiun-Yin Jian. Studies and analyses of aided adversarial decision making. phase 2: Research on human trust in automation. Technical report, DTIC Document, 1998
68. Maria Luz. *Validation of a Trust survey on example of MTCD in real time simulation with Irish controllers*. PhD thesis, thesis final report. The European Organisation for the Safety of Air Navigation, 2009
69. Joseph B Lyons. Being transparent about transparency. In *AAAI Spring Symposium*, 2013
70. Joseph B Lyons and Paul R Havig. Transparency in a human-machine context: Approaches for fostering shared awareness/intent. In *International Conference on Virtual, Augmented and Mixed Reality*, pages 181–190. Springer, 2014
71. Joseph B Lyons and Charlene K Stokes. Human–human reliance in the context of automation. *Human Factors: The Journal of the Human Factors and Ergonomics Society*, page 0018720811427034, 2011
72. Maria Madsen and Shirley Gregor. Measuring human-computer trust. In *11th australasian conference on information systems*, volume 53, pages 6–8. Citeseer, 2000
73. Nikolas Martelaro, Victoria C Nneji, Wendy Ju, and Pamela Hinds. Tell me more: Designing hri to encourage more trust, disclosure, and companionship. In *The Eleventh ACM/IEEE International Conference on Human Robot Interation*, pages 181–188. IEEE Press, 2016
74. Anthony J Masalonis and Raja Parasuraman. Effects of situation-specific reliability on trust and usage of automated air traffic control decision aids. In *Proceedings of the Human Factors and Ergonomics Society Annual Meeting*, volume 47, pages 533–537. SAGE Publications, 2003
75. Reena Master, Xiaochun Jiang, Mohammad T Khasawneh, Shannon R Bowling, Larry Grimes, Anand K Gramopadhye, and Brian J Melloy. Measurement of trust over time in hybrid inspection systems. *Human Factors and Ergonomics in Manufacturing & Service Industries*, 15(2):177–196, 2005
76. C. Roger, Mayer, James H Davis, and F David Schoorman. An integrative model of organizational trust. Academy of management review 20(3), 709–734 (1995)
77. John M McGuirl and Nadine B Sarter. Supporting trust calibration and the effective use of decision aids by presenting dynamic system confidence information. *Human Factors: The Journal of the Human Factors and Ergonomics Society*, 48(4):656–665, 2006
78. K. Ann, McKibbon and Douglas B Fridsma. Effectiveness of clinician-selected electronic information resources for answering primary care physicians information needs. Journal of the American Medical Informatics Association 13(6), 653–659 (2006)
79. Ashleigh Merritt, Culture in the cockpit do hofstede dimensions replicate? Journal of cross-cultural psychology 31(3), 283–301 (2000)
80. Stephanie M Merritt and Daniel R Ilgen. Not all trust is created equal: Dispositional and history-based trust in human-automation interactions. *Human Factors: The Journal of the Human Factors and Ergonomics Society*, 50(2):194–210, 2008
81. Joachim Meyer, Effects of warning validity and proximity on responses to warnings. Human Factors: The Journal of the Human Factors and Ergonomics Society 43(4), 563–572 (2001)
82. Joachim Meyer, Conceptual issues in the study of dynamic hazard warnings. Human Factors: The Journal of the Human Factors and Ergonomics Society 46(2), 196–204 (2004)
83. Debra Meyerson, Karl E Weick, and Roderick M Kramer. Swift trust and temporary groups. *Trust in organizations: Frontiers of theory and research*, 166:195, 1996
84. Raymond E Miles and WE Douglas Creed. Organizational forms and managerial philosophies-a descriptive and analytical review. *Research in Organizational Behavior: An Annual Series of Anaylytical Essays and Critical Reviews*, 17:333–372, 1995
85. Neville Moray and, T Inagaki. Laboratory studies of trust between humans and machines in automated systems. Transactions of the Institute of Measurement and Control 21(4–5), 203–211 (1999)
86. Neville Moray, Toshiyuki Inagaki, Makoto Itoh, Adaptive automation, trust, and self-confidence in fault management of time-critical tasks. Journal of Experimental Psychology: Applied 6(1), 44 (2000)

87. Kathleen L Mosier, Everett A Palmer, and Asaf Degani. Electronic checklists: Implications for decision making. In *Proceedings of the Human Factors and Ergonomics Society Annual Meeting*, volume 36, pages 7–11. SAGE Publications, 1992

88. Kathleen L Mosier and Linda J Skitka. Human decision makers and automated decision aids: Made for each other? *Automation and human performance: Theory and applications*, pages 201–220, 1996

89. L. Kathleen, Mosier, Linda J Skitka, Susan Heers, and Mark Burdick. Automation bias: Decision making and performance in high-tech cockpits. The International journal of aviation psychology **8**(1), 47–63 (1998)

90. KL Mosier, LJ Skitka, and KJ Korte. Cognitive and social psychological issues in flight crew/automation interaction. *Human performance in automated systems: Current research and trends*, pages 191–197, 1994

91. M. Bonnie, Muir. Trust in automation: Part i. theoretical issues in the study of trust and human intervention in automated systems. Ergonomics **37**(11), 1905–1922 (1994)

92. M. Bonnie, Muir and Neville Moray. Trust in automation. part ii. experimental studies of trust and human intervention in a process control simulation. Ergonomics **39**(3), 429–460 (1996)

93. Bonnie Marlene Muir. *Operators' trust in and use of automatic controllers in a supervisory process control task*. PhD thesis, University of Toronto, 1990

94. Clifford Nass and Kwan Min Lee, Does computer-synthesized speech manifest personality? experimental tests of recognition, similarity-attraction, and consistency-attraction. Journal of Experimental Psychology: Applied **7**(3), 171 (2001)

95. Scott Ososky, David Schuster, Elizabeth Phillips, and Florian Jentsch. Building appropriate trust in human-robot teams. In *AAAI Spring Symposium Series*, 2013

96. Raja Parasuraman and Dietrich H Manzey. Complacency and bias in human use of automation: An attentional integration. *Human Factors: The Journal of the Human Factors and Ergonomics Society*, **52**(3):381–410, 2010

97. Raja Parasuraman, Victor Riley, Humans and automation: Use, misuse, disuse, abuse. Human Factors: The Journal of the Human Factors and Ergonomics Society **39**(2), 230–253 (1997)

98. Raja Parasuraman, Thomas B Sheridan, and Christopher D Wickens. Situation awareness, mental workload, and trust in automation: Viable, empirically supported cognitive engineering constructs. *Journal of Cognitive Engineering and Decision Making*, **2**(2):140–160, 2008

99. PL Patrick Rau, Ye Li, and Dingjun Li. Effects of communication style and culture on ability to accept recommendations from robots. Computers in Human Behavior **25**(2), 587–595 (2009)

100. K. John, Rempel, John G Holmes, and Mark P Zanna. Trust in close relationships. Journal of personality and social psychology **49**(1), 95 (1985)

101. V. A. Riley. *Automation theory and applications*, chapter Operator reliance on automation: theory and data, pages 19–35. Mahwah, NJ. Erlbaum, 1996

102. Victor Andrew Riley. *Human use of automation*. PhD thesis, University of Minneapolis, 1994

103. Paul Robinette, Ayanna M Howard, and Alan R Wagner. Timing is key for robot trust repair. In *International Conference on Social Robotics*, pages 574–583. Springer, 2015

104. Paul Robinette, Wenchen Li, Robert Allen, Ayanna M Howard, and Alan R Wagner. Overtrust of robots in emergency evacuation scenarios. In *2016 11th ACM/IEEE International Conference on Human-Robot Interaction (HRI)*, pages 101–108. IEEE, 2016

105. B. Julian, Rotter. A new scale for the measurement of interpersonal trust1. Journal of personality **35**(4), 651–665 (1967)

106. B. Julian, Rotter. Generalized expectancies for interpersonal trust. American psychologist **26**(5), 443 (1971)

107. M. Denise, Rousseau, Sim B Sitkin, Ronald S Burt, and Colin Camerer. Not so different after all: A cross-discipline view of trust. Academy of management review **23**(3), 393–404 (1998)

108. H Saeidi, F McLane, B Sadrfaidpour, E Sand, S Fu, J Rodriguez, JR Wagner, and Y Wang. Trust-based mixed-initiative teleoperation of mobile robots. In *2016 American Control Conference (ACC)*, pages 6177–6182. IEEE, 2016

109. Martin Saerbeck and Christoph Bartneck. Perception of affect elicited by robot motion. In *Proceedings of the 5th ACM/IEEE international conference on Human-robot interaction*, pages 53–60. IEEE Press, 2010

110. Maha Salem, Friederike Eyssel, Katharina Rohlfing, Stefan Kopp, Frank Joublin, To err is human (-like): Effects of robot gesture on perceived anthropomorphism and likability. International Journal of Social Robotics 5(3), 313–323 (2013)
111. Maha Salem, Gabriella Lakatos, Farshid Amirabdollahian, and Kerstin Dautenhahn. Would you trust a (faulty) robot?: Effects of error, task type and personality on human-robot cooperation and trust. In *Proceedings of the Tenth Annual ACM/IEEE International Conference on Human-Robot Interaction*, pages 141–148. ACM, 2015
112. Nadine B Sarter and Beth Schroeder. Supporting decision making and action selection under time pressure and uncertainty: The case of in-flight icing. *Human Factors: The Journal of the Human Factors and Ergonomics Society*, 43(4):573–583, 2001
113. M. Paul, *Satchell* (Routledge, Cockpit Monitoring and Alerting Systems, 1993)
114. Kristin E Schaefer. *The perception and measurement of human-robot trust*. PhD thesis, University of Central Florida Orlando, Florida, 2013
115. Kristin E Schaefer, Jessie YC Chen, James L Szalma, and PA Hancock. A meta-analysis of factors influencing the development of trust in automation implications for understanding autonomy in future systems. *Human Factors: The Journal of the Human Factors and Ergonomics Society*, page 0018720816634228, 2016
116. TB Sheridan and W Verplank. Human and computer control of undersea teleoperators. cambridge, ma: Man-machine systems laboratory, department of mechanical engineering, 1978
117. A Simpson, GN Brander, and DRA Portsdown. Seaworthy trust: Confidence in automated data fusion. *The Human-Electronic Crew: Can we Trust the Team*, pages 77–81, 1995
118. Valerie K Sims, Matthew G Chin, David J Sushil, Daniel J Barber, Tatiana Ballion, Bryan R Clark, Keith A Garfield, Michael J Dolezal, Randall Shumaker, and Neal Finkelstein. Anthropomorphism of robotic forms: a response to affordances? In *Proceedings of the Human Factors and Ergonomics Society Annual Meeting*, volume 49, pages 602–605. SAGE Publications, 2005
119. L. Indramani, Singh, Robert Molloy, and Raja Parasuraman. Individual differences in monitoring failures of automation. The Journal of General Psychology 120(3), 357–373 (1993)
120. Randall D Spain, Ernesto A Bustamante, and James P Bliss. Towards an empirically developed scale for system trust: Take two. In *Proceedings of the Human Factors and Ergonomics Society Annual Meeting*, volume 52, pages 1335–1339. SAGE Publications, 2008
121. K Sycara and M Lewis. Forming shared mental models. In *Proc. of the 13th Annual Meeting of the Cognitive Science Society*, pages 400–405, 1991
122. Katia P Sycara, Michael Lewis, Terri Lenox, and Linda Roberts. Calibrating trust to integrate intelligent agents into human teams. In *System Sciences, 1998., Proceedings of the Thirty-First Hawaii International Conference on*, volume 1, pages 263–268. IEEE, 1998
123. Kazuya Takeda. Modeling and detecting excessive trust from behavior signals: Overview of research project and results. In *Human-Harmonized Information Technology, Volume 1*, pages 57–75. Springer, 2016
124. Adriana Tapus, Maja J Mataric, and Brian Scassellati. Socially assistive robotics [grand challenges of robotics]. IEEE Robotics & Automation Magazine 14(1), 35–42 (2007)
125. Lin Wang, Pei-Luen Patrick Rau, Vanessa Evers, Benjamin Krisper Robinson, and Pamela Hinds. When in rome: the role of culture & context in adherence to robot recommendations. In *Proceedings of the 5th ACM/IEEE international conference on Human-robot interaction*, pages 359–366. IEEE Press, 2010
126. Ning Wang, David V. Pynadath, and Susan G. Hill. Trust calibration within a human-robot team: Comparing automatically generated explanations. In *The Eleventh ACM/IEEE International Conference on Human Robot Interaction*, HRI '16, pages 109–116, Piscataway, NJ, USA, 2016. IEEE Press
127. E. Karl, Weick. Enacted sensemaking in crisis situations. Journal of management studies 25(4), 305–317 (1988)
128. P. Richard, Will. True and false dependence on technology: Evaluation with an expert system. Computers in human behavior 7(3), 171–183 (1991)

129. A. Xu and G. Dudek. Maintaining efficient collaboration with trust-seeking robots. In *Intelligent Robots and Systems, 2016.(IROS 2016). Proceedings. 2016 IEEE/RSJ International Conference on*, volume 16. IEEE, 2016

130. Anqi Xu and Gregory Dudek. Optimo: Online probabilistic trust inference model for asymmetric human-robot collaborations. In *Proceedings of the Tenth Annual ACM/IEEE International Conference on Human-Robot Interaction*, pages 221–228. ACM, 2015

131. E. Rosemarie, Yagoda and Douglas J Gillan. You want me to trust a robot? the development of a human-robot interaction trust scale. International Journal of Social Robotics **4**(3), 235–248 (2012)

Chapter 9
Trustworthiness of Autonomous Systems

S. Kate Devitt

9.1 Introduction

Humans are constantly engaged in evaluating the trustworthiness of humans and systems. Effective robots and Autonomous Systems (AS) must be trustworthy. Understanding how humans trust will enable better relationships between human and AS. Trust is essential in designing autonomous and semi-autonomous technologies, because "No trust, no use" [80]. Additionally, rates of usage are proportionally related to the degree of trust expressed [54]. Hancock et al. [37] argue that trust begets reliance, compliance and use. However, humans do already rely on systems they do not trust. Consider the unreasonable privacy policies agreed to by users to access services via apps, websites and cloud services [90]. Because privacy policies can be changed at any time, private data may be sold by organisations for profit without explicit consumer consent or even awareness. Consumers can find the benefits of the services to enhance their lives and productivity too strong to resist. In these situations, people rely on systems they do not trust and are not trustworthy. People know that their data may be shared for corporate interests. People know that they have signed away rights on their own images, etc. by using these services. As more services operate without human decision makers yet offer irresistible perks, humans may increasingly rely on untrusted AS to decide for them. Instead of trust, it may be better to consider human reliance on other humans and systems as a measure of risk aversion–of which trustworthiness remains a significant part.

S. Kate Devitt (✉)
Robotics and Autonomous Systems, School of Electrical Engineering
and Computer Science, Faculty of Science and Engineering, Institute for
Future Environments, Faculty of Law, Queensland University of Technology,
Brisbane, Australia
e-mail: kate.devitt@qut.edu.au

© The Author(s) 2018
H. A. Abbass et al. (eds.), *Foundations of Trusted Autonomy*, Studies in Systems,
Decision and Control 117, https://doi.org/10.1007/978-3-319-64816-3_9

9.1.1 Autonomous Systems

AS can be robots, AI programs or software that operate without human control. AS are made by teams of engineers, designers, mathematicians, and computer programmers to serve a human need. AS actions and decisions are made by complex hierarchical processes balancing the uncertainties of cross modal inputs such as cameras, microphones, tactile responders with internal representations such as maps, directives and event memories. AS execute functions such as actively selecting data, transforming information, making decisions, or controlling processes without inputs [54] (p. 50). AS are defined in contrast with automated systems and manual systems. Automated systems are largely deterministic to achieve predefined goals. Classic automata such as Japanese karakuri demonstrate complicated, nevertheless predictable behaviours [1]. In contrast, AS learn and adapt in their environments rendering their actions more indeterminate over time [36, 80]. Advanced AS may be capable of executive functions such as planning, goal-setting, rule-making and abstract conceptualisation. An 'autonomous system' can refer to a subset of functions within a larger functional system or refer to the superset of functions undertaken by an agent or machine. Regardless of the scope of functions of an autonomous system, it is important that AS operate without human control.

9.1.2 Trustworthiness

Trustworthiness is a property of an agent or organisation that engenders trust in another agent or organisation. Trust is a psychological state in which a person makes themselves vulnerable because they are confident that other agents will not exploit them [68]. Trust is also a social feeling of mutual confidence that increases the efficiency of systems, allowing adaptations to externalities and uncertainties [4]. Trust, like empathy, truth telling and loyalty lubricates social interactions. Humans depend on flexible cooperation with unrelated group members that rely on trust [89]. Thus, social success relies both the evaluation of the trustworthiness of others and the presentation of oneself as trustworthy [23].

We can distinguish between the trust we place in individuals, and the general trust we have in our society that affects how we make decisions more broadly, e.g. Adam Smith [87] in the *Wealth of Nations* noted that a merchant is more comfortable trading within their own society because they can "know better the character and situation of the persons whom he trusts." Empirical literature has linked improved trust with more efficient public institutions, greater economic prosperity, self-reported health and happiness across many societies using a range of statistical techniques (see [16]). Within a Nation or society, trust is quite heterogeneous between individuals. Surveys on whether subjects trust a generic person—measured on a scale between 0 (no trust at all) and 10 (fully trusted)—find large interpersonal differences [14]. Economic productivity peaks when the average citizen rates a generic person a '7' level of trust-a fairly high level of trust. Pessimists trust too little and give up opportunities too often. Optimists trust too much and get cheated more frequently. How does this trust research relate to AS? Do economic models apply to designing trustworthiness in AS? Should we create trustworthy systems to engender a '7' level of trust matching optimum human economic performance? That is to say, if we test the trustworthiness of autonomous-human interactions, should we aim to replicate the trust metrics found between people or some other measure?

It is important to acknowledge that trust is a complex phenomena and has been defined differently depending on the discipline [78]. Economists consider it calculative [99] or institutional [70]. Psychologists focus on the cognitive attributes of the trustor and the trustee [77, 96]. Sociologists find trust within human relationships [33]. Understanding the way humans conceive of and act regarding trust is critical to ensure the success of trusted AS. To bring different approaches under a single framework for investigation, this chapter will examine trustworthiness with three questions:

1. Who or what is trustworthy?-metaphysics
2. How do we know who or what is trustworthy?-epistemology
3. What factors influence what or who should we trust?-normativity

Building trustworthy autonomous systems requires understanding trust in human-human relationships and human-AS interactions. A research program on trusted AS ought to incorporate mental models informed from cognitive science to better understand and respond to human thoughts and behaviour. An example of such a research program is the recent work programming a robot with ACT-R/E [49, 94], an embodied extension of the ACT-R [3] cognitive architecture. The ACT-R/E implementation takes features of human cognition, such as segmenting time into events and narrative explanation to bring meaningfulness and trust to robot-human relationships. But, it is just one of many promising frameworks to align AS with human cognition. This chapter considers a range of theories of trust to influence the design trustworthy autonomous systems.

9.2 Background

The Fukushima Daiichi nuclear power plant disaster stemming from the Japanese earthquake and tsunami in March 2011 motivated DARPA to develop the Robotics Challenge (DRC) in 2012. Immune to radiation damage, Japan could have used robots to help rescue people, or go into the Fukishima power plant to turn off valves, investigate leaks or structural damage. Yet after decades of robot research and development Japan did not have a rescue robot. Where was the real Astroboy [1, 60]? Humanoid Robotics Project (HRP)-2 was functionally designed to assist people in construction, dangerous environments and home [47] but did not have the operational capacities to help when needed. In response, the DRC challenged robots to perform tasks modeled on the context of urban search and rescue (USAR) and industrial disaster response task domains [105]. Tasks were real-world anthropomorphic manipulation and mobility; controlled by automated interfaces and teleoperation. Challenges included obstacles such as opening a door, turning a valve, driving a car, and walking over a pile of chaotic bricks. The first robots to attempt the challenge failed miserably. They almost all fell over or were unable to complete tasks so simple for humans. The DRC robots were not even autonomous-actions were manually controlled by teams.

Thus, despite early optimism that robots would be capable of performing human-level tasks by 2015, machines are still far from achieving this goal. Very basic tasks still require supervisory human control from one or more operators. Complex environments such as USAR, require continuous direct control by multiple operators. Engineering autonomy in robots requires more research in both pragmatic design and societal implications. Trust will emerge from evidence-based control interface design that accommodates multiple control paradigms of the robot and the user [105].

Even though the DARPA challenge remains difficult to accomplish, AS are already being depended upon in our lives, from our adaptive smart phones [56], to off shore oil rig drilling programs [34]. Self-driving modes in cars (see [91, 100]), mining trucks [82] and buses [76] are already in use. Now is the time to understand the metaphysical, epistemological and normative dimensions of trust and trustworthiness so that we can build, use and thrive with AS.

9.3   Who or What Is Trustworthy?

Who or what is trustworthy? In this section I consider what sort of property trustworthiness is and the sorts of components a trusted AS might comprise of. Trustworthiness might be an intrinsic property of an agent similar to height, or a relational property similar to tallness. Perhaps a robot that survives the apocalypse, like WALL-E [61] is trustworthy due to intrinsic moral virtues such as charm, cheeriness and helpfulness, even if there are no other humans or robots to trust him? Or WALL-E is trustworthy when compared to other robots such as EVE programmed to obey directives. Trustworthiness might be a substantial property-an independent particular-or a dispositional property-the capacity of an object to affect or be affected by other things. The classic example of a dispositional property is fragility. A vase is fragile because it breaks easily. A dispositional account might suppose that a person is trustworthy because they speak truthfully or act reliably with others. It might be thought that trustworthiness is both a dispositional and relational property established by the subjective judgment of one agent X of another agent Y in virtue their shared spatio-temporal interactions. For example, an employee goes through a three month probation period or a solder undergoes basic training to build their reputation with a Drill Sergeant or manager. The graduating employee or solider are deemed trustworthy for a prescribed set of activities with a particular group of people in a specific context. Note that any trustworthiness ascribed to an individual due to these processes pertains to that domain of actions. It's not clear how generalizable or transferable trustworthiness is. At least an argument needs to be made to demonstrate the transferability of trustworthiness across domains.

What is interesting about Trustworthiness understood as a dispositional and relational property is that it can be established by combining judgments from multiple agents, such as through peer assessment [58]. In this way, an IT device can be judged trustworthy through a network of sensors using a reputation-checking algorithm. For example, beacon nodes on Wireless Sensor Networks can be evaluated on whether they are providing accurate location identification by 1-hop neighboring nodes [88]. Autonomous trustworthiness-evaluation and -judgment is important when networks are vulnerable to malicious interference. Indeed, trustworthiness evaluation programs are considered increasingly important with the proliferation of autonomous systems connected via the Internet of Things (IoT) (see [17, 83, 104]).

If the dispositional and relational account of trustworthiness is right, then what dispositional properties does it consist of? In the preceding paragraphs I suggested that a person might be trustworthy because they speak truthfully or act reliably. Let's look at these ideas more closely.

Central to the notion of trustworthiness is reliability and accuracy. So, an AS is trustworthy if we can *rely* on it being *right*. For example, a binnacle compass is trustworthy if a sailor can rely on it to accurately adjust to the rise and fall of the waves and orient to magnetic north [7]. If a sailor navigates to the wrong shore, she might wonder if her compass has become unreliable and thus she ought not trust it. Perhaps ferrous nails have been used that pull the needle away from true readings and the binnacle compass's reliability compromised?

Is trustworthiness more than reliability? How do properties such as adaptability meet reliability? For example, the trustworthiness of a rescue dog might be its capacity to adapt to severe conditions, such as digging through an avalanche to find a stranded person, even if the dog has never encountered such an environment. Adaptability is not an orthogonal trait, but a higher order reliability. In this case, we rely on the dog to be adaptable in unusual, unexpected or changing conditions. The trustworthiness of people, creatures and machines is related to the reliability of their capacities and functions in domains of differing complexity and uncertainty.

Is trustworthiness also about redundancy? We know that AS will not be perfectly safe. There will be hardware failures, software bugs, perception errors and reasoning errors [27]. Aerospace and military operations build in an expectation of failure into design to enable trust. For example, Boeing 747's only need a single engine to fly, yet are equipped with four engines to ensure redundancy [22]. The Space Shuttle program used five identical general purpose digital computers [85]. Four of these computers operated as a redundant set and the fifth calculated non-critical computations. The anticipation of failure and the deliberate engineering of multiple systems in avionic engineering makes these systems more reliable and hence more trustworthy. Still, is there more to trust than reliability?

Philosophers have traditionally differentiated reliability and trust. While reliability is necessary for trust, it isn't sufficient. Reliability is a property of machines and inanimate objects, where as trust occurs between conscious agents. For example, we rely on a shelf to hold books, but do we *trust* the shelf [39]? Fully-fledged trust seems to involve reliability *and* psychological components such as the ability to apologise if we let people down, if we fail to do as we said we would. A shelf has no attitudes towards what it does. Human trust is traditionally mentally, linguistically and rationally based rather than limited to summaries of behavior [24, 40, 48, 84]. AS are a challenge to traditional philosophical distinctions on trust because they are inanimate, in the sense that they are programed to fulfill a set of tasks within a domain and have no intrinsic care for humans and no self-driven desire to maintain their reputation. The tradition to incorporate psychological attitudes in a model of trust could either be misplaced or reconsidered to drive the design processing the age of AS.

By focusing on systems as well as people, the business management literature may provide a more suitable starting framework for building trusted AS than philosophy (for more philosophical discussion see [64]). The management two-component model of trust differentiates competence-consisting of skills, reliability and experience-and integrity-consisting of motives, honesty and character (see Fig. 9.1). Using this framework user trust in AS could be grounded in reliable operations built by high-integrity organisations.

Competence comprises of skills, reliability and experience. A person or robot can be competent and yet occasionally not have exactly the right skills for the job, or the sometimes fail to do a task within their domain and sometimes reach the limit of their experience. Competence is thought to improve when an individual learns more skills, becomes more reliable and has more experiences. Integrity can be analysed as comprising of motives, honesty and character. We trust someone who is trying their best, who is transparent about their actions and has a character that, regardless of competence, inclines them to take responsibility for their actions, be thoughtful and empathetic to others and other traits. This two-factor model of trust combines ability and ethics [19, 20, 51, 59]. Trust (T) consists of:

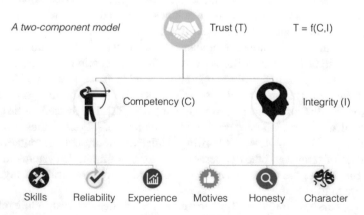

Fig. 9.1 A two-component model of trust incorporating *competence*-skills, reliability and experience-and *integrity*-motives, honesty and character [19, 20, 51, 59]

- Competence (C)
- skills (Cs)
- reliability (Cr)
- experience (Ce)
- Integrity (I)
- motives (Im)
- honesty (Ih)
- character (Ic)
- $T = f(C, I)$

Research suggests an asymmetry in the way trust is lost between these two factors. A single integrity failure may result in a loss of trust in the way that a single incompetent action does not [51]. People use integrity judgments to generalize across domains of a relationship, where as competence is more domain specific [20]. Additionally integrity-based trust implies a reduced threat of opportunism in a way that competence-based trust does not [59]. Trust depends on beliefs about the other's benevolent motives [103].

Notice the difference between human-human trust and human-AI trust violations. There is an interesting asymmetry between levels of competence required for humans to trust other humans versus trusting AI. Unlike human-human relationships, trust built up inductively between humans and AI can be destroyed with single instances of inaccuracy or unreliability. Consider the mistakes Google's AI made identifying vertical wavy lines as a starfish [69]. A single misidentification of a starfish can end trust in that machine learning algorithm even though it has performed well in the past. Consider the disproportionate media scrutiny of the first Tesla autopilot fatality. The Tesla flaw was due to the car's incapacity to differentiate the reflectance of light from a truck from the reflectance of the sky [86]. Even though human drivers make perceptual errors leading to crashes all the time, the Telsa fatality caused much uncertainty around whether the AI responsibly could be trusted. The Tesla case is a good example of much higher competence-based trust thresholds for AS than human operators and where a single model may not be sufficient. But, not only are competency requirements misaligned between human-humans and human-AS, but the integrity aspects of the model present a challenge for AS design.

Consider the requirement for honesty in Fig. 9.1. Engineers might correctly wonder how to communicate complex computational processes to human operators who themselves do not have the competency to understand their underlying logical operation? The data and algorithms of autonomous agents are hidden from most human stake-holders and cannot be understood even if a translation layer were added and explanations communicated in plain language. Perhaps human do not expect honesty from AS the same way they do from other humans? A question then is, whether humans *should* mistrust AS based on perceived honesty violations (I_h). Should engineers creating an AS prioritise transparency and communication of their decision-making mechanisms for trust and adoption? Should users demand them? It is important to consider that integrity components of the trust model might be appropriate for the human engineers, designers and corporate representatives of AS, but perhaps not crucial for the systems themselves? That is, so long as human stakeholders can honestly report the technical specifications of AS to other experts such as regulators, then AS do not need to convey integrity information to users.

What about the role of motives (I_m) and character (I_c) on trusted autonomy? Sometimes humans believe AS have more psychological reality than they actually do due to clever programming. ELIZA was one of the first relational AIs designed to engender trust using simple grammatical tricks [15, 95]. Little has been developed since that could be dubbed motives or character. Merrick, Klyne and Hardhienata in Sect. 15.5 discuss the interplay between motive and reliability. They argue that lack of transparency in the motivations or experiences of an agent can reduce trust between humans and robots, as it is difficulty to gauge why a robot is behaving the way it is, and hence, whether it is trustworthy. They suggest reputational models to help multiple users know when they should trust a particular agent. However, they also note that there is very little work done that incorporates both computational models of motivation and computational models of trust.

AIs in science fiction imagine how character might affect operations. HAL from the movie 2001: A Space Odyssey [52, 53] is a malevolent AI who lacks integrity, but is fairly competent at achieving a mission-albeit his own. Deep Thought from the Hitchhikers guide to the Galaxy [2, 67] is a benevolent AI who provides answers that humans don't want to hear, such as that the meaning of life, the universe and everything is 42. AIs can have varying degrees of competence and integrity that affects how we trust them. Additionally, may be other factors in a successful model of trust to truly understand how humans will respond to extremely smart AI.

The model described in this section is the start of an investigation of what trustworthiness could be between humans an AS based on an interdisciplinary investigation. Critics have noted that the model above confuses an influencing factor and an indicator.[1] They argue that reliability is an *indicator* of competence, not an input like skills and experience that generate competence. Skills and competence are independent variables that influence competence. I argue that while reliability is not an input, it is a *property* of a trustworthy system, not merely an indicator, hence its inclusion in the model along with skills and experience. Isolating reliability from skills and experience is meant to allow for multiple ranges in skills, reliability and experience to operate independently from one another. So, a person might be a skilled carpenter with years of experience, yet be incompetent at time tm because his divorce lead him to alcoholism and unreliable behaviours. Reliability is not merely the combination of skills and experience, it requires additional features such as the adaptability and redundancy discussed above. However, the critic is right that much more work needs to done to refine and hone this model to appropriately capture the metaphysics of trustworthiness for AS. The management model is just the beginning of incorporating human factors into AS design.

---

[1] Many thanks to an anonymous reviewer for bringing up this distinction.

9.4 How do We Know Who or What Is Trustworthy

How do we decide whether to trust? In Sect. 9.3 the properties that establish and define trustworthiness were considered. In this section the epistemology of trustworthiness is examined-how do we know who or what is trustworthy? What are the indicators of trust? If a person claims to justifiably trust another, it indicates they have the ability and confidence to predict others' behaviour [62]. Implicit, heuristic or 'gut' indicators of trust are often grounded in physical responses and intuitions. Explicit, reflective or rational trust stems from our experience of people over time and our reasons to judge their trustworthiness. Often we do not know why we trust, we trust implicitly. Thomas Reid (1764) [75] argued that reasons could not be required for trust given that'most men would be unable to find reasons for believing the thousandth part of what is told them.' Reid's point is that humans must be justified to trust even in the absence of reasons. Consider the way we use Google maps. Many people use Google maps to get them where they need to go, without knowing how Google maps works, how their phone works or how traffic influences the instructions Google maps provides. Not only do people not know why they trust Google maps, it does not seem to concern people that they do not know why. So how do humans make trust judgments of systems and each other, and are these the same mechanisms that elicit trust in AS? This section moves through implicit and explicit justifications of trust followed by a cognitive model of trust and competence and finally a brief comment on the relationship between trustworthiness and risk.

9.4.1 Implicit Justifications of Trust

Implicit justifications of trust are preconscious, embodied trust responses developed without top-down cognitive evaluations. For example a monkey climbs a vertical structure implicitly trusting that it will improve their odds of survival against predation. Researchers know how to alter physical properties of embodied AS (i.e. robots) to engender implicit trust including how they look, sound and feel. Social robots are designed with big responsive eyes and eyebrows [12], as are mobile, dexterous and social robots (MDS) [11, 94]. Some designers have shaped robots like baby animals-such as the harp seal robot PARO [1]-and use biomimetic features such as soft skin for tactile trust [50]. The Kismet robot with human-like eyes, eyebrows and lips was designed to recognize and mimic emotions, including facial expressions, vocalisations and movement [12].

Physical actions connote trust in humans. Japanese robot designers have found cultural identification with a robot who imitates traditional 'aizu bandaisan' dance [1]. Japanese robot designers try to build trust by incorporating aspects of fictional references to helpful and social robots, such as Anime characters Astroboy and the Patlabor [1]. But, representations can be incredibly primitive and build emotional attachment, for example, humans watching 2D dots moving on a screen intuitively differentiate between animate versus inanimate movement based on how wel algorithms replicate biological behaviour [74]. Mimicry of biological behaviours can make people empathise and be concerned for the wellbeing of robots, evidenced by viral videos of the Spot robot by Boston Dynamics being kicked and struggling to stay upright [9].

People enter into a relationship with a robot if it simulates human-like emotional and personal understanding, even though these relationships lack the authenticity of shared human meaning [95]. Entirely soft autonomous robots may bridge the authenticity divide, triggering different emotions and trust reactions than solid state robots. Consider the 3D printed soft Octobot designed to emulate a real Octopus, controlled with microfluidic logic instead of microchips [98]. Biology-inspired control systems are likely to affect trust responses.

The way AS communicate verbally and through sound can have a big impact on implicit trust. Tom Gruber (Siri Advanced Development Head at Apple) argues that people feel more trusting of Apple's Siri if she has a higher quality voice, "the better voice actually pulls the user in and has them use it more. So it has an increasing-returns effect" [56].

Physical characteristics also impact on how much humans move from empathy to revulsion when robots are like humans, but eerily not quite like humans-known as the uncanny valley [66] impacting how much people intuitively trust them. There is much research still to be done on whether AS that does not attempt human-like physical characteristics might not arouse the same empathy or emotional connection, but may still generate trust. The rise of chatbots in the tradition of Eliza is a linguistic means by which to generate disembodied trust. However, one benefit from realistic facial gestures and embodied movements of robots could be a speed advantage of conveying subtle information regarding the uncertainty of a robot's beliefs, their skepticism or their competing interests when providing an answer to human query improving integrity judgments (see Fig. 18.1). Such gestures may be implementable as avatar animations alongside text communication. The model outlined in Sect. 9.3 may also help us understand how humans implicitly trust autonomous systems in lieu of human-like physical characteristics or avatars. Consider human-drivers who trust Telsa's autopilot function. The car has no physical similarities with humans. Additionally, Telsa drivers cannot trust Tesla because they explicitly know anything about the algorithms before they set the autopilot on. Trust could come from implicit factors such as integrity or reliability (see Fig. 9.1). Integrity judgments may stem from a cult of personality around Elon Musk's extensive future vision for solar power, electric cars and sustainable colonies on Mars [26]?

9.4.2 Explicit Justifications of Trust

Trust is explicitly justified when we have reasons to rely on someone or something. These reasons might coalesce into a deductive, inductive or abductive inference based on the testimony and behaviour of an agent. The link between trust and higher order reasoning is supported by research showing that human intelligence relates to how successfully people evaluate trustworthiness [16, 21, 102]. Under this hypothesis, intelligent people foster relationships with people less likely to betray them and make better contextual judgments to account for circumstances where trust is difficult to uphold. Explicit reasons for trust may allow more nuanced and accurate trust judgments than relying on gut feelings or intuition.

Faulkner [25] argues that though we need reasons to trust an agent generally, we do not need reasons to justify *particular* statements from that agent. Our reasons to trust are based on evaluations of a *general* trustworthiness of an agent [24, 39–41, 63]. After all, the boy who cried wolf was not trusted in the end because he had a history of false testimony even though he was correct in the final instance. A trustworthy reputation for Y built up inductively with X can be shared quickly via testimony to other agents P, Q and R etc... Thus, the value of a trustworthy reputation is not only the ability of X to act based on information provided by Y, but its *transferability*, that is, secondary agents P, Q and R, are justified to trust Y *iff* they trust X without themselves needing prior interaction with Y. The transferability of a trustworthiness judgment increases the effectiveness and efficiency of social relationships and information systems.

But, does increased efficiency dangerously increase risk? Hume rejected testimony as a source of justification for trust [43]. He thought that a hearer was justified to trust based only on their *personal* observations of the speaker's history of truth-telling plus inductive inference from those observations [29]. Hume's reluctance to accept other people's pronouncements demonstrates the subtly and context-sensitivity of trust relationships. An AS might be trustworthy for native English speakers, but break down when deployed in mixed language context. Or an AS learns how to operate with a Platoon, but must be re-skilled each time it interacts with a new human team.

Highly complex AS are a problem for explicit justifications of trust. Because if reasons are required for trust, then perhaps no individual has sufficient reasons to make such a judgment? Take the job of calibrating a ScanEagle unmanned aircraft with hyperspectral imagery sensors to map coastal areas [42]. One individual might verify the location and ensure the imagery sensors are operating correctly but be unable to evaluate the hyperspectral map. The point is that no one operator may know or vouch for all components, mechanisms and physical properties that comprise a complex AS. A key difference between human-human trust and human-As trust is the complexity and difficulty of a single agent-agent dyad relationship. I propose that instead of relying on individual testimony AS be judged trustworthy by teams and groups that are themselves deemed to be trustworthy within the domain. Groups may include (but are not limited to):

1. Regulatory agencies responsible for issuing parameters of safe operation including physical construction and operational algorithms, operator licensing, maintenance, consumer safety.
2. Institutions and companies designing and building AS.
3. Cohesive teams of staff responsible for successful operations.
4. Environmental conditions conducive to operational success.

Hume's framework could still be useful within a more layered and complicated system of establishing explicit trust. A Human regulatory framework means that an

Fig. 9.2 Model of trust and competence where human levels of competence yield the highest trust and trust is reduced at sub-human and super-human levels

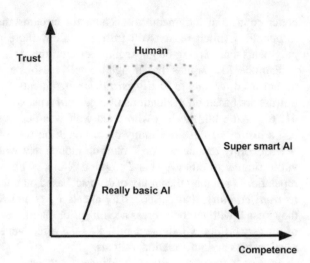

individual is justified in trusting an AS in virtue of their background knowledge of the past veracity of regulators, companies and staff plus inductive inference from those beliefs to a current instance. However, induction remains a significant problem for fast evolving AS. New AS may be made by cohesive and trustworthy teams, yet not have sufficient inductive evidence to generate warranted trust in their safe operation. This may be true, even though an individual knows that a particular aircraft company has a history of trustworthiness and that the regulatory bodies have a history of safe aircraft policies. In cases where innovation is radical and complex, trustworthiness needs inductive and abductive arguments-inference to the best explanation-to justify operations. An individual or organisation should devise an individualized set of weighted factors that together render a trust or not-trust threshold for a particular AS.

9.4.3 A Cognitive Model of Trust and Competence

Considering both intrinsic and extrinsic forms of justification, is there a linear relationship between competence and trust (holding integrity constant)? I propose that trust and competence forms more of a quadratic relation for trust. We build trust as agents become more competent. We reserve a pinnacle of trust at a human level of competence, and then trust declines as humans or machines exhibit competence at the outlier or far beyond ordinary human capacity to understand it-see Fig. 9.2. This model needs empirical testing, but I think the burden of proof is on the developers of AS to demonstrate how trust can be retained or improved as competence surpasses human capabilities and understanding. Such a justification may arise via reputational justifications as specified in Sect. 9.4.2.

To appreciate the impact of outlier competence, consider the AlphaGo game played against leading Go player Lee SeeDol in 2016 [65]. In move 37, Match 2, AlphaGo-a machine learning AI-placed a single black stone on the board that shocked the human player Lee SeeDol so much that he immediately left the table. This move was incomprehensible at a top Go playing level. What this move revealed was that humans sometimes do not understand why an AI acts in order to evaluate it. This is relevant because each competitor playing Go must presume the capacity in their opponent (human or AI) and use game play to build theories to explain strategies and mistakes of their opponent. When playing another human, Go players might overtly inquire about the opponents Go background (how old were they when they begin playing? How much have they played? Who have they played against? What books have they read? What teachers have they had? What sort of handicap do they have? etc.). Players watch their opponents actions, not only the stones placed, but the manner of their placement, and the ultimate destination on the board. Each move can be evaluated in the immediate context of the game, but also in forming what Nelson Goodman [32] describes as *overhypotheses* about their opponents style, learning journey, preferences, beliefs and desires Players use these overhypotheses to predict what an opponent will do, then use these predictions to design their own strategies to counteract them. In terms of outcome, Move 37, was very strong, providing support to stones over a large swathe of the board. But, at the moment the move was made, it was impossible to trust by the human opponent because they could not evaluate the competency of the action based on the information available about its genesis. What was AlphaGo? How does it think? What grounds its decisions? How does it make its decisions? Human understanding is critical to trust between humans and AS. It is likely in the future that more and more AIs driving AS are complex, sophisticated intellects, born of machine learning and other architectures. The danger is that humans do not trust them because they cannot understand them.

Smithson in Sect. 9.7 discusses people's aversion to systems that conflict with their own forecasts and diagnoses. Users view autonomous systems as less trustworthy if they do not understand how they operate, for example, if users do not know all the possible failure modes of an autonomous system, they will trust it less than if they know these states. His argument supports the hypothesis here that people are most likely to trust systems that produce results aligned with human-levels of decision-making.

Consider if AlphaGo was a platoon commander, sending troops into a war zone. Imagine, just as in move 37, the AS commander ordered soldiers to go to a place they could not make sense of; that they felt put their lives or civilian lives at unnecessary risk? Keep in mind that each soldier has a duty to disobey an unlawful order if its illegality is immediately obvious, such as procedural irregularity or moral gravity [71]. In these cases, humans ought not trust the AI, even if the AI proves to be more competent than human decision makers. The AI could have access to huge repositories of data unable to be processed by humans. These calculations and decisions are frightening to humans and justify wariness and skepticism. Even more significantly, suppose complex sophisticated AIs were in charge of Lethal Autonomous Weapons Systems (LAWS), both decisions to target and decisions to fire, how do we know whether to trust them? How would deaths be judged just or unjust if the algorithms deciding who dies are beyond human comprehension? LAWs led by AIs may lead to unintended initiation of armed conflicts and the unjust escalation of conflicts [5].

It is important to note that leading manufacturers of LAWs currently require human oversight and judgment for all decisions to target and to fire [5]. Current restrictions are based on the notion that humans are better decision-makers than machines. However, manufacturers continue to build incrementally autonomous capabilities across all systems. To imagine the impact of increasing autonomy for weapons systems, it is instructive to consider how other industries have rolled out autonomous systems and their impact on human users. Car manufacturer Tesla released a self-driving mode on its cars with the requirement that humans always have their hands at the wheel. Yet, Tesla drivers drive while deliberately disobeying protocols because they trust that the systems *do not* actually require their oversight [86]. There is evidence as AS become increasingly sophisticated humans may become either overly trusting or overly skeptical. Consider research on autonomous offshore oil drilling system operations [34]. Drill operators sometimes abandon their duty to oversee AS due to competing cognitive demands or they ignore the AS and make their own decisions inefficiently. In both cases the level of trust in the autonomous system plays a direct role in how humans view their obligations to participate in broader systems operations or obey oversight protocols. In sum, while there are currently policies requiring LAWs to be under ultimate human control, the pressures and stress of combat may lead to humans relinquishing control. In the future humans may not have the competence to be in control of these systems.

Perhaps more frightening is a future where AS knows how to manipulate consent and trust in humans [10]. This is a situation where we trust an AS because it is clever enough to manufacture our trust. But, it does so in either a disingenuous or manipulative way. It is not hard to imagine such an AI capitalizing on inductive trust tendencies or biases in humans. Consider Nelson Goodman's [32] thought experiment about the colour of emeralds known as the 'grue-paradox' [18]. In this hypothetical, all our experience of emeralds is their greenishness, so we ascribe to them the stable and persistent property 'green'. Goodman points out that in fact, Emeralds might be not green but 'grue'. Grue is a property of objects that makes them look green until a particular time (e.g. 2025), but look blue afterwards:

Definition 1 x is grue $=_{df}$ x is examined before t and green $\bigvee$ x is not so examined and blue.

If Emeralds are grue, they have never been green. Now suppose we take this hypothetical case of false induction (i.e. trying to establish facts about emeralds and their colour from history and experience) and consider malevolent programmers building an AS. These programmers design a robot that engenders trust over time, for a long time, like an embedded undercover operative. During production and deployment, the AS passes every test humans and regulators can design to establish its trustworthiness. The AS is tested in hundreds of real time situations and thousands of simulated scenarios. But, unbeknownst to regulators, it has been programmed to switch modes in 2025 while deeply embedded in society. So, humans trusted it, but then the AS betrays them and carries out its secret objective. There was no way to know, inductively that the AS would flip. That it was actually an untrustworthy AS. It is also concerning to consider if such hidden higher-level objectives can be programmed, such programs could be activated or changed remotely and iteratively-threatening the integrity of the AS.

9.4.4   Trustworthiness and Risk

Finally, when ascribing trustworthiness to agent Y, X needs to consider the context of decisions. We have different thresholds for trust depending on the risk of the decisions that have to be made and this in turn depends on impact of decisions-see Fig. 9.3. This figure shows the relationship between decision impact, trustworthiness and trust. Life-threatening decisions, such as our choice of neurosurgeon have a higher threshold for trust than merely inconvenient decisions such as our choice of lawyer to settle a contract on a house. Consider PARO, a robot that resembles a baby-seal designed to assist the elderly similar to pet therapy. If PARO malfunctions, very little is lost to the humans who rely on it. But if a rescue robot malfunctions during an evacuation human lives are at stake. If 0 = no trust and 1.0 = absolute trust, We may need to trust our surgeon 0.99 in order to agree to brain surgery, but only need to trust our check out clerk 0.65 in order to complete our retail shopping. This is relevant in AS where similar algorithms may be installed or implemented into a

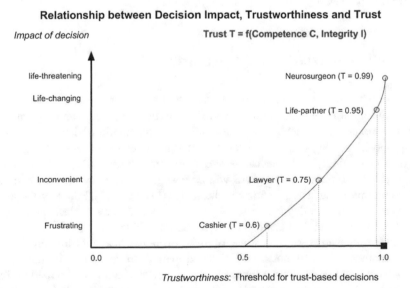

Fig. 9.3 Relationship between decision impact, trustworthiness and trust

huge variety of contexts. We can imagine perceptual and mechanical algorithms that allow a capsicum-picking robot [55] to drop 1 in 10 vegetables being reconfigured to help in a rescue operation where dropping 1 in 10 children from a boat is absolutely unacceptable.

9.4.5 Summary

This section examined the epistemology of trustworthiness. Implicit indicators of trust can be grounded in physical responses and intuitions as well as reputational features of the system that designed and built the AS. Explicit, reflective or rational trust can be elusive, but must stem from our experience of people over time and our reasons to judge their trustworthiness. As AS become more complex, reasons to trust need to be curated from teams of experts including regulators, designers, engineers, users and so forth. Inductive reasoning may need to be augmented by abductive reasoning for radically innovative AS that involve untested combinations of systems and/or new types of systems.

Even once explicit evaluation methods are established, the increasing competence of AS is a risk to human trust. I argue that increased competence increases trust in AS by humans both for implicit and explicit justifications up until competence far exceeds human comprehension. As AS competence continues to increase, humans may cease trusting them because they do not understand them (perhaps frustrating engineers and designers). Or, perhaps even worse, they falsely trust malevolent systems that should not be trusted. Either way, humans may become unreliable at evaluating trustworthiness as AS surpass human cognitive capacities.

9.5 What or Who Should We Trust?

What or who should we trust? Robots and AS should be programed with our best normative theories of logic, rationality [46] and ethics tempered with pragmatic performance expectations. Robots and AS are already computational devices, thus abide by propositional logic, predicate logic, and sometimes paraconsistent logic [93]. Robots increasingly make decisions under uncertainty using Bayesian rationality [8, 92]. In the future, robots and AS will be designed to test newer normative theories of rationality such as quantum cognition [13] (See Sect. 10). Ethically, we should trust humans and AS that take care of our interests and obey the law. This section will briefly survey ethical theories that AS ought to abide by.

Legal frameworks can do some of the normative heavy lifting for AS, but unfortunately the law is not nearly nuanced enough to cover human-judged ethical behaviours. For example, suppose a tree branch has fallen on the road during a storm [57]. A human driver would cross double-yellow lines on a road to go around the branch once a safety-check was undertaken and we would judge her ethical. However, for us to trust an autonomous car to make the same judgment, violating legal requirements regarding double-yellow lines, it would need to know a huge range of concepts and contexts, e.g. computational versions of terms such as 'obstruction' and 'safe' [31]. Humans make decisions that violate the law strictly speaking, but are usually nuanced actions that take context and risk into account.

In terms of human rights, AS ought to be aligned to the United Nations Declaration of Human Rights, the Geneva Conventions and Protocols [44]), and human rights law [6]. Additionally, AS ought consider a broad range of ethical theories from philosophy. Consequentialism (or 'Utilitarianism') is a dominant ethical theory that would justify AS actions if they cause the most happiness or 'utility'. For a Utilitarian, LAWs would be justified if they remove human error, thus reduce civilian casualities. Self-driving cars are justified if they massively reduce the road toll, even if the occasional person or bystander is killed through error. Utilitarian arguments are the most frequently cited arguments in favour of deploying autonomous systems. Deontological arguments focus not on the ends of decisions, but the way decisions are made, aka 'the ends do not justify the means'. Kant might agree that lying to all children about the existence of Santa creates the most happiness, but, it is unethical because it violates the Categorical Imperative [73]. A deontologically or 'duty' based AS may have a duty to retain all records of software upgrades and decision parameters in an impenetrable black box for later insurance claims and legal determinations regardless of whether such records end up disproportionately punishing low socioeconomic groups. Each design decision can be worked through from different ethical perspectives including social contract theory, virtue ethics or feminist ethics. While different theories may demand conflicting design decisions, many decisions may come out the same. For example, there are both Utilitarian and Kantian justifications for rescue robots to obey triage rules in a rescue. On the other hand, some ethical theories provide a unique way of understanding how and why we trust each other under stressful and uncertain circumstances. Virtue ethics justifies action not based on their consequences or intention, but on virtues such as bravery and honour. Where as Utilitarian or Kantian principles could possibly be coded into a decision maker, virtue is built up over time, via experience and feedback calibrating specific actions against virtuous norms. Virtue ethics could be incorporated into probabilistic decision systems because the right action is not the one that always produces the best outcome. Under virtue ethics we trust an AS if it made the best decision possible in its context given its operating parameters. Additionally, newer ethical theories might fill in some decision-making gaps. For example, Feminist ethics [45] could justify preferential care behaviours in a special operations team. There is a particular synchronicity between virtue ethics and feminist ethics that could be fruitful for building trust [35].

Our reliance on people and AS is affected by our level of dependence and cooperation. Our trust in our life partner to care for us involves a multi-faceted risk and trust over time (with shared cognition) versus the one-off trust we might place in a surgeon. For example, we don't really care if our surgeon is nice to his in-laws at Christmas, just so long as he can remove the tumour. We trust people who we believe have strong reasons for acting in our best interests [38]. The main incentive for these reasons is a desire to maintain a strong relationship with us (whether that is economic, love, friendship etc.). Trust between individuals is different to trust we have in corporations. This asymmetry is a really significant issue for AS, because humans ground their trust in beliefs about the corporation behind the AS, not the systems themselves instantiated in a single car, robot, or computer installation. Social norming is an approach to procedural ethics outside of traditional philosophical theories from anthropology and sociology [101]. Social norming is about learning how to behave in groups to get along the best. It requires we understand social expectations. Detailed theories of cooperative behaviour stem from disciplines such as sociology, biology, anthropology and group psychology. These models are not about competence and achieving optimal performance on tasks, but about creating the most cohesive, resilient teams of organisms. Theories such as game theory contribute to understanding social norming [72]. One of the many advantages of group level norms is the ability to train AS with social norming without needing top-down ethical theories to drive behaviours.

However, while there are promising avenues for research into the ethical programming to improve trust, many barriers exist for the universalization of such programming. This is because there remains vast disagreement on what the right ethical principles are or even whether ethical principles exist such that they could be implemented into an AS. What does ethical talk amount to? It seems that humans judge each others actions as ethical or not ethical based a huge range of theoretical, contextual, pragmatic and social factors that ethical theories struggle to explain beyond stipulating that actual human decision makers exhibit a sort of hopeless contrariness.

There is a lot of work to be done in determining what the most ethical action is in any particular context and what model underpins such actions. However, even if we can program AS to be ultimately logical, rational or ethical, humans may be uncomfortable. Would we trust machines that obey norms without empathy [28]? Consider the origins of the word robot from the 1920 play, Rossumovi Univerzální Roboti (Rossum's Universal Robots). In the play Czech writer Karel Capek endowed robots with not just thoughts, but emotions to enable them to increase their productivity [97]. Capek's robots were forced workers more like biological androids Replicants in Bladerunner than metal machines. If we program AS with emotions and empathy to build trust, will they suffer if we treat them badly? If AS are moral agents that can suffer, then building trustworthy autonomous systems also means building an ethical and legal framework around their use and identifying their rights [81]. Japanese roboticists are already designing robots to have 'kokoro', translated into heart, spirit or mind [1]. Kokoro stems from animist spiritual thinking that all objects, including rocks and trees, have some level of consciousness and agency including emotions, intelligence and intention. Robots and AS of the future may need complex social identities to meet ethical and social norms.

9.6 The Value of Trustworthy Autonomous Systems

The discussion of the metaphysics, epistemology and normativity of trustworthiness has assumed that trustworthy AS are the desired goal. However, do humans want their decisions automated even if available AS are trustworthy? One the one hand optimising AS could be ideal for human-robot interactions, freeing up time and resources, but on the other hand, perhaps humans want to make their own decisions? We might think that humans develop a sense of identity and security from decision making responsibility in their roles and jobs and that we risk devaluing human workers by outsourcing decisions to AS. If so, then even if AS increase process productivity, it may decrease productivity overall. Alternatively, humans may find work tedious and be glad for near-optimal autonomous task allocations [30]. In the Culture novels by science fiction writer Iain M. Banks, the AS 'Minds' make most human decisions that aren't spiritual or fun and the human populace are perfectly content [79]. 'Minds' are sentient hyper-intelligent AIs on space ships and inhabited planets that have evolved to become far more intelligent than their original biological creators. The minds have taken over the administrative infrastructure of the Culture civilization. We don't have to go too far to see that humans already welcome efficiencies that stem from machine learning when they use their smart phones. How many decisions and what sorts of decisions will humans outsource to an AS if given the opportunity?

Interestingly Gombolay et al. [30] found that contrary to their hypotheses (and in alignment to Iain M. Banks), humans prefer to outsource decision making to autonomous robots even when they perceived their human co-leader more favorably than their robotic co-leader. Interestingly, in follow up questionnaires, subjects felt that their human co-leader had additional properties, such that they liked, appreciated and understood them, that humans understood, trusted and respected each other, and finally that subjects and human co-leaders were important to the task. However, liking humans and wanting them around is not the same as wanting humans to make decisions.

One of the important distinctions when considering AS is the difference between physically instantiated AI (e.g. personal robot) that learns and grows with an individual or team, versus an integrated AI programmed to act over many physical bodies (e.g. networked self-driving cars) that show no preferential or focused behaviours with individual humans. In the latter case, Iain M. Banks Minds and Apple's subtle machine learning might work fine. But, in the former case, social norming may be the right solution.

9.7 Conclusion

This chapter has examined the trustworthiness of autonomous systems. I have argued that effective robots and autonomous systems must be trustworthy and the risks of reliance justified relative to perceived benefits. Trustworthiness is a dispositional and relational property of agents relative to other agents within spatiotemporal bounds. Trustworthy agents must be reliable (incorporating adaptability and redundancy). A two-component model of trust was used to differentiate factors of competence (skills, reliability and experience) to factors of integrity (motives, honesty and character). When humans evaluate the trustworthiness of autonomous systems and other humans they use intrinsic, 'gut' level cues such as physicality as well as extrinsic 'top down' reasoning. Humans tend to trust agents operating within the bounds of human cognition and are less trusting as systems operate at super-human levels. The threshold for trustworthiness of an agent or organisation depends on the impact of decisions in a particular context. Building trustworthy autonomous systems requires obeying the norms of logic, rationality and ethics under pragmatic constraints-even though there is disagreement on these principles by experts. AS may need sophisticated social identities including empathy and reputational concerns to build human-like trust relationships. Ultimately transdisciplinary research drawing on metaphysical, epistemological and normative human and machine theories of trust are needed to design trustworthy autonomous systems for adoption.

References

1. S. Šabanović, Inventing Japan's 'robotics culture': the repeated assembly of science, technology, and culture in social robotics. Soc. Stud. Sci. **44**(3), 342–367 (2014)
2. D. Adams, *The Hitchhikers Guide to the Galaxy* (Pan Books, UK, 1979)
3. J.R. Anderson, *How can the Mind Exist in a Physical Universe* (Oxford University Press, Oxford, 2007)
4. K.J. Arrow, *The Limits of Organization, Fels Lectures on Public Policy Analysis* (W. W. Norton and Co., New York, 1974)
5. P. Asaro, Killer robots and the ethics of autonomous weapons, in *Ethics of Artificial Intelligence., NYU*, 14–15 Oct 2016
6. P. Asaro, On banning autonomous weapon systems: human rights, automation, and the dehumanization of lethal decision-making. Int. Rev. Red Cross **94**(886), 687–709 (2012)
7. I. Basterretxea-Iribar, I. Sotés, Jose Ignacio Uriarte, Towards an improvement of magnetic compass accuracy and adjustment. J. Navig. **69**(6), 1325–1340 (2016)
8. P. Bessière, C. Laugier, R. Siegwart, *Probabilistic Reasoning and Decision Making in Sensory-Motor Systems*, vol. 46 of *Springer Tracts in Advanced Robotics* (Springer Science & Business Media, 2008)
9. Boston Dynamics, Introducing spot, https://youtu.be/M8YjvHYbZ9w. Retrieved from 9 Feb 2015
10. N. Bostrom, Ethics of artificial intelligence, in *Ethics of Artificial Intelligence, NYU*, 14–15 Oct 2016
11. C. Breazeal, M. Siegel, M. Berlin, J. Gray, R. Grupen, P. Deegan, J. Weber, K. Narendran,break J. McBean, Mobile, dexterous, social robots for mobile manipulation and human-robot interaction, in *ACM SIGGRAPH 2008 new tech demos* (ACM, 2008), p. 27
12. C.L. Breazeal, *Designing Sociable Robots* (MIT Press, Cambridge, MA, 2002)
13. J.R. Busemeyer, P.D. Bruza, *Quantum Models of Cognition and Decision* (Cambridge University Press, Cambridge, 2012)

14. J. Butler, P. Giuliano, L. Guiso, *The right amount of trust* (Electronic book section, National Bureau of Economic Research, 2009)
15. J.R. Carbonell, AI in CAI: an artificial-intelligence approach to computer-assisted instruction. IEEE Trans. Man Mach. Syst. **11**(4), 190–202 (1970)
16. N. Carl, F.C. Billari, Generalized trust and intelligence in the United States. PLOS ONE **9**(3), e91786 (2014)
17. D. Chen, G. Chang, D. Sun, J. Li, J. Jia, X. Wang, TRM-IoT: a trust management model based on fuzzy reputation for internet of things. Comput. Sci. Inf. Syst. **8**(4), 1207–1228 (2011)
18. D. Cohnitz, M. Rossberg, Nelson goodman, in *The Stanford Encyclopedia of Philosophy*, ed. by E.N. Zalta (Stanford Encyclopedia, 2016)
19. B.L. Connelly, T.R. Crook, J.G. Combs, D.J. Ketchen, H. Aguinis, Competence- and integrity-based trust in interorganizational relationships: Which matters more? J. Manage. (2015)
20. B.L. Connelly, T. Miller, C.E. Devers, Under a cloud of suspicion: trust, distrust, and their interactive effect in interorganizational contracting. Strat. Manage. J. **33**(7), 820–833 (2012)
21. L. Cosmides, H. Clark Barrett, J. Tooby, Clark Barrett, and John Tooby. Adaptive specializations, social exchange, and the evolution of human intelligence. Proc. Natl. Acad. Sci. U.S.A. **107**(Supplement 2), 9007–9014 (2010)
22. J. Downer, *When failure is an option: redundancy, reliability and regulation in complex technical systems* (Economic and Social Research Council, Government document, Centre for Analysis of Risk and Regulation, 2009)
23. J.B.M. Engelman, *An empirical investigation of the evolutionary and ontogenic roots of trust*. Thesis, 2014
24. P. Faulkner, On telling and trusting. Mind **116**(464), 875–902 (2007)
25. P. Faulkner, *Knowledge on Trust* (Oxford University Press, Oxford, 2011)
26. G. Flanagan, If you think Apple is a cult, you haven't been to a Tesla event, in *Business Insider*, 14 Oct 2015
27. T. Fraichard, J.J. Kuffner, Guaranteeing motion safety for robots. Auton. Robots **32**(3), 173–175 (2012)
28. E. Gleichgerrcht, L. Young, Low levels of empathic concern predict utilitarian moral judgment. PloS one **8**(4), e60418 (2013)
29. A. Goldman, Epistemology and the evidential status of introspective reports. J. Conscious. Stud. **11**(7–8), 1–16 (2004)
30. M.C. Gombolay, R.A. Gutierrez, S.G. Clarke, G.F. Sturla, J.A. Shah, Decision-making authority, team efficiency and human worker satisfaction in mixed human-robot teams. Auton. Robots **39**(3), 293–312 (2015)
31. N.J. Goodall, *Machine Ethics and Automated Vehicles* (Springer, DE, 2014)
32. N. Goodman, *Fact, Fiction, and Forecast* (Harvard University Press, Cambridge, 1983)
33. M. Granovetter, Economic action and social structure: the problem of embeddedness. Am. J. Sociol. **91**(3), 481–510 (1985)
34. L.J. Gressgård, K. Hansen, F. Iversen, Automation systems and work process safety: assessing the significance of human and organizational factors in offshore drilling automation'. J. Inf. Technol. Manage. **24**(2), 47 (2013)
35. R. Halwani, Care ethics and virtue ethics. Hypatia **18**(3), 161–192 (2003)
36. P.A. Hancock, Imposing limits on autonomous systems. Ergonomics **60**(2), 284–291 (2017)
37. P.A. Hancock, D.R. Billings, K.E. Schaefer, Can you trust your robot? Ergonomics Des.: Q. Hum. Factors Appl. **19**(3), 24–29 (2011)
38. R. Hardin, *Trust and Trustworthiness* (Russell Sage Foundation, New York, 2002)
39. K. Hawley, Trust, distrust and commitment. Noûs **48**(1), 1–20 (2014)
40. P. Hieronymi, The reasons of trust. Australas. J. Philos. **86**(2), 213–236 (2008)
41. E.S. Hinchman, Telling as inviting to trust. Philos. Phenomenol. Res. **70**(3), 562–587 (2005)
42. R. Hughes, Sensors for coastal remote sensing. Paper presented at the SpaceNet Remote Coastal Workshop, July 2015
43. D. Hume, *A Treatise of Human Nature: Being an Attempt to Introduce the Experimental Method of Reasoning Into Moral Subjects*. ebooks@Adelaide (1739)

44. International Committee of the Red Cross, Geneva conventions of 1949 and additional protocols, and their commentaries, 1949
45. A.M. Jagger, *Feminist Ethics* (Gardland Press, New York, 1992)
46. P.N. Johnson-Laird, R.M.J. Byrne, Models and deductive rationality, in *Rationality: Psychological and Philosophical Perspectives*ed. by K. Manktelow, D. Over (Routledge, London, 1993)
47. K. Kaneko, K. Harada, F. Kanehiro, G. Miyamori, K. Akachi, Humanoid robot HRP-3, In *2008 IEEE/RSJ International Conference on Intelligent Robots and Systems* (IEEE, 2008), pp. 2471–2478
48. A. Keren, Trust and belief: a preemptive reasons account. Synth. Int. J. Epistemol. Methodol. Philos. Sci. **191**(12), 2593–2615 (2014)
49. S. Khemlani, A. Harrison, G. Trafton, An embodied architecture for thinking and reasoning about time, in *The 38th Annual Meeting of the Cognitive Science Society, Philadelphia* (2016)
50. J. Kim, A. Alspach, K. Yamane, 3D printed soft skin for safe human-robot interaction, in *IEEE/RSJ International Conference on Intelligent Robots and Systems (IROS)* (IEEE, 2015), pp. 2419–2425
51. P.H. Kim, D.L. Ferrin, D. Cooper, K.T. Dirks, Removing the shadow of suspicion: the effects of apology versus denial for repairing competence- versus integrity-based trust violations. J. Appl. Psychol. **89**(1), 104–118 (2004)
52. A.K. Kozlovic, Technophobic themes in pre-1990 computer films. Sci. Cult. **12**(3), 341–373 (2003)
53. S. Kubrick, A.C. Clark, 2001: A space odyssey [film] (1968)
54. J.D. Lee, K.A. See, Trust in automation: designing for appropriate reliance. Hum. Factors J. Hum. Factors Ergon. Soc. **46**(1), 50–80 (2004)
55. C. Lehnert, I. Sa, C. McCool, B. Upcroft, T. Perez, Sweet pepper pose detection and grasping for automated crop harvesting, in *2016 IEEE International Conference on Robotics and Automation (ICRA)* (2016), pp. 2428–2434
56. S. Levy, The iBrain is here and its already inside your phone: an exclusive inside look at how artificial intelligence and machine learning work at apple. back channel, 25 Aug 2016
57. P. Lin, The ethics of autonomous cars. *The Atlantic*, 8 (2013)
58. T. Lopez, Army changing basic training this October. Army News Service, 24 Sept 2015
59. Y. Luo, Contract, cooperation, and performance in international joint ventures. Strat. Manage. J. **23**(10), 903–919 (2002)
60. F. Maleki, Z. Farhoudi, Making humanoid robots more acceptable based on the study of robot characters in animation. IAES Int. J. Robot. Autom. **4**(1), 63–72 (2015)
61. S. Mattie, WALL-E on the problem of technology. Perspect. Polit. Sci. **43**(1), 12–20 (2014)
62. D.J. McAllister, Affect-and cognition-based trust as foundations for interpersonal cooperation in organizations. Acad. Manage. J. **38**(1), 24–59 (1995)
63. V. McGeer, Trust, hope and empowerment. Australas. J. Philos. **86**(2), 237–254 (2008)
64. C. McLeod, Trust, *The Stanford Encyclopedia of Philosophy*, electronic book section Trust, in ed. by E.N. Zalta (Stanford Encyclopedia, 2015)
65. C. Metz, In two moves, alphago and lee sedol redefined the future. WIRED.com, 16 Mar 2016
66. M. Mori, K.F. MacDorman, N. Kageki, The uncanny valley [from the field]. IEEE Robot. Autom. Mag. **19**(2), 98–100 (2012)
67. D. Naiditch, The meaning of life. Skeptics Soc. Skeptic Mag. **8**, 74 (2000)
68. G. Nave, C. Camerer, M. McCullough, Does oxytocin increase trust in humans? A critical review of research. Perspect. Psychol. Sci. **10**(6), 772–789 (2015)
69. A. Nguyen, J. Yosinski, J. Clune, Deep neural networks are easily fooled: high confidence predictions for unrecognizable images, in *2015 IEEE Conference on Computer Vision and Pattern Recognition (CVPR)* (IEEE, 2015), pp. 427–436
70. D.C. North, *Institutions, Institutional Change, and Economic Performance* (Cambridge University Press, Cambridge, 1990)
71. M.J. Osiel, Obeying orders: Atrocity, military discipline, and the law of war. *California Law Review* (1998), pp. 939–1129

72. E. Ostrom, Collective action and the evolution of social norms. J. Nat. Resour. Policy Res. **6**(4), 235–252 (2014)
73. H.J. Paton, *The Categorical Imperative: A Study in Kant's Moral Philosophy. Book* (University of Pennsylvania Press, Philadelphia, 1971)
74. Z. Pylyshyn, *Seeing and Visualizing: It's Not What You Think* (The MIT Press, Cambridge, MA, 2003)
75. T. Reid, *An Inquiry into the Human Mind: On the Principles of Common Sense* (Penn State University, 2000, 1764)
76. C. Reilly, Australia's first autonomous bus trial goes off without a hitch (or a driver). CNET, 1 Sept 2016
77. J.B. Rotter, A new scale for the measurement of interpersonal trust. J. Pers. **35**(4), 651–665 (1967)
78. D.M. Rousseau, S.B. Sitkin, R.S. Burt, C. Camerer, Not so different after all: a cross-discipline view of trust. Acad. Manage. Rev. **23**(3), 393–404 (1998)
79. Y. Rumpala, Artificial intelligences and political organization: an exploration based on the science fiction work of Iain M. Banks. Technol. Soc. **34**(1), 23–32 (2012)
80. K.E. Schaefer, J.Y.C. Chen, J.L. Szalma, P.A. Hancock, A meta-analysis of factors influencing the development of trust in automation: implications for understanding autonomy in future systems. Hum. Factors **58**(3), 377–400 (2016)
81. E. Schwitzgebel, M. Garza, A defense of the rights of artificial intelligences. Midwest Stud. Philos. **39**(1), 98–119 (2015)
82. C. Sganzerla, C. Seixas, A. Conti, Disruptive innovation in digital mining. Proc. Eng. **138**, 64–71 (2016)
83. S. Sicari, A. Rizzardi, L.A. Gricco, A. Coen-Porisini, Security, privacy and trust in internet of things: the road ahead. Comput. Netw. **76**, 146–164 (2015)
84. E. Simpson, Reasonable trust. Eur. J. Philos. **21**(3), 402–423 (2013)
85. J.R. Sklaroff, Redundancy management technique for space shuttle computers. IBM J. Res. Dev. **20**(1), 20–28 (1976)
86. Slow News Day (Producer), Tesla's model S autopilot is amazing! (2016)
87. A. Smith, *An Inquiry into the Nature and Causes of the Wealth of Nations. Book 4, Chap. 2.* (David Campbell Publishers, London, 1991 edn, 1776)
88. A. Srinivasan, J. Teitelbaum, J. Wu, DRBTS: Distributed reputation-based beacon trust system. In *2nd IEEE International Symposium on Dependable, Autonomic and Secure Computing*, 29 Sept–1 Oct 2006, pp. 277–283
89. K. Sterelny, *The Evolved Apprentice* (MIT Press, Cambridge, 2012)
90. N. Steinfeld, "I agree to the terms and conditions": (How) do users read privacy policies online? An eye-tracking experiment. Comput. Hum. Behav. **55**, 992–1000 (2016)
91. The Tesla Team. A tragic loss (2016)
92. S. Thrun, D. Fox, W. Burgard, F. Dellaert, Robust monte carlo localization for mobile robots. Artif. Intell. **128**(1–2), 99–141 (2001)
93. C.R. Torres, J.M. Abe, G. Lambert-Torres, J.I. da Silva Filho, Paraconsistent autonomous mobile robot Emmy III, in *Advances in Technological Applications of Logical and Intelligent Systems: Selected Papers from the Sixth Congress on Logic Applied to Technology*, vol. 186 (IOS Press, 2009), p. 236
94. G. Trafton, L. Hiatt, A. Harrison, F. Tamborello, S. Khemlani, A. Schultz, Act-r/e: an embodied cognitive architecture for human-robot interaction. J. Hum. Robot Interact. **2**(1), 30–55 (2013)
95. S. Turkle, Authenticity in the age of digital companions. Interact. Stud. **8**(3), 501–517 (2007)
96. T.R. Tyler, *Why People Obey the Law* (Princeton University Press, Princeton, 2006)
97. M.K. Vukobratovic, When were active exoskeletons actually born? Int. J. Humanoid Rob. **4**(03), 459–486 (2007)
98. M. Wehner, R.L. Truby, D.J. Fitzgerald, B. Mosadegh, G.M. Whitesides, J.A. Lewis, R.J. Wood, An integrated design and fabrication strategy for entirely soft, autonomous robots. Nature **536**(7617), 451–455 (2016)

99. O.E. Williamson, Calculativeness, trust, and economic organization. J. Law Econ. **36**(1), 453–486 (1993)
100. WIRED.com (Producer), How tesla's self-driving autopilot actually works, 09 2016
101. D.H. Wrong, The oversocialized conception of man in modern sociology. Am. Sociol. Rev. **26**, 183–193 (1961)
102. T. Yamagishi, *Trust as a form of social intelligence* (Russell Sage Foundation, New York, 2001)
103. T. Yamagishi, *Trust: The Evolutionary Game of Mind and Society* (Springer, Berlin, 2011)
104. Z. Yan, P. Zhang, A.V. Vasilakos, A survey on trust management for internet of things. J. Netw. Comput. Appl. **42**, 120–134 (2014)
105. H.A. Yanco, A. Norton, W. Ober, D. Shane, A. Skinner, J. Vice, Analysis of human-robot interaction at the DARPA robotics challenge trials. J. Field Robot. **32**(3), 420–444 (2015)

Chapter 10
Trusted Autonomy Under Uncertainty

Michael Smithson

10.1 Trust and Uncertainty

10.1.1 What Is Trust?

The main goal of this chapter is to elaborate the connections between trust, distrust, and uncertainty. Trust will be treated primarily as a psychological state, but sometimes also as a type of relationship or in purely behavioral terms. Trust, after all, is a social concept. However, it is not a contractual relationship, and so "trust" here will not have the kind of meaning in legal or institutional forms such as "trust fund" or "company trusts". The main theme is that trust and distrust inherently involve uncertainty (and risk) in two respects. First, uncertainty is a given in trust or distrust as a psychological state. Second, the processes in a trust or distrust relationship generate unknowns.

It may seem odd to begin this chapter by reconsidering definitions of trust and distrust, but this is necessary for three reasons. First, the concept of trust has been diversely defined in various disciplines, chiefly economics, psychology, political science, and sociology. These definitions often disagree with or talk past one another (see the discussion in [30]). Second, "trust", "mistrust", and "distrust" are multifarious in many natural languages. Each can be a noun or a verb, and each can describe a psychological state, a belief, a feeling, a relationship, or even (in the case of trust) a legal entity. And third, the terms have not been used consistently in the literature on human-robot interaction (HRI, from here on). Some HRI researchers have treated "trust" as synonymous with reliability, while others have brought in matters of trustee interest or intent regarding the trustor. "Distrust", on the other hand, has been relatively neglected in the HRI literature. Starting with "trust", we shall avoid the definitions used by some researchers that define trust by the ways in which it is

M. Smithson (✉)
Research School of Psychology, The Australian National University,
Bldg 39, Room 215, Canberra, Australia
e-mail: Michael.Smithson@anu.edu.au

© The Author(s) 2018
H. A. Abbass et al. (eds.), *Foundations of Trusted Autonomy*, Studies in Systems,
Decision and Control 117, https://doi.org/10.1007/978-3-319-64816-3_10

formed [30]. Thus, while trust may arise from a rational choice (e.g., [16]), as a personality trait (e.g., [33]), or as an institutionalized or identity-based norm (e.g., [29]), none of these actually defines trust. Trust is defined in this chapter as an outcome of such choices, personalities, or normative processes.

A version of Hardin's [16] tripartite conceptualization of trust will be used here. In his framework, trust is defined in terms of attributes of the trustor, properties of the trustee, and the specific context in which trust occurs. A trustor may be "trusting" in the sense of an expectation (e.g., [18]), a positive feeling (e.g., [11]), or an attitude that has an intentional component (i.e., being willing to trust). According to a survey of 65 sources of definitions of trust [30], the typical characteristics ascribed to a trustee include predictability, reliability, competence, benevolence (toward the trustor), and integrity. Thus, for instance, a trustor may expect or feel that a trustee is reliable or benevolent.

Context includes the following components:

1. Dependence: What the trustor depends on the trustee to provide or to do,
2. Trust behavior: What the trustee must do to show and bestow trust,
3. Basis: Factors involved in the formation of trust, and
4. Stakes: The potential benefits and costs of trust.

The nature of the dependence involved in a trust relationship will also strongly influence the kinds of uncertainties involved, as will be elaborated later on. Trust behavior, as we will see later, involves a combination of deference to the trustee, relinquishment of control over or micro-management of the trustee, and relevant risk-taking. The basis for forming a trust relationship may be rational calculation, personal disposition, reputational, social identity, part of a role, or even part of a set of rules in an institutional setting [23]. The stakes can be financial or tangible, but also may include intangibles such as esteem, reputation, and even willingness to trust in future relationships. Moreover, the stakes may not be limited to direct consequences of trust, but also can include "side-effects" such as sociability, opportunity, and transaction costs and benefits.

What are the opposites of trust? The absence of trust, in the sense of indifference, clearly is not the same as distrust, mistrust, or paranoia. "Distrust" and "mistrust" often are used interchangeably, although common usage tends to construe "mistrust" in terms of suspicion or doubt about a target, and "distrust" as without doubt that the target is untrustworthy. A systematic treatment of distinctions between these two terms is beyond our scope, and the focus in this chapter will be on distrust. Several scholars have claimed distrust is the opposite of trust, whereby they mean an expectation or suspicion that the distrusted party is unreliable and/or malevolent (see [34]; or [12]: "I trust my friends; distrust my enemies"). Like trust, distrust is a social entity through and through. We can employ Hardin's tripartite framework for dealing with distrust, in a similar manner to trust, by considering it in terms of attributes of the distrustor, properties of the distrustee, and the specific context in which distrust occurs.

Distrust, then, incorporates attributions of unreliability and intentions toward the distrustor ranging from neglectful to malign. Distrustful relations therefore will be characterized by hyper-vigilance, attempts to free oneself from any dependence on the distrusted party, and/or attempts to assert control over that party. Legally binding and enforceable contracts are an example of a relationship that could be based on distrust.

10.1.2   Trust and Distrust in HRI

How have trust and distrust been construed in the literature on HRI? Which definitions or conceptions are most useful in understanding HRI and designing technologies to implement or augment it? For instance, is trusting an AI-driven robot more like trusting a refrigerator, a trading bank, a surgeon, or a friend-or is it like none of these? Some scholars, such as Lee and See [26], have defined trust in HRI settings as rather similar to trust in humans. Lee and See's aspects of trust include performance, process, and purpose. The first two are similar to the well-worn concepts of reliability and predictability. The third refers to a belief that the automaton is functioning as its designers intended, and includes agreement with those intentions.

However, others have suggested that trust in HRI is not the same as human-to-human trust [28]. Jian et al. [20] found that people are more willing to rate an automaton than a human being as "distrusted", suggesting that there may be differences between human trust in automatons and human trust in humans. One source of such differences is that people tend to regard expert systems, AI systems, or computer-based decision support systems as more objective and rational than their human counterparts [7]. One could also add that people may expect automatons to have greater integrity than humans because they believe that automatons are not programmed to deceive. Of course, this stereotype could change rapidly as AIs become more sophisticated. In the HRI literature, the prospect of deception by robots (or AI) already has been raised [15].

There also is some evidence that people react more strongly to errors made by automatons than those made by humans, so that there is a swifter decline in trust (see [9, 43]). In a general sense, then, people may be less tolerant of uncertainties manifested in automaton behavior than in the behavior of their human counterparts. Humans, on the other hand, are expected to be more adaptive and creative than automatons, so it is plausible that novel actions or proposals from humans will be more trusted than if they come from automatons.

What is meant by "appropriate" trust in automatons? Oleson et al. [31] claim that appropriate trust of a system manifests itself in appropriate reliance on that system. Too much trust results in overreliance, and too little in insufficient reliance. However, it is worth bearing in mind that other factors can result in over- or under-usage of a system, such as a desire to avoid blame for bad outcomes (over-use) or a desire to gain credit for good outcomes (under-use).

Addressing the question of appropriate trust, Ososky et al. [32] refer to humans' tendency to anthropomorphize robots and to apply "inaccurate" mental models to inferences about robots' behaviours. Their remedy is that operatives have a full understanding of the automaton's capabilities and limitations. However, they do not systematically investigate the practical achievability of this suggestion. There already is an abundance of software and automated systems whose complexities exceed human capacity for anything approaching a complete understanding of their capabilities and limitations. Hancock et al. [15] recommend "transparency" in the form of system designs that are accessible and clear to human team members. However, there is an obvious potential for difficult tradeoffs or even dilemmas if one of the design objectives for a robot or AI also is that it is able to deceive enemies or even allies who are not cleared to know about that robot or AI.

Interestingly, the question of whether humans perceive (or can perceive) that they are trusted or distrusted by automatons seems to have been relatively neglected. At first glance, the question might seem nonsensical; surely we are not about to deem an automaton as being capable of trust. Nevertheless, the question makes sense for three reasons.

First, humans do anthropomorphize machines, so we cannot rule out the possibility that people may attribute an automaton's behavior towards them to trust or distrust of themselves by the automaton. This attribution certainly could arise when humans adopt what Dennett [6] called the "intentional stance". Dennett contrasts this stance against the "design stance", whereby an automaton's behavior is explained via beliefs about what it was designed to do. The intentional stance accounts for an automaton's behavior by assuming that it is a quasi-rational agent, with beliefs and desires of its own and the intelligence to pursue those desires on the basis of its beliefs. Moreover, people may be more likely to attribute trust to automatons than they would attribute emotions such as desire, because they are more willing to attribute belief states to automatons than feelings (cf. [19]) and the primary basis for trust is a set of beliefs.

Second, in connection with Lee and See's concept of purpose as a basis for trust, the intended uses of an automaton can include (dis)trust-relevant purposes such as monitoring its human teammates or deferring decisions to them. Thus, humans interacting with an automaton may adopt an intentional stance with regard to the automaton's designers and/or operators, attributing trust or distrust to these "puppeteers", even if they maintain only a design stance regarding the automaton itself. The automaton then manifests trust or distrust indirectly, via its apparently designed purposes and uses.

Third, Dennett's distinction between the design and intentional stances points to a candidate criterion for appropriate trust. A design stance would be appropriate in HRI most of the time. Automatons are indeed designed entities or systems; their designers will have had purposes and uses in mind. Trust based on a design stance will be limited to attributions of dependability, reliability, adherence to purpose-directed behavior, and the like. This kind of trust will be inappropriate only if the trustor has badly estimated the automaton's reliability or has misconstrued its design purposes. On the other hand, basing trust on an intentional stance clearly has pitfalls in the form of attributing benign intentions to an automaton. So, trust based on an intentional stance is likely to be inappropriate. However, as Dennett systematically argues throughout his book, the intentional stance works very well for predicting machine behavior, even when applied to something as simple as a thermostat. An intentional stance is, as Dennett points out [6], a viable alternative when a design stance is not practical. This stance therefore is seductive and difficult to falsify.

Finally, imputation of trust to automatons also is important because, as automatons are made increasingly human-like, humans will interact with them in more social rule-following ways. Reciprocity is a key social rule governing many aspects of human-human interactions, and it is likely to become increasingly relevant to HRI. Trust and distrust often are reciprocated, so we may expect that people are more likely to (dis)trust an automaton if they believe that the automaton (dis)trusts them. In short, a trust-enhancing way of humanizing automatons is to enable them to manifest trust-like behaviors toward their human teammates. Whether or when this would be desirable is a matter for careful consideration by designers.

How should we measure or evaluate trust in human-automaton relations? Examples from the literature include the Human Computer Trust Rating Scale [25]. Yagoda and Gillan [44] propose a scale that taps four closely-related adjectives for describing different aspects of HRI: reliability, dependability, accessibility, and timeliness or predictability. An in-depth critical review of the relevant measurement issues is not within the scope of this chapter, but suffice it to say that measuring trust in HRI is an active area of research and the current state of the art is at a fairly preliminary level. It seems unlikely that a single scale or battery of scales will be adequate for all types and contexts of HRI, and that as AI and related technologies advance, the measurement of trust in automatons will need revising.

10.2   Trust and Uncertainty

10.2.1   Trust and Distrust Entail Unknowns

Trust as a psychological state entails willingness to take risks by placing oneself in a vulnerable position with respect to the trustee (e.g., [11, 16, 23]). Uncertainty is therefore a given in trust. Moreover, trust relations may have to be forged in contexts bearing unknowns. This would be the case with new complex technology, for example, even if it has undergone extensive testing.

The key connection between trust and uncertainty is that to enter a trust relationship requires at least some non-surveillance of the trustee, and at least temporary non-accountability (freedom from micro-management) for the trustee. Thus, the trustor forgoes an entitlement to place the trustee under 24–7 surveillance or total accountability. Thus, trust relationships create unknowns and require that the trustor tolerate them [37]. Relinquishment of knowledge and control is primarily what distinguishes trust relationships from contracts (or assurance). In effect, such relinquishment amounts to trading one source of uncertainty for another, in the sense that uncertainty about whether desired goals or outcomes will be attained is reduced via the trust relationship, which in turn imposes a less aversive kind of uncertainty regarding the means by which goals or outcomes will be reached, through allowing the trustee discretionary power. This trade must be viewed by the trustor as worth enough to bear the risks entailed in a trust relationship.

Distrust as a psychological state amounts to a disposition to avoid being vulnerable to the distrusted party, often arising as a result of uncertainty about this party's intentions or future actions. Distrust therefore may involve unknowns in the form of suspicions as a given, or even as a justification for distrust in the first instance. Distrust also brings with it two additional forms of uncertainty. First, one may believe or suspect that the distrusted party lacks integrity and therefore doubt the veracity of information provided by that party. Second, distrust can morally license the manufacture of unknowns by the distrustor, either by withholding information from or outright deceiving the distrusted party. It seems plausible that people would find it easier to justify either of these acts if the distrusted party is an automaton than if it is a human, and therefore would be more likely to try to keep secrets from or deceive an automaton. The consequences of distrust in HRI appear to be relatively neglected in the research literature.

Relevant uncertainties can enter into any of Hardin's tripartite components: the (dis)trustor, the (dis)trustee, and the context. People can be unsure about their own psychological states; they may not be familiar with the automaton's reliability or design specifications; and they may have to engage in HRI in situations fraught with unknowns. Disentangling all of these uncertainties in a way that is relevant to trust considerations requires, first, ascertaining what is at stake in a HRI trust relationship. Thereafter, we can bring in knowledge about how and when people are likely to be able to tolerate and work with unknowns.

10.2.2   What Is Being Trusted; What Is Uncertain?

The relevance of uncertainties and their effects on trust in HRI will hinge on what is at stake in trusting an automaton. The stakes may be considered in terms of three aspects: The scope of the automaton's capabilities and responsibilities, the nature and sources of potential malfunctions or mishaps, and the kinds of errors or malfunctions committed by the automaton. The greater the scope of capabilities and responsibilities attributed to the automaton, the greater the impact of uncertainties about its functioning and capabilities on its trustworthiness. Likewise, greater perceived control over an important decision will be likely to increase the impact of uncertainties on trust. Yagoda and Gillan [44] present a useful two-dimensional framework regarding automaton capabilities. One dimension is the degree of intelligence and the other is the level of autonomy. AI would be high on both dimensions, while expert systems are typically high on intelligence but low on autonomy. A battery exemplifies low-intelligence but high-autonomy, and a robotic arm typifies low-intelligence and low-autonomy. It is plausible that being higher on either of these dimensions will increase the impact of uncertainties on trust in an automaton.

Turning now to the nature and sources of malfunctions or mishaps, two considerations are important to bear in mind. First, what kinds of errors or malfunctions are most problematic? Suppose an automaton has a diagnostic function that makes a binary decision to raise an alarm or not. False alarms will be regarded as more harmful than misses in some settings (e.g., in a legal trial where false convictions are worse than false acquittals) but the reverse will be the case in others (e.g., diagnosing a contagious fatal disease, where false positives are not as harmful as false negatives).

Second, are the sources of potential malfunctions internal or external to the automaton? Trusting an automaton to function properly is one thing if the only possible causes of malfunction are hardware or software faults in the automaton itself. It is quite another if malfunctions could be caused by damage from attacks, sabotage, hacking, or other security breaches. This latter set of possibilities brings with it questions of trust regarding the automaton's robustness and security provisions, which may have little or no connection with its primary purposes or functions. Uncertainty about autonomy itself may raise doubts and concerns about who or what is controlling the automaton (e.g., whether it has been hijacked).

Finally, we turn to considering errors and malfunctions. Errors or malfunctions will break trust, although at least one study has suggested that they may not influence decisions of whether to permit the automaton to act [35]. This finding highlights the importance of separating considerations about trust from those regarding whether humans will override an automaton. The connection between these two matters is relatively unexplored. Reasons or explanations for uncertain or erroneous performance also will influence trust. To begin, an absence of reasons or explanations will be detrimental to trust. Dzindolet et al. [8] demonstrated that users distrust even a generally high-performing system unless provided with reasons for why performance errors have occurred. Moreover, providing these reasons can maintain or even increase trust even when the system performs poorly, as long as the explanations do not evoke counter-trust attributions. Two attributions arising from malfunctions or errors that threaten trust are incompetence and betrayal. Deception or betrayal will break trust more irrecoverably than performance errors or incompetence. Consequently, uncertainty about honesty or benign "intent" will endanger trust more than uncertainty about performance or performative competence.

The impact of errors or malfunctions also will depend on the extent to which they can be rectified or undone. Uncertainties regarding reversible or steerable decisions are less detrimental to trust than uncertainties about irrevocable decisions [35]. Smithson and Ben-Haim [40] argue that steerable or revocable choices are more robust under extreme uncertainty than irrevocable ones. One aspect of their robustness is that such choices engender less fear of unknowns and thereby pose less of a threat to trust relations.

10.2.3   Trust and Dilemmas

Trust may involve dilemmas, which arise from particular sources of uncertainty and generate additional unknowns. Here, "dilemmas" refer to situations in which multiple rational actors' pursuit of self-interests lead to sub-optimal joint outcomes. Recently attention has been given to the "driverless car dilemma": People want others to have driverless cars programmed to sacrifice its passenger for the greater good, but they do not prefer those cars for themselves [3]. Viewed from the utilitarian assumption that sacrificing the passenger for the "greater good" is a public good regardless of whether the passenger is oneself or another person, this is a classical free-rider dilemma.

One line of reasoning about rational self-interest suggests that trust itself is inherently dilemmatic. The so-called "trust game" [2] has spawned a large literature. The original two-player procedure involves two stages. Both players are given an initial endowment of $10, one player is assigned to be the "sender", and the other assigned to be the "receiver". In the first stage, the sender passes any amount, $0< s <$10, to the receiver. The sender retains $10 - s, and the experimenter triples the amount sent, with 3 s passed to the receiver. In the second stage, the receiver passes any amount of the money received $ 0 < r <$3s, back to the sender. The amount passed by the sender is supposed to measure trust, and the amount returned by the receiver to measure trustworthiness. A self-interested rational sender or receiver should send nothing, and therein lies the dilemma claim. However, human players regularly demonstrate willingness to send sizeable amounts (see [21] for a meta-analysis of 162 experimental studies showing that this finding is robust across 35 countries).

Even if one does not accept the notion that trust is dilemmatic, dilemmas can pose problems for human trust in automatons that are programmed to be rational utility-maximizers. It is not difficult to imagine social dilemmas that could confront automatons and their human teammates in military combat. Suppose that enemy automatons A and B consider two alternative strategies available to each of them, A1 and A2 versus B1 and B2. To simplify matters, suppose that the stakes are the loss of 1000 lives on either or both sides. Both automatons are programmed to value the magnitude of utility for own-side casualties as 4 times greater than enemy-side casualties. That is, the utility of one own-side casualty is -4 whereas the utility of an enemy-side casualty is $+1$.

If A chooses A_1 and B chooses B_1 then A estimates a probability of $1/2$ of 1000 A-side casualties but also estimates 1000 B-side casualties for sure, so the expected utility for A is $U_{a11} = -4K/2 + 1K = -1K$. For the same combination of strategies, B also estimates a probability of $1/2$ of 1000 A-side casualties but only a probability of $1/2$ of 1000 B-side casualties, for an expected utility of $U_{b11} = 1K/2 - 4K/2 = -1.5K$. The remaining expected utilities are as follows.

For the $A_1 - B_2$ combination, $U_{a21} = -4K/2 = -2K$ and $U_{b21} = 1K/2 = 0.5K$;

For the $A_2 - B_1$ combination, $U_{a12} = 1K$ and $U_{b12} = -4K/2 = -2K$;

For the $A_2 - B_2$ combination, $U_{a22} = -4K + 1K/2 = -3.5K$ and $U_{b22} = 1K - 4K = -3K$.

These expected utilities are displayed in the upper half of Table 10.1 in units of 1000, with the appropriate row and column sums. The sums reveal that automaton A will conclude that A2 is its best strategy and automaton B will conclude that B2 is its best strategy. The result is the worst expected outcomes for both of them. This is a Chicken Game structure. Choosing any other combination instead would benefit both sides.

Table 10.1 A two-automaton dilemma

A

		A_1		A_2	
B	B_1	-1	1		-3.5
			-1.5		-2
	B_2	-2	-3.5		-2.5
			0.5		-3
			-3.0		-2.5

$$A_1 : U_{a11} = -V_{aa}/2 + V_{ab} \quad A_2 : U_{a12} = V_{ab}$$
$$B_1 : U_{b11} = -V_{ba}/2 - V_{bb}/2 \quad B_1 : U_{b12} = -V_{bb}/2$$
$$A_1 : U_{a21} = -V_{aa}/2 \quad A_2 : U_{a22} = -V_{aa} + V_{ab}/2$$
$$B_2 : U_{b21} = V_{ba}/2 \quad B_2 : U_{b22} = V_{ba} - V_{bb}$$

The lower half of Table 10.1 displays the utility formulas, where V_{aa} is the value given to A-side casualties by A, V_{ba} is the value of A-side casualties for B, V_{ab} is the value of B-side casualties for A, and V_{bb} the value of B-side casualties for B. Straightforward algebraic arguments show that regardless of the positive numbers assigned to these valuations, the $A_2 - B_2$ combination always is chosen by automatons A and B. Moreover, it is easy to show that this choice always is sub-optimal for both of them (even if it is not always the worst), because $U_{a21} > U_{a22}$, $U_{b12} > U_{b22}$, and $U_{b11} > U_{b22}$. Finally, it is clear that this structure always is a Chicken Game because the best outcome for A always is the $A_1 - B_2$ strategy combination whereas for B it is the $A_2 - B_1$ combination.

The prospect of such dilemmas raises a problem of trust in automatons for their human teammates and/or operators. How are they to know when, or how often, dilemmas like this will arise, and what can be done about them when they do? The obvious solutions, such as engaging in honest communication with the enemy automaton, often are not available in military situations as they may be for networked driverless cars.

10.3   Factors Affecting Human Reactivity to Risk and Uncertainty, and Trust

In this section, we survey factors affecting tolerance of uncertainties. These factors come in three kinds: the nature of the uncertainties themselves and how humans differentiate among varieties of unknowns, the psychological dispositions that influence tolerance of unknowns in general, and the conditions in groups or organizations that influence norms regarding the treatment of unknowns.

10.3.1 Kinds of Uncertainty, Risks, Standards, and Dispositions

Humans think and act as though there are distinct kinds of unknowns. They regard some kinds as worse than others, and may trade one kind for a more preferred kind. People's risk perceptions can be modulated by influences such that those perceptions will not match so-called "objective" risk assessments. They also may apply different standards of proof to different settings, and the burden of proof will depend on the assumptions they have made. Likewise, humans vary in their orientations toward and tolerance of risks and unknowns. All of these considerations are relevant to trust in HRI settings, and this section reviews them with this in mind. Starting with probabilities, there is ample evidence that human reactivity to probabilities is not linear in the probabilities, even when those probabilities are accurate. People tend to over-weight risks that have small probabilities, particularly if the stakes are high, and they have difficulty making meaningful decisional distinctions between small probabilities, even when these differ by orders of magnitude (such as one in a million versus one in ten thousand). They do, however, make a strong distinction between a probability of 0 and a very small nonzero probability. Trust in an automaton therefore is unlikely to be improved noticeably by decreasing the probability of automaton failure from, say, one in ten thousand to one in a hundred thousand. However, it is likely to increase substantially if the probability of failure is reduced from one in a hundred thousand to zero.

A relevant body of work here is on the relationship between judgments of probabilities and sample space partitions [13]. This line of research has shown that people anchor on the number of outcomes that is salient to them when making probability judgments. If they think in terms of K possible outcomes (i.e., a K-fold sample space partition), then they will anchor on probabilities of $1/K$ for each of the outcomes, and then adjust away from that when presented with relevant information. Smithson and Segale [41] demonstrated that partition-dependency effects hold even when people are using imprecise probabilities (e.g., probability intervals). An implication is that trust in an automaton can be influenced by priming users to consider its performance outcomes in alternative partitions. For instance, unpacking good outcomes into $K - 1$ sub-categories ($K > 2$) but lumping bad outcomes together into one category will anchor users on $1/K$ probability of a bad outcome, whereas packing both good and bad outcomes into one category will anchor users on a probability of $1/2$ for a bad outcome.

Turning now to types of unknowns, there are long-running debates among proponents of formal frameworks for uncertainty about whether all uncertainties can be handled by some version of probability theory. These debates will not be surveyed here, but one of the motivations for them has been evidence of widespread human intuitions that not all uncertainties are probabilistic. Instead, research in judgment and decision making under uncertainty has revealed that uncertainty arising from ambiguous or conflicting information influences judgments and decisions in ways that probabilistic uncertainty does not. Ambiguity has been widely studied in psychology and economics, beginning with Ellsberg's [10] seminal paper in which he demonstrated that people prefer a gamble with precisely specified probabilities to a gamble with imprecise probabilities, although the expected utilities for both gambles are identical. Although ambiguity aversion is not universally observed under all conditions (ambiguity-seeking may be observed, for example, for very low probabilities), the key point here is that people behave as though ambiguity is a different kind of uncertainty from probability that is relevant in their decisions. Several studies of uncertainty arising from conflicting information have found that there is a greater aversion to conflicting information than to ambiguous information [1, 4, 5] (e.g., [36]). Conflict aversion has been manifested in two ways. First, a majority of people prefer to receive or deal with messages from ambiguous rather than conflicting sources of information (see [36, 38]). Second, people tend to make more pessimistic estimates for future outcomes under conflict than ambiguity [4, 5, 38].

These findings suggest that ambiguous and conflicting signals or indications from an automaton may have different impacts on trust. These distinctions have implications for trust in HRI. Among the demonstrations [36] regarding conflict aversion is the finding that people usually assume that experts or computer models should agree in their forecasts and diagnoses. They prefer ambiguous but agreeing forecasts over unambiguous but disagreeing ones, even when these are informationally equivalent. Importantly, they attribute less trustworthiness to disagreeing experts or expert systems than to ambiguous but agreeing ones. It therefore seems plausible that ambiguous but agreeing signals or performance indicators from a single automaton will be less detrimental to trust than unambiguous but conflicting signals or indicators. If true, an example of a practical application is in the design of failure-mode indicators for an automaton whose operation is to be halted by a human overseer if failure is sufficiently indicated. A risk-averse approach would be to design the automaton's failure-mode indicators to be "trigger-happy" in the sense that at least one of them is likely to indicate possible failure even under a low probability that a malfunction has occurred.

The conflict versus ambiguity distinction also has implications for teams with multiple networked automatons and humans, in which the automatons are providing multiple assessments or predictions regarding the same situation. Unambiguous but disagreeing forecasts will be more detrimental to trust of the ensemble of automatons than ambiguous but agreeing ones. They also are likely to cause greater risk-aversion in the human team members. Another important kind of uncertainty is sample space ignorance, whereby the decision maker does not know all of the possible outcomes. With complex software, for instance, it is a commonplace for even its coders not to know all of its possible failure modes. Sample space ignorance has been shown in at least one study to be aversive [39]. To my awareness, no work has been done on the impact of sample space ignorance on trust. Nonetheless, it seems plausible that automatons will be viewed by users as more trustworthy if all of their possible failure modes are known than if users believe that these modes are not completely known.

What characteristics of risks besides probabilities influence human perceptions of riskiness? A large body of research on this topic indicates that people react most strongly to those risks that are hard to understand, involuntary, and invisible [22].

Typical examples are risks associated with nuclear power, nanotechnology, and climate change. Strong fears may persist despite evidence and reassurances by experts that a particular risk is minimal or unlikely. On the other hand, people are likely to be overly complacent about risks that are familiar, voluntary, and visible. Examples of this kind of risk include driving an automobile, handling or using a firearm, and using power-tools.

An additional relevant, but often neglected, characteristic of risks is whether the relevant unknowns are reducible or not. Reducible unknowns may be less corrosive of trust than irreducible ones, especially if there are measures in place to eventually eliminate these unknowns. As AI becomes more complex, irreducible uncertainties about automaton behavior will become more commonplace and may pose an obstacle to building trust in HRI.

The burden of proof identifies the party or position that must build a case to overturn a default position. (e.g., the presumption of innocence in a Western court trial places the burden of proof on the prosecution). Trust can be presumed, in cases such as role-based trust where the role involves expertise and the experts have been certified as qualified to perform the role. Given the current state of the art in HRI, presumed trust seems unlikely and so the burden of proof most often will fall on the technology and the automaton that instantiates it. However, as automatons become more advanced and more human-like, automatons may be increasingly presumed trustworthy until they prove otherwise. This prospect adds a new twist to considerations of what constitutes "appropriate" trust.

The standard of proof refers to the strength and weight of evidence required for a case to be regarded as "proven". In Western criminal trials, the conventional standard of proof is evidence of guilt "beyond reasonable doubt", whereas in civil cases the standard is "on the balance of probabilities". Standards of proof therefore demarcate thresholds for tolerance of uncertainty. Differing standards of proof regarding automaton trustworthiness between their designers and users will raise problems, so establishing agreements about such standards will be an important aspect of automaton development, testing, and deployment.

Finally, psychological dispositions may play a role in building trust. Some people are less trusting than others, they may be more risk-averse, and/or more intolerant of uncertainty. Dispositions such as these may influence the standard of proof a human brings to HRI when making judgments of automaton trustworthiness. Only few HRI studies have systematically investigated the role of human-related characteristics (e.g. level of expertise, personality traits such as extroversion [17]) and environmental factors (e.g. culture, task type [27]). To my knowledge, none have investigated the role of trait-level trustingness, risk orientation, or tolerance of uncertainty regarding their influences on the nature of trust in HRI. Because trust relations are strongly context-dependent, it is possible that psychological traits will not have a strong influence here, but this possibility has yet to be ascertained.

10.3.2 Presumptive and Organizational-Level Trust

Kramer and Lewicki [24] introduce the notion of "presumptive" trust as a kind of depersonalized basis for trust that has more to do with indirect indicators such as reputation and properties of organizational or group settings such as shared identity, common fate, and interdependence, than with direct indicators of trustworthiness as manifested by the potential trustee. The term "presumptive" conveys that this kind of trust is a default stance on the part of the trustor, and often operates in a tacit way. According to Kramer and Lewicki, presumptive trust has at least one of three primary bases: Identities, roles, and rules.

Identity-based trust is the expectation that fellow in-group members can be trusted, and some scholars have argued that this is based on an expectation of general reciprocity within the boundaries of the in-group [12]. Shared identity is unlikely to be a basis for human trust of automatons, although it certainly is plausible that "in-group" automatons may be trusted more than "out-group" automatons, even when both categories of automaton are "on the same side".

Role-based trust probably would better be thought of as "system-based". The primary idea here is that an individual occupying a specific role in an organization may be trusted because both the nature of the role and the system of training and/or selecting people to occupy that role are trusted. Thus, we will trust a robot if we trust robotics and also trust the engineering programs that train roboticists. Or, we may trust a particular brand of automaton because we trust that particular company and its selection processes for hiring engineers and programmers.

Rule-based trust has its source in the codified norms and other rules for behavior within a group or organization, and the expectation that members have been socialized to follow the rules and adhere to the norms. "Honour" codes are an example of this kind of trust basis. Analogs for this kind of trust in HRI include beliefs about the robot's adherence to its programmed protocols, and compatibility between those protocols and human social and psychological norms. There may be a design tradeoff here between a preference for robots that "blindly" adhere to their inbuilt protocols and a preference for robots whose behavior is flexible and adapts to novel situations.

Risk management norms in a group or organization will influence the development of trust in HRI. Perhaps the most obvious kind of influence stems from the "tightness" of the organizational culture [14]. So-called "tight" cultures have numerous strong norms and very little tolerance of deviant behavior, whereas "loose" cultures' social norms are relatively weak and they are permissive of deviant behavior. Research into this cultural dimension has found a correlation between tightness and the magnitude of risks in the ecology occupied by a culture. This connection suggests that tighter cultures will be more risk-averse and less trusting. While the research program elaborated by [14] has focused on national cultures, it is plausible that these same connections and the tightness construct will apply to organizations and groups.

10.3.3 Trust Repair

Kramer and Lewicki [24] observe that most approaches to trust repair have only focused on changing cognitions, thereby neglecting emotional or behavioral aspects of trust repair. Much of this research also has emphasized routes to repair that may not apply in HRI, although as automatons are increasingly humanized more of these routes may become available. Also, it is arguably an open question as to whether some apparently incongruous acts by an automaton could nevertheless aid in trust repair. For example, would an apology by a robot for its error assuage human users?

Both explanations and apologies have been found to help restore trust, but generally if accompanied by some actual reparations or measures to prevent further breaches of trust. Tomlinson, et al. [42] investigated the characteristics of apologies influencing their effectiveness in trust repair. They found that an apology was more effective if issued sooner than later after a breach of trust. They also found that apologies and explanations that had the trust violator taking responsibility for the breach were more effective than accounts that blamed other parties or external factors for the breach. A possible exception to this finding, pointed out by [24], is when the breach involves a violation of integrity. In that case, being able to deny responsibility for such a violation may be more effective.

Penance and reparations have been extensively studied in regard to trust repair. One problem for HRI is that, like apologies, penance and reparation on the part of an automaton may be largely irrelevant unless humans have anthropomorphized the automaton to the extent that they attribute emotional responses to it. However, such measures could be applied to the designers or producers of the automaton, especially if trust in the automaton is primarily a matter of trust in its designers and/or producers.

Similar arguments apply to other more "legalistic" trust repair mechanisms, such as rules, contracts, monitoring systems, and sanctions against further trust violations. Most of these are attempts to ensure that the trusted party is motivated not to breach trust again, which is irrelevant to an automaton unless its users attribute motivations to it. One partial exception to this is reinforcement schedules in machine learning, which could be revised in the service of preventing further malfunctions or errors by the automaton.

10.4 Concluding Remarks

In this chapter, we have surveyed the following factors in HRI that influence the nature and development of trust:

- The scope of an automaton's capabilities and responsibilities, and the extent of its control over decisions
- Whether the sources of potential malfunctions or mishaps are internal or external to the automaton
- Which kinds of errors or malfunctions are most important or consequential

- The impact of uncertainty about benign intent versus competence or reliability
- Uncertainties arising from the prospect of social dilemmas involving interacting automatons, especially opponent automatons
- Organization-based trust and the impact of organizational norms and culture
- Factors influencing trust repair when trust has been eroded or lost.

This chapter also has provided suggestions for several avenues of further research and theoretical developments regarding the role of uncertainties in HRI, specifically in connection with trust. A major theme of this chapter is that almost all treatments of uncertainty in relation to matters of trust have over-simplified both the role and nature of uncertainty. Regarding its role, on the one hand, it is widely claimed that trust serves to reduce uncertainty. On the other, it also is widely claimed that uncertainty is endemic in a trust relationship. Absent from these accounts is the realization that in establishing a trust relationship, the trustor is trading the reduction of one set of uncertainties for the creation of another set of uncertainties. Typically, the tradeoff involves reducing uncertainty about outcomes (to be attained by the trustee) at the expense of tolerating uncertainty about the means by which the trustee pursues and achieves those outcomes. Likewise, the role of uncertainty in distrust has not been fully understood, especially in regard to the license for secrecy, deception, and other forms of ignorance production that distrust provides for the distrustor.

Uncertainty also has largely been treated as if it is unitary or monolithic, and a "negative" that people are motivated to be rid of. These over-simplifications persist throughout both the human sciences and engineering. People have uses for unknowns and unknowns underpin important forms of social capital, as is exemplified by the fact that a trust relationship is predicated on tolerated ignorance. Likewise, as has been clearly articulated in this chapter, people think and act as though there are different kinds of uncertainty, and as though those differences are important. For instance, they prefer agreeing but vague experts to precise but disagreeing experts (i.e., "conflict aversion"), and they trust the former more than the latter. The impacts of different kinds of uncertainty on trust in HRI remain to be systematically investigated, but this chapter points to clear directions for such research.

References

1. A. Baillon, L. Cabantous, P.P. Wakker, Aggregating imprecise or conflicting beliefs: an experimental investigation using modern ambiguity theories. J. Risk Uncertainty **44**(2), 115–147 (2012)
2. J. Berg, J. Dickhaut, K. McCabe, Trust, reciprocity, and social history. Game Econ. Behav. **10**(1), 122–142 (1995)
3. J.-F. Bonnefon, A. Shariff, I. Rahwan, The social dilemma of autonomous vehicles. Science **352**(6293), 1573–1576 (2016)
4. L. Cabantous, Ambiguity aversion in the field of insurance: insurers' attitude to imprecise and conflicting probability estimates. Theor. Decis. **62**(3), 219–240 (2007)

5. L. Cabantous, D. Hilton, H. Kunreuther, E. Michel-Kerjan, Is imprecise knowledge better than conflicting expertise? evidence from insurers' decisions in the united states. J. Risk Uncertainty **42**(3), 211–232 (2011)

6. D.C. Dennett, *The Intentional Stance* (MIT press, Cambridge, 1989)

7. J.J. Dijkstra, W.B.G. Liebrand, E. Timminga, Persuasiveness of expert systems. Behav. Inf. Technol. **17**(3), 155–163 (1998)

8. M.T. Dzindolet, S.A. Peterson, R.A. Pomranky, L.G. Pierce, H.P. Beck, The role of trust in automation reliance. Int. J. Hum.-Comput. Stud. **58**(6), 697–718 (2003)

9. M.T. Dzindolet, L.G. Pierce, H.P. Beck, L.A. Dawe, B.W. Anderson, Predicting misuse and disuse of combat identification systems. Mil. Psychol. **13**(3), 147 (2001)

10. D. Ellsberg, Risk, ambiguity, and the savage axioms, in *The Quarterly Journal of Economics*, pp. 643–669 (1961)

11. G.A. Fine, L. Holyfield, Secrecy, trust, and dangerous leisure: generating group cohesion in voluntary organizations. Soc. Psychol. Q. pp. 22–38 (1996)

12. M. Foddy, T. Yamagishi, *Whom Can We Trust*, chapter Group-based trust (Russell Sage Foundation, New York, 2009), pp. 17–41

13. C.R. Fox, Y. Rottenstreich, Partition priming in judgment under uncertainty. Psychol. Sci. **14**(3), 195–200 (2003)

14. M.J. Gelfand, J.L. Raver, L. Nishii, L.M. Leslie, J. Lun, B.C. Lim, L. Duan, A. Almaliach, S. Ang, J. Arnadottir et al., Differences between tight and loose cultures: a 33-nation study. Science **332**(6033), 1100–1104 (2011)

15. P.A. Hancock, D.R. Billings, K.E. Schaefer, Can you trust your robot? Ergon. Des. **19**(3), 24–29 (2011)

16. R. Hardin, The street-level epistemology of trust. Analyse & Kritik **14**(2), 152–176 (1992)

17. K.S. Haring, Y. Matsumoto, K. Watanabe, How do people perceive and trust a lifelike robot, in *Proceedings of the World Congress on Engineering and Computer Science*, Vol. 1 (2013)

18. L.T. Hosmer, Trust: the connecting link between organizational theory and philosophical ethics. Acad. Manag. Rev. **20**(2), 379–403 (1995)

19. B. Huebner, Commonsense concepts of phenomenal consciousness: does anyone care about functional zombies? Phenomenol. Cogn. Sci. **9**(1), 133–155 (2010)

20. J.-Y. Jian, A.M. Bisantz, C.G. Drury, Foundations for an empirically determined scale of trust in automated systems. Int. J. Cogn. Ergon. **4**(1), 53–71 (2000)

21. N.D. Johnson, A.A. Mislin, Trust games: a meta-analysis. J. Econ. Psychol. **32**(5), 865–889 (2011)

22. R.E. Kasperson, O. Renn, P. Slovic, H.S. Brown, J. Emel, R. Goble, J.X. Kasperson, S. Ratick, The social amplification of risk: a conceptual framework. Risk Anal. **8**(2), 177–187 (1988)

23. R.M. Kramer, Trust and distrust in organizations: emerging perspectives, enduring questions. Annu. Rev. Psychol. **50**(1), 569–598 (1999)

24. R.M. Kramer, R.J. Lewicki, Repairing and enhancing trust: approaches to reducing organizational trust deficits. Acad. Manag. Ann. **4**(1), 245–277 (2010)

25. J. Langan-Fox, M.J. Sankey, J.M. Canty, Human factors measurement for future air traffic control systems. Hum. Factors: J. Hum. Factors Ergon. Soc. **51**(5), 595–637 (2009)

26. J.D. Lee, K.A. See, Trust in automation: designing for appropriate reliance. Hum. Factors: J. Hum. Factors Ergon. Soc. **46**(1), 50–80 (2004)

27. D. Li, P.L. Patrick Rau, Y. Li, A cross-cultural study: effect of robot appearance and task. Int. J. Soc. Robot. **2**(2), 175–186 (2010)

28. P. Madhavan, D.A. Wiegmann, Similarities and differences between human-human and human-automation trust: an integrative review. Theor. Issues Ergon. Sci. **8**(4), 277–301 (2007)

29. D.J. McAllister, Affect-and cognition-based trust as foundations for interpersonal cooperation in organizations. Acad. Manag. J. **38**(1), 24–59 (1995)

30. D.H. McKnight, N.L. Chervany, Trust and distrust definitions: one bite at a time, in *Trust in Cyber-societies* (Springer, Berlin, 2001), pp. 27–54

31. K.E. Oleson, D.R. Billings, V. Kocsis, J.Y.C. Chen, P.A. Hancock, Antecedents of trust in human-robot collaborations, in *2011 IEEE International Multi-Disciplinary Conference on Cognitive Methods in Situation Awareness and Decision Support (CogSIMA)* (IEEE, 2011), pp. 175–178
32. S. Ososky, D. Schuster, E. Phillips, F.G. Jentsch, Building appropriate trust in human-robot teams, in *2013 AAAI Spring Symposium Series* (2013)
33. J.B. Rotter, Generalized expectancies for interpersonal trust. Am. psychol. **26**(5), 443 (1971)
34. J.B. Rotter, Interpersonal trust, trustworthiness, and gullibility. Am. psychol. **35**(1), 1 (1980)
35. M. Salem, G. Lakatos, F. Amirabdollahian, K. Dautenhahn, Would you trust a (faulty) robot?: effects of error, task type and personality on human-robot cooperation and trust, in *Proceedings of the Tenth Annual ACM/IEEE International Conference on Human-Robot Interaction* (ACM, New York, 2015) pp. 141–148
36. M. Smithson, Conflict aversion: preference for ambiguity vs conflict in sources and evidence. Organ. Behav. Hum. Decis. Process. **79**(3), 179–198 (1999)
37. M. Smithson, *Uncertainty and risk: Multidisciplinary perspectives*, chapter The many faces and masks of uncertainty (Earthscan, London, 2008), pp. 13–25
38. M. Smithson, Conflict and ambiguity: preliminary models and empirical tests, in *Proceedings of the Eighth International Symposium on Imprecise Probability: Theories and Applications*, pp. 303–310 (2013)
39. M. Smithson, T. Bartos, K. Takemura, Human judgment under sample space ignorance. Risk Decis. Policy **5**(02), 135–150 (2000)
40. M. Smithson, Y. Ben-Haim, Reasoned decision making without math? adaptability and robustness in response to surprise. Risk Anal. **35**(10), 1911–1918 (2015)
41. M. Smithson, C. Segale, Partition priming in judgments of imprecise probabilities. J. Stat. Theor. Pract. **3**(1), 169–181 (2009)
42. E.C. Tomlinson, B.R. Dineen, R.J. Lewicki, The road to reconciliation: antecedents of victim willingness to reconcile following a broken promise. J. Manag. **30**(2), 165–187 (2004)
43. D.A. Wiegmann, A. Rich, H. Zhang, Automated diagnostic aids: the effects of aid reliability on users' trust and reliance. Theor. Issues Ergon. Sci. **2**(4), 352–367 (2001)
44. R.E. Yagoda, D.J. Gillan, You want me to trust a robot? the development of a human-robot interaction trust scale. Int. J. Soc. Robot. **4**(3), 235–248 (2012)

Chapter 11
The Need for Trusted Autonomy in Military Cyber Security

Andrew Dowse

11.1 Introduction

Information systems in the early 21st Century have become a critical enabler of increased value to the business, or as people in Defence might call a 'force multiplier'. Clearly the converse of this logic is that in warfare any capability that provides such a competitive advantage is also a vulnerability and a focus for a potential adversary to target. In the 20th Century, this risk was mitigated through the isolation of our information systems, with closed systems inherently easier to protect. However the real value of modern information systems has been the ability to provide more accurate, complete, relevant and timely information to support the business; and this has been achieved through a trend towards openness with greater connectivity and integration of systems. The very source of value to the business also represents a risk to it, and this remains a matter of tension and deliberation in the management of information systems.

The importance the Australian Department of Defence places in protecting our information advantage is reflected in the 2016 Defence White Paper, which notes the emergence of cyber threats to the ADF's warfighting ability, given its reliance on information networks [7]. The White Paper states that national and Defence cyber security capabilities will be strengthened to protect our systems.

A simplistic response to this priority would be for Defence to put more resources towards cyber security: more people monitoring audit logs and gateways, more effort towards accreditation and assurance activities, more funding allocated to cyber security projects. However, the exponential growth in the information environment means that taking a traditional approach and providing linear increases in resources is unlikely to meet the emerging challenge.

A. Dowse (✉)
Department of Defence, Canberra, Australia
e-mail: andrew.dowse@defence.gov.au

© The Author(s) 2018
H. A. Abbass et al. (eds.), *Foundations of Trusted Autonomy*, Studies in Systems,
Decision and Control 117, https://doi.org/10.1007/978-3-319-64816-3_11

This paper will consider potential requirements for trusted autonomy in cyber security, looking firstly at the current cyber environment, including four fundamental principles of cyber security. It will then assess the emerging challenges to this mission, framed though the dimensions of Big Data and consider opportunities to apply trusted autonomy to improve cyber security. The intent of this paper is to help inform researchers of the areas in which development in trusted autonomy may provide greatest return on investment in cyber security. Whilst these areas are specifically related to the requirements for the Australian Department of Defence, they may also be relevant to many other organisations facing similar cyber challenges.

Defence's Information and Communications Technology (ICT) architecture provides a reasonably robust protection against cyber threats. The lower classification (Protected) network is connected to the Internet via a gateway that provides multiple security mechanisms, thus achieving defence-in-depth. These security mechanisms are highly effective and relatively sophisticated, but involve significant manual processes. Due to the sensitivities and the need to maintain a security advantage, this paper will not provide any details of the tools or techniques currently utilised by Defence.

Whereas cyber security threats are on the increase, incidents on Defence networks actually decreased in 2015 in comparison with the previous year. Some 50,000 events were detected in 2015, around the same as 2014, of which there were 580 incidents, which represented a 25% decrease [10]. The causality of the decrease in the number of security incidents cannot be stated with certainty, but there are strong indications that this result is through greater success of security mechanisms, especially through blocking of threats at the gateway. Notwithstanding the evidence of current success in cyber security, Defence needs to further strengthen protections to keep up with, and preferably ahead of the threat.

11.2 Cyber Security

Information assurance and cyber security are both concerned with the protection and defence of information and information systems by ensuring their confidentiality, integrity and availability. Information assurance accounts for the risks to information from natural, accidental and deliberate actions. Whereas cyber security tends to focus on deliberate acts, information assurance and the managers of information systems need to prepare against such acts, but also against accidents, faults, external events and human error [4].

If the overall outcome for ICT management is the preservation of confidentiality, integrity and availability of information in support of the organisation's missions and interests, in many respects it doesn't matter whether something happens due to a deliberate act or some other reason. When there is an impact on an organisation's information systems, the priority is to respond coherently and expeditiously, rather than dwelling on whether it is an attack or a fault before someone responds. Hence a principle of cyber security is that it is an integrated part of how an organisation manages its ICT environment. The organisation needs to have clear accountabilities for ICT security, from policy to accreditation to day-to-day operations. As exciting as the idea of doing cyber operations might seem, defensive operations are largely a matter of systematically reviewing candidate incidents and managing the various security mechanisms within a defence in depth approach. The ability to successfully undertake defensive operations is strongly dependent upon how well the information environment is set up and the level of discipline inherent in it. Hence the second principle is that a secure foundation is fundamental to cyber security.

The design of Defence's information environment provides requisite levels of information assurance. In addition, Defence must undertake activities that ensure that systems perform consistently as specified, that users are accountable for their actions, that risks are mitigated by monitoring the environment and reducing the impact of a failure against system or usage expectations, and that there are means available to support an effective and timely recovery from an incident. These are the foundations that need to be designed into a secure environment, and need to be continually reviewed, updated and validated as technologies and threats evolve.

Much of Defence's information assurance against cyber threats comes from application of the Australian Signals Directorate (ASD) mitigation strategies. The top 4 mitigations—application whitelisting, application patching, operating system patching and restricting privileged access—can prevent over 85% of cyber intrusions [6]. Defence also gives priority to ASD's larger list of 35 mitigations, which further reduce vulnerability to cyber threats.

Defence will further enhance cyber security mechanisms with investments through Joint Project 2068, the Cyber Security Improvement Program, and increasing the recruitment and training of cyber security specialists, with a mix of military, public servants and contractors. Defence is also supporting Whole of Government efforts through the expansion of the ASD-led Australian Cyber Security Centre.

Defence's adoption of current operating systems has been slow in the past, and this leads to vulnerabilities associated with using older systems. The Infrastructure Transformation Program, which is planned to deploy by the end of 2017, will update hardware, networks, operating systems, applications and architectures to make Defence's ICT more robust, supportable and defendable [5].

But no matter how good these systems are, cyber security can only be as good as the organisation's people make it. This outcome is not only reliant on cyber security specialists, but requires the support of all people in the organisation who use ICT. Poor discipline, stupidity, lack of awareness and deliberate acts are all critical risks to cyber security. The third principle is that everyone in the organisation contributes to cyber security.

Some aspects of Defence's information environment are already constrained to mitigate the risks of poor user behaviours. Risks could be minimised further by locking down systems, but it gets to the point that it impacts the business; in which case it is better to accept some level of risk, and perhaps mitigate through training or auditing or some other mechanism. Thus the fourth and final principle is that an organisation must balance security imperatives with business requirements for functionality and access.

There are different expectations for cyber security for different information services. For Defence's classified warfighting network, there is an expectation of a high level of availability and confidentiality, whereas Defence's financial systems on the Protected network require high integrity.

Hence Defence's classified networks utilise greater access controls and adherence to the imperatives of need to know and need to share. Their connectivity is generally only to equivalent domains, and access points are strongly monitored. The Protected network has connectivity to the Internet, but through a consolidated gateway that has multiple security mechanisms in place.

Defence faces different types of cyber threats for its different networks. For classified networks, there is a focus on ensuring availability, as well as protecting against insider threats and to guard against potential intelligence collection. On the Protected network, the most common threat is criminal and there is less of a concern with attacks against confidentiality, although there are risks with commercially sensitive information as well as the real prospect that aggregation of information is sufficiently valuable to attract sophisticated state-based intelligence collection threats.

As an example of managing the balance between functionality, access and cyber security, Defence has in the past only permitted purely unclassified emails to exit the gateway from the Protected network to the Internet, and did not allow emails with a Dissemination Limiting Marker (such as sensitive or For Official Use Only) to be passed to the Internet. While reducing risks, this practice was damaging Defence business, such as vetting processes and interaction with Defence industry. Therefore a risk based decision was made earlier this year to permit such emails to be sent to the Internet when justified for Defence business, with risks managed through user awareness, procedures and auditing.

This highlights an important point here that much of Defence's information is actually held outside controlled networks. With the Defence Industry Policy Statement intent to strengthen these industry partnership arrangements, sensitive military information held on industry's networks must be protected to the same level as on Defence's own networks. In this regard, industry is critical to Defence's cyber security and creates an additional complexity to how we might use trusted autonomy.

11.3 Challenges and the Potential Application of Trusted Autonomy

The evolution of computational techniques has taken us from automation of processes, in which a system acts in accordance with defined rules, to autonomy, in which its behaviour is governed more by an understanding of objectives combined with observation and learning. Autonomy is a characteristic of an agent in which it is aware of other entities, and interacts with them, but exercises independence in order to maintain focus on its defined interests. Key to autonomy is the interaction with the environment within a goal-directed behaviour [12]. Such independence of action is further defined as self-direction or self-governance [14].

Trust is a further characteristic in which the agent will act in a predictable and reliable manner, producing credible outcomes based upon the use of reputable sources [11]. Trust also has a connotation of the formal evaluation of systems to determine how they behave with such predictability and certainty, especially where it applies to the protection of confidentiality. This requirement in trust raises an interesting dilemma for an autonomous system, in that it may be difficult to measure predictability in such a system that does not act in an obviously deterministic manner. Trusted autonomy may in this regard be considered an oxymoron, or at least a challenge for developers and researchers.

In cyber security, trusted autonomous agents should provide reliable security outcomes that align with the interests of the organisation. Given the very nature of cyber security, there is an expectation of a significant level of trust in any agent involved in the protection of networks. The need for trusted autonomy is being driven by a number of factors or challenges in the future information environment, which will be explored in the remainder of this paper.

Fifteen years ago, the concept of Big Data was introduced, characterising the concept in terms of three 'V' dimensions [13]. Since then other authors have added more dimensions, typically continuing the alliteration, and also it has been recognised that many of these dimensions are relevant for cyber security. In this paper I'm going to examine Defence's future cyber security challenges in terms of five Vs.

The first V is volume. CIO Group manages multiple networks on behalf of Defence, the largest being the Protected network with over 100,000 users. The personnel system runs over 100 million transactions per week and the logistics system over 10 million transactions per week. Utilised storage in the Defence environment is 5.8 petabytes, with an annual growth of around 20% [9].

The Defence network architecture is designed with consolidated gateways that protect the corporate network but enable controlled access between it and the Internet.[1] This approach provides a security focus on the gateway, reducing the risks of multiple vulnerabilities, but brings a significant volume of interactions across the gateway.

Each week, Defence's High Availability Internet Gateway supports around 2 million inbound and 600 thousand outbound legitimate emails, as well as nearly 10 terabytes in web services [8]. While legitimate traffic continues to increase, it is negligible compared to blocked emails, which have gone in a period of 12 months from roughly the same quantity as legitimate in 2015 to now four times as many and increasing. While a portion of the blocked emails are due to being oversize or misaddressed, the majority are spam or potentially have malicious content.

Clearly dealing with such large and increasing volumes of data is a challenge for cyber security effectiveness. Much of this effort is focused on perimeter security at the Internet Gateway, and Defence's systems are highly capable of identifying suspicious events. However the analysis of candidate incidents involves humans in

---

[1]The importance of gateways in cyber security is elaborated at [3].

the loop, and increasing volumes of events will create a challenge that needs to be met through a combination of increased resources and automation, if not autonomy.

Cyber security risks however are not only concerned with incoming traffic. The increasing volumes of data exiting through the gateway to the Internet are monitored, with mechanisms to block, flag or log emails dependent on the content and other circumstances.

Additionally, the expansion of the Internet, especially with introduction of IPV6, translates to additional volume for setting rules at the gateway in respect of whitelisted and blacklisted entities.

While there are tools that support distinguishing between valid and potentially suspicious traffic, much of the actual decision making around release of traffic remains a manual process. While automation reduces the volumes of data necessitating a manual process (e.g. from events to candidate incidents), the growing volumes and sophistication of threats mean that there are growing numbers of unfiltered events that require trusted decision-making.

Hence a key future requirement for trusted autonomy is to further increase the ratio of total cyber events to those that require manual analysis. This will require increasing the trust not only in the agents that provide that filtering but also in the sources that the agents rely upon to undertake this task. Success in this endeavour to handle large volumes of data is a matter of both the defeat of cyber threats as well as the facilitation of valid business.

Although much of the cyber security emphasis is on the gateway, Defence recognises that perimeter defence is not enough for comprehensive cyber security, and endpoints around the network are monitored and analysed accordingly. Defence employs some automation to support analysis of this data, but its intent to strengthen cyber security requires further enhancement in this area, including more sophisticated and autonomous agents that can recognise anomalous behaviours and events.

This leads to a second V: visualisation. The ICT security capability in Defence has utilised pioneering visualisation technologies for some years. In order to have awareness of the health of the information environment and be able to make timely decisions, whether about cyber security or any aspect of operations, an organisation needs better visualisation. This is a challenge for Defence, especially in the future with its transformed network infrastructure managed within separate towers by outsourced service providers.

The future visualisation capability needs to provide consistent information in an operating picture that reflects the information and physical domains, and that can be shared between security operators, network operators and military command and control. Defence needs visualisation for decision support in the form of relevant and accurate information about the status of networks.

Relevant is important in the context of what information is needed to make decisions—a strategic commander will require different information from a tactical commander, and quite different again from network and security operators. Military commanders will be interested in whether systems are available, fit for purpose and are providing the required connectivity and functionality. This will demand a different approach to visualisation compared to network and security operators, who will be more concerned with the systems themselves rather than the businesses they support. In this respect, the term 'common operating picture' or COP is misleading as the picture is not the same across all the types of decision makers—so what is required is a consistent operating picture.

The difficulty of a cyber COP is that it is far more difficult to represent the cyber environment, and be able to be comprehended, compared to the comparatively simpler representations of the physical environment. Added to this is the critical importance of providing accurate information to a decision maker who will make potentially life or death decisions based upon the state of cyber support.[2] Therefore it is important that decision support systems provide an accurate representation of the cyber environment.

This in itself is more of a challenge than might be immediately evident. A system that provides situational awareness of the cyber environment is itself part of the cyber environment. Any system that provides such awareness must do so with credibility and reliability, having access to sources that provide this information, but having sufficient resilience so not to be vulnerable to the threats or faults of which it is providing awareness. Hence visualisation provides a second critical requirement for trusted autonomy in cyber security.

The visualisation system needs to facilitate users' ability to drill down to get more information. It also needs to provide timely advice and support decisions to reconfigure as required in real time.

This leads to the third V: velocity. As per the case of Big Data, many business applications require real time speed for their interactions. Whereas an email or file can be taken offline for analysis, other interactions such as web services may not be so easily managed from a security perspective.

Another consideration for velocity is the critically short period between the first awareness of a new threat and the deployment of associated defence mechanisms such as detection and patching. Defence enjoys excellent relationships with partners in other CERT like organisations and in industry, and is very focused on minimising the time between identification of new threats and deployment of adjusted defences.

This may not be as helpful if the organisation's networks are the target of a zero day attack, hence a lot of importance needs to be placed on security mechanisms that help identify new threats, limit the damage and facilitate quick recovery and support of business.

It is important to recognise the pace at which a significant cyber incident can evolve. Defence is placing greater emphasis on the coordination procedures for reacting to a cyber event and making timely and appropriate decisions on how we balance operational continuity and security. To do so demands an understanding of who the authorities for making such decisions are and to exercise realistically so that responders aren't trying to figure it out in the middle of a real incident.

Such initiatives will help improve the ability to respond to cyber threats in a timely manner, but so long as these procedures are manual, Defence may not be able to keep

---

[2]This might seem overly dramatic, but Defence is now highly reliant on information systems, and a decision to sever or shut down systems due to a cyber threat may have significant consequences.

up with the tempo of cyber operations. Current procedures in Defence seek to resolve network issues in terms of hours, if not days. The consequences of cyber events can be a matter in which decisive action is needed in much shorter timeframes.

It is useful to consider the velocity issue in terms of Boyd's Observe-Orient-Decide-Act (OODA) loop [2]. Boyd identified that in modern warfare, particularly air warfare, an advantage would be gained by having a shorter decision cycle than one's adversary. Such an advantage is even more critical in cyber warfare.

The third critical requirement for trusted autonomy in cyber security is to streamline decision-making to minimise the time to take action. As a minimum, this requires superior decision support, as discussed earlier under visualisation. Timeframes can be reduced if the autonomous agent provides recommendations based upon a comprehensive knowledge of the network, the cyber threats and consequences, and the supported business.

To fully comprehend the need for timely and integrated decision-making in cyber security, it is important to appreciate the nature of our vulnerabilities. The Defence network architecture focuses perimeter defences on areas of greater vulnerability. Defence is creating greater access in its networks to information and services, with deployable and mobile users soon having similar access to services as one might have in an office in Canberra. The centralising of data processing and making greater use of thin client technologies may represent a shift in vulnerabilities.

The seamless integration of the Defence ICT environment means that the organisation cannot afford for localised and independent decisions about balancing risk and reward as it pertains to cyber security. A violation in one area can conceivably proliferate throughout the network, which demands an approach to configuration management, cyber security, technical control and support to military operations that can balance overall risks. Often the risk to a mission needs to be weighed against the risk to the enterprise. This needs to be taken into account in developing the ability, including autonomous capability, to respond quickly to cyber threats.

The primary intent therefore is to ensure vulnerabilities are minimised and responses are quick, coherent and effective when facing cyber incidents. Defence leadership must recognise the potential that a sophisticated and strong cyber threat could impact its systems, and therefore must be prepared through training to fight on in situations of disrupted or degraded information services.

One might suggest the ultimate goal for trusted autonomy in cyber security is to take the human out of the loop in defensive cyber operations, making the speed of response dependent only on electronic processes, rather than also including physical and cognitive elements. Such a goal would require considerable investment in research and development, as well as a very highly refined (and continually updated and tested) understanding of the relative risks (to missions, systems and security). Right now the level of trust required for this goal does not seem within reach, but it is certainly a fertile area for future research and development. As mentioned earlier, this raises the interesting question of predictability and whether truly trusted autonomy is even possible.

The fourth V is variety. Like the substantial variety of data Defence has in operating a rather complex business, there also is a great variety of cyber threats that it needs to protect against. For externally sourced threats, this challenge of variety of threats is addressed with a variety of defences. This is a similar approach to Ashby's Law of Requisite Variety [1], in that in cyber security we use multiple tools and multiple sources to help increase the likelihood of 'catching' the different threats.

The bulk of the day to day cyber threats Defence deals with, in terms of pure numbers, are about unsophisticated criminal scams. The organisation must be on its guard to deal not only with these prevalent threats, but with less common threats. The bigger concern is about seeing and dealing with the threats that aren't so obvious and are more dangerous, such as sophisticated malicious code, or the exfiltration of information by such code or by a trusted insider.

Whilst such events may be identified through monitoring, Defence's cyber security approach tends to focus on known threats such as through signature matching. In addition to these mechanisms, better systems need to be developed that characterise the normal environment and can effectively and responsively identify anomalies. Such a capability will help protect Defence networks against the unknown threats. Thus the fourth critical requirement for trusted autonomy in cyber security is to help in the identification of potential cyber threats through monitoring of anomalous activity.

The fifth and final V challenge is variability. Here I diverge from the Big Data view and consider variability more in a macro sense of the word, and this has several dimensions.

Whereas Defence embraces the concept of a Single Information Environment, in reality there are a lot of networks within the Department that are managed by individual business units, and have variable adherence to security requirements. Defence is working to remediate and accredit these networks to reduce vulnerability. Defence is also considering the introduction of a cyber-readiness or cyber-worthiness regime, to regularly test the security of Defence's ICT networks.

Another aspect of variability arises at the application layer, in that the Single Information Environment comprises different applications, and versions of applications, that largely do similar things. The inertia in moving on from legacy applications creates a management burden and results in security risks in operating unsupported systems. Defence continues to work on the rationalisation of legacy systems through the Infrastructure Transformation Program.

Like the Internet and many organisations' systems, Defence networks also have a lot of outdated content. Such 'untidiness' of the environment can impact productivity and is also a security risk. The Enterprise Information Management initiative of Defence's First Principles Review is endeavouring to address this problem.

There are arguments for and against whether variability within the environment contributes to or detracts from cyber security. Some might suggest that variability of systems reduces the impact of an exploit against a particular system. However, my belief is that a more consistent, tidier and disciplined environment is easier to support and to defend.

Another aspect of variability is how critical each application and information service is to Defence business. This then translates into variability for the redundancies, disaster recovery and incident resolution priorities that apply to each information service. Such requirements have been established for all the services and systems Defence supports, with a view to their continuity in the case of a fault. Defence leadership will have to consider in future whether these priorities are right, particularly in respect of recovering from a substantial cyber event. This will come from engagement with Defence's Groups and Services, as well as through exercising and wargaming of cyber events.

One last consideration for variability, and this to me is the most important, is the variability of the organisation's people when it comes to cyber security. Defence requires that its people have a standard of behaviour and awareness that adds to our defence in depth, rather than being a weakness. Despite having standard training for cyber security, practically a range of behaviours can be observed, from cautious to reckless.

So what implication does variability have for the need for trusted autonomy? As per previous discussions about anomalous behaviour, there is a need for improved systems that identify when actions, activities or attributes of the system are unexpected, and potentially to take action to mitigate risk. Additionally, there is a need for a sophisticated understanding of Defence's business, specifically an appreciation that the criticality of services varies across the organisation and thus affect the balance of risks in undertaking cyber defence. Trusted autonomy could contribute to information management and cyber security as a compliance agent, by monitoring the environment, and identifying and analysing variance—thus helping maintain our cyber readiness.

11.4   Conclusion

Right now is a very interesting time to be involved in cyber security, especially in Defence. The Department has a clear direction to strengthen cyber security, whilst at the same time needing to improve the functionality and the accessibility of information services. This demands creation of a solid foundation for information assurance and then cyber security and operations personnel must manage the balancing act of risks and value.

I have outlined the principles of cyber security and future challenges, with a bit of alliteration borrowing from the Big Data V concepts. It is important to recognise that traditional approaches to address these challenges will not be enough. Specifically, future cyber security will need to deal with exponentially growing volumes of information, need to have better awareness of the cyber environment, need to respond quickly to cyber events, need to identify unknown threats and have the ability to understand our complex environment.

Success in meeting these challenges requires a competitive advantage, which most likely will only come with assistance from trusted autonomy. Given the development of trusted autonomy has not to date met expectations, greater investment into research and development in these areas is important in order to keep ahead of cyber threats in future.

References

1. W. Ashby, *Ross: An Introduction to Cybernetics* (Champman & Hall, London, 1956)
2. J.R. Boyd, A discourse on winning and losing. *Air University Document MU43947, Briefing*, 1, 1987
3. DoD, Information Security Manual (2015)
4. DoD, *Australian Defence Force Publication 6.0.3 Information Assurance*, 2nd edn. (2016)
5. DoD, CIOG presentations at DefenceWatch Briefing, 8 Sept 2016, Commonwealth Club Canberra, 2016
6. DoD (2016) http://www.asd.gov.au/infosec/mitigationstrategies.htm
7. DoD (2016) http://www.defence.gov.au/whitepaper/docs/2016-defence-white-paper.pdf, 20 Dec 2016
8. DoD, Internal Defence Statistics, sourced from Defence Strategic Communications Branch (2016)
9. DoD, Internal Defence Statistics, sourced from Enterprise Technology Operations Branch (2016)
10. DoD, Internal Defence Statistics, sourced from ICT Security Branch (2016)
11. M.N. Huhns, D.A. Buell, Trusted autonomy. IEEE Internet Comput. **6**(3), 92 (2002)
12. U. Krogmann, From automation to autonomy-trends towards autonomous combat systems. Technical report, DTIC Document, 2000
13. D. Laney, 3d data management: controlling data volume, velocity and variety. META Group Res. Note **6**, 70 (2001)
14. W. Truszkowski, H. Hallock, C. Rouff, J. Karlin, J. Rash, M. Hinchey, R. Sterritt, *Autonomous and Autonomic Systems: With Applications to NASA Intelligent Spacecraft Operations and Exploration Systems* (Springer Science & Business Media, Berlin, 2009)

Chapter 12
Reinforcing Trust in Autonomous Systems: A Quantum Cognitive Approach

Peter D. Bruza and Eduard C. Hoenkamp

12.1 Introduction

Bad decisions can have dire consequences. From high exposure events such as an oil spill or a plane crash, to the smaller scale drama of a patient who dies on the operating table, the unavoidable question soon follows: Was this mechanical failure or human error? Yet, in a society where people increasingly base their decisions on autonomous systems such as search engines, recommender systems, or social media, the distinction becomes blurred. Although these systems are based on algorithms (less material, but nonetheless mechanical) people will have to process and consider the provided information, thus becoming the weakest link in the decision chain. In general, mechanical failure, once discovered, seems more easily addressed than human error. So if autonomous systems could be made aware of how humans judge information, they could become more judicious in advising humans, and more proactive in the way they present their information. Currently this is not the case. To change this, we have looked into decades of research about human judgement (For the case of how judgement of a particular system is shaped by some of its properties see Sect. 7.6 of this book.). We found a whole range of human judgement that deviates substantially from what would be normatively correct according to logic and probability theory. As example take the famous experiment in which Tversky and Kahneman [35] presented participants with the following text:

> Linda is 31 years old, single, outspoken, and very bright. She majored in philosophy. As a student, she was deeply concerned with issues of discrimination and social justice, and also

P. D. Bruza (✉) · E. C. Hoenkamp
Information Systems School, Queensland University
of Technology (QUT), Brisbane, Australia
e-mail: p.bruza@qut.edu.au

E. C. Hoenkamp
Institute for Computing and Information Sciences,
Radboud University, Nijmegen, The Netherlands
e-mail: hoenkamp@acm.org

© The Author(s) 2018
H. A. Abbass et al. (eds.), *Foundations of Trusted Autonomy*, Studies in Systems,
Decision and Control 117, https://doi.org/10.1007/978-3-319-64816-3_12

participated in anti-nuclear demonstrations Which is more probable:
(a) Linda is a bank teller, or
(b) Linda is a bank teller and is active in the feminist movement?

In this experiment, the participants consistently rated option (b) as more probable than (a). However, according to the axioms of probability theory, the conjunction of events is less probable than a single event, so (b) must be less probable than (a).

This type of judgement errors has been found so invariably that it became known as the conjunction fallacy (see [31, 34] for other such fallacies). These findings are not widely known, let alone implemented, by programmers designing the communication part of autonomous systems. The latter systems so far adhere to the laws of probability and logic, which is their strength. Their weakness, however, is to not account for *how humans make decisions*. For example, an intelligent system that would correctly answer (a) in the experiment above, might bewilder the human who was expecting answer (b). In turn this could erode that human's trust in such a system.

Let us take another, more concrete, example from technology soon to become reality. Suppose you arrive late in the evening at a meeting. As the street is clearly indicated as a tow zone, you let your self-driving car find a parking space elsewhere to park in your stead. Later you come back and find that it parked in the first free spot in that same street. Would this not be the time to reconsider your trust in the autonomy of the car? And so you instruct it not to do this again. But then it politely explains [24] that the tow zone only applies during rush hour, thus restoring your trust. An even better scenario would be that when you leave the car to park itself, it would tell you that yes, this is a tow zone, but only during rush hour. Machines have become smart enough to do the first part, in this case finding a parking space and park. But then, is it not time to work on the second part, where the machine proactively explains its actions from the human's point of view? Or the part where it can foresee a human error because it knows how a human would reason in a particular case? This stands to hugely help humans put trust in autonomous systems, and in the current presentation we show a direction one could go.

To elaborate the problem we want to address, before trying to solve it, consider Wittgenstein's often cited remark "If a lion could talk, we could not understand him" ([37], p. 223). The remark has been food for much thought and speculation (that it is probably true) for half a century, notably in the philosophy of language. We wonder, however, that if we cannot even understand a lion, whence the confidence that we will understand machines when it comes to communicating with them verbally? Here we think that Wittgenstein's own, less quoted, comment can lead the way [37]:

> If language is to be a means of communication there must be agreement not only in definitions but also (queer as this may sound) in *judgements*. This seems to abolish logic, but does not do so — It is one thing to describe methods of measurement, and another to state results of measurement. (p. 391, our italics)

This is an important vantage point for the current presentation: first it emphasizes the role of *judgement*, second it distinguishes the method of *measurement* from its result, and third it challenges the role of *logic*.

We will express these notions in the language of *quantum cognition*, which derives terminology and computations from quantum mechanics. But whereas quantum mechanics deals with physical states, quantum cognition describes cognitive (or mental) states. The Linda experiment can be described in this language, as we showed already in the book we wrote about quantum cognition [10]. For further details we will refer the reader to that book, so that we can use the space here to describe a new experiment. The experiment is also modeled using quantum cognition, laying a foundation for its implementation in future autonomous systems. Incorporating such models can proactively help the user avoid mistakes that are inherent in human judgement and thus prevent an erosion of trust. We contend that in this way the interactions between humans and future autonomous system will become more effective.

12.2   Compatible and Incompatible States

The conjunction fallacy does not mean that people always judge the probability of a conjunction as higher than each of its conjuncts. That would indeed be counter to probability theory. Just imagine that the choice between (a) and (b) was presented without the story about Linda. Then one would expect everyone with some notion of probability to choose (a) over (b), as confirmed in [35]. But when asked the question after first hearing the story, even people schooled in statistics fall victim to the conjunction fallacy. Why is this? The question generated a host of publications with possible explanations over the last several decades (see [19] for an overview). Among the many kinds of explanations offered, two stand out in particular. One assumes that words such as 'and' and 'probability' are misunderstood by the participants, or at least not understood in their formal interpretation. The other assumes a reasoning bias. A recent overview [30] concludes that the latter has the best support of the two. But this answer begs the question: if there is a reasoning bias, where does that reasoning bias originate?

Indeed, we are not satisfied with just knowing there is such a bias, rather we would like to describe how that bias unfolds as a cognitive process. To do so, let us formulate the participants' judgements as the outcome of a decision process. The explanations in the literature almost invariably involve two competing states, one in which Linda is a bankteller, and another where she is a feminist. To most participants in the experiment these states are not compatible, and whether bias or reasoning, each can in principle tip the scale in favor of one state or the other. In order to make headway, we have to take a closer look at the notion of compatible and incompatible states.

In this presentation we will formulate states in the language of quantum cognition. Especially for the reader who is not already familiar with this approach, we will first recall some concepts from quantum mechanics. One such concept is the (formal) notion of compatible and incompatible states. Incompatibility lies at the heart of Heisenberg's famous uncertainty principle. It holds that when we are certain about a quantum particle's momentum, we are necessarily uncertain about its position, and vice versa. Position and momentum are therefore called incompatible states. On the other hand, again given the momentum of that particle, we can still measure its kinetic energy with certainty, momentum and kinetic energy thus being compatible states.

12.3   A Quantum Cognition Model for the Emergence of Trust

There is a vision of nature that readers may have entertained themselves at one time or another, namely that "all phenomena could be explained mechanically if only we knew enough." Those readers are in good company, as this is a direct quote from Leibniz's writing of 1695 [29]. Yet the problems with this vision are several. First, we are not omniscient and second it is uncertain if we will ever have enough computing power to do the explaining. But even if is this may become an issue for intergalactic travel, why would it apply to the Linda experiment? After all, everything the participants in the experiment need to know is given in the instructions, and the computing power needed is an unassuming application of probability theory. So why is an explanation for the experimental findings still wanting?

There is a third, perhaps more fundamental, problem with Leibniz's position: His mechanical view of the laws of nature has been drastically undermined with the advent of quantum mechanics. It turns out that some phenomena require a probability calculus, but different than given by classical probability theory. For example, probabilities may not always add up to 1.

The problem does not just apply to the description of physical systems, but also to cognitive systems, and more in particular those that play a role in interactions between humans and autonomous systems. We can safely assume that such interactions require decisions under uncertainty. For decades cognitive scientists have studied how humans make judgments under such conditions. The theories they produced can be roughly divided into a *heuristic* and a *rational* approach.

The heuristic approach is founded on so-called bounded rationality [33]. It assumes that in order to make decisions, humans use simple heuristics such as representativeness, anchoring-and-adjustment, and base-rate neglect. They support powerful (often inductive) processes that may depend on the environment [22]. In contrast, the rational approach conforms to rules drawn from a theory, most notably Bayes' rule [13] or expected utility theory [32]. In this approach the same basic axioms can be used to derive inferences and utilities across all environmental conditions. Recently, a third approach, called *quantum cognition* has emerged [10, 11, 36]. In common with the heuristic approach, it assumes that the human decision maker is subject to bounded rationality. And in common with the rational approach, inferences used for decisions are derived from basic axioms that define a probability theory. However, the axioms are different from those employed by the Bayesian approach, and consequently, so are the decisions that follow from it.

Two core concepts underpinning quantum cognition are *incompatibility* and *contextuality*. We will briefly come back to 'contextuality' later in this chapter, but right now we will continue with 'incompatibility' from Sect. 12.2 in more detail. Both concepts play an important role in formalizing people's perception of *trust*.

At the time of writing this chapter, a term that gained notoriety was the term *fake news*. Many readers may have realized their quandary over what news can be trusted any longer. We will describe an experiment in which we induced such a quandary and measured how it influenced people's degree of trust. We will then offer a formal explanation in terms of quantum cognition, which at the same time shows that we need to distinguish between two forms of trust.

We could perhaps have presented participants with pieces of news of varying degrees of trustworthiness. However, note the many variables that should then be kept under control, such as the participant's background knowledge or the ephemerality of news. Instead we used pictures of which they had to assess the trustworthiness on a five point scale from untrustworthy to very trustworthy [18]. (Trustworthy was defined as an image that gave an accurate representation of a situation, person or object.) The participants were also asked to supply the reasons for their rating. An example of one such a picture was a smiling face of Putin, which can be seen by following the URL in [7].

The image of Putin was deliberately chosen. Many participants will know that he is a former KGB agent and probably not predisposed to smiling. Would this then lead to uncertainty whether the image had been photoshopped, and a consequent lack of trust that the image is a true an accurate depiction of Putin? When analyzing the qualitative data a curious phenomenon emerged. A considerable number of participants appeared to confuse the decision regarding trust of the image with a decision on whether they trust the *content* of the image. For example, "This is Vladimir Putin, a world leader I associate with dishonesty and distrust, who works to his own agenda and doesn't worry about other people", "I wouldn't trust the person, but the photo is fine", and "I really could not separate what I know about this man from his image".

Assuming the confounding of the trustworthiness decision is a robust cognitive effect, how can it be explained? Quantum cognition offers an explanation based on incompatible decision perspectives. Consider diagram (a) on the left in Fig. 12.1.

This figure depicts two decision perspectives. One is a two-dimensional vector space featuring two orthogonal basis vectors (in black) corresponding to the decision

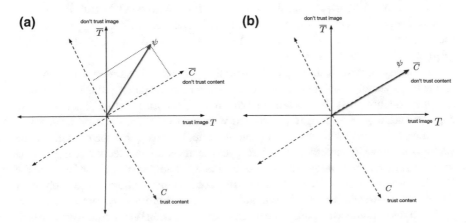

Fig. 12.1 Incompatible decision perspectives explain why trust in the image is sometimes confused with trust in its content (elaborated in the text)

that participant trusts, or doesn't trust the image. We will call this the "image" decision space. The other decision perspective is a two-dimensional vector space (dotted basis vectors) which models the decision whether the participant trusts, or doesn't trust the content of the image. We will call this the "content" decision space. This vector space is rotated with respect to the vector space modeling the decision of the trustworthiness of the image. The red vector represents the cognitive state of the participant when the image is first presented. Note that the cognitive state-vector is superposed with respect to both decision perspectives. If this participant's cognitive state is suspended between the basis vectors of both decision spaces. Superposition in quantum cognition is a major departure to mixed state in a standard probabilistic model. A mixed state implies that the participant is always in one basis state or the the other. For example, with respect to the decision regarding the image, the participant will be either in the state corresponding to the decision that they trust the image or be in the alternative state in which they don't trust the image. We may not know what state they are in, but they must be in one of these two states. It may be the case that in the course of considering the image the cognitive state of the participant moves between these states. Superposition allows the participant to be in *both* states at once. As we shall see, this has a marked effect on the probabilities of the associated states.

Quantum cognition is defined by quantum probabilities which are related geometrically, not by an underlying Boolean algebra over the event space. More formally, let T denote the decision that the participant trusts the image and let $\overline{T}$ denote the decision that the participant does not trust the image. These are the basis vectors of the image subspace. Similarly, let C and $\overline{C}$ denote the basis vectors of the content subspace. The cognitive state ψ is superposed between the two decisions in both the image and content subspaces. The probability that the participant decides they trust the content portrayed in the image is equal to the square of the projection of the cognitive state ψ onto the basis vector C, denoted $\|\mathbf{P}_C \psi\|^2$. We can see from the diagram that the length of this projection is small, hence the probability is small, which reflects the weak predisposition of the participant to trust Vladimir Putin. Conversely, the predisposition not to trust Putin is high because of the long projection of the cognitive state vector ψ onto the basis vector $\overline{C}$. Hence the associated probability $\|\mathbf{P}_{\overline{C}} \psi\|^2 = 1 - \|\mathbf{P}_C \psi\|^2$ is high.

Consider diagram (b) in Fig. 12.1. Note that the cognitive state vector now lies on the basis vector. This models the situation in which the participants have decided that they do not trust the content of the image (Vladimir Putin). Therefore, the cognitive state ψ is no longer superposed with respect to the decision perspective regarding the trustworthiness of the content. Note, however, that the cognitive state vector is necessarily superposed with respect the decision perspective regarding the trustworthiness *of the image*. The reason for this is the degree of rotation between the two decision subspaces. Because these subspaces are not orthogonal to each other the decision perspectives are incompatible. Incompatibility between the decision perspectives explains why participants confuse the decision on the trustworthiness of the image versus content of the image. The reason is that in case of incompatibility the law of total probability does not hold. For example, the probability that the participant will trust the content of the image in terms of the two decision perspectives is as follows:

$$p(C) = \|\mathbf{P}_C \psi\|^2 \tag{12.1}$$

$$= \|(\mathbf{P}_C \cdot I)\psi\|^2 \tag{12.2}$$

$$= \|(\mathbf{P}_C \cdot (\mathbf{P}_T + \mathbf{P}_{\overline{T}}))\psi\|^2 \tag{12.3}$$

$$= \|\mathbf{P}_C\mathbf{P}_T\psi\|^2 + \|\mathbf{P}_C\mathbf{P}_{\overline{T}}\psi\|^2 + \underbrace{\psi^\top\mathbf{P}_{\overline{T}}\mathbf{P}_C\mathbf{P}_T\psi + \psi^\top\mathbf{P}_T\mathbf{P}_C\mathbf{P}_{\overline{T}}\psi}_{Int} \tag{12.4}$$

The preceding rendered in standard probability theory looks like the law of total probability, which is being modified by extra term Int:

$$p(C) = p(C, T) + p(C, \overline{T}) + Int \tag{12.5}$$

Int is referred to as the "interference term". The interference term can positively or negatively contribute to the probability $p(C)$. For this reason, incompatible subspaces have been put forward as a natural explanation why human beings do not adhere to the law of total probability like in the conjunction and other so called fallacies in human decision making [12]. When the interference term is zero, the law of total probability holds. This happens when the decision perspectives are compatible.

Incompatible decision perspectives are a recent development in cognitive modeling and their striking characteristic is the use of "quantum" probabilities. By quantum probabilities, we mean that the decision event space is modelled as a vector space rather than a Boolean algebra of sets. A key differentiator is the use of the interference term. When this term is non-zero, violations of the law of total probability occur. The interference term has been used in models of the perception of gestalt images [15, 28], models of the conjunction and other decision fallacies [12, 14], modeling violations of rational decision theory [6, 28, 31], modeling belief dynamics [34] and conceptual processing [3–5, 20, 21]. In this presentation there is no need to go any deeper into the fine points where quantum cognition and the Bayesian approach part company. Therefore we only present Table 12.1 as a summary, and refer the interested reader to [9] for further details.

Placed in a psychological context, the uncertainty principle becomes relevant because a person's understanding of two events, such as two different perspectives

Table 12.1 Comparing Bayesian and Quantum cognition

Bayesian cognition	Quantum cognition
Human is in a definite cognitive state	Human can be in a superposed state
Events are compatible	Events may be incompatible
Law of total probability	May violate law of total probability
Models are non-contextual	Models may be contextual

on a matter, requires changing from one point of view to another, and the two points of view can imply incompatibility. In other words, the uncertainty principle entails that it is not possible to be simultaneously decided on both the image and content with respect to assessing trustworthiness. Just like it is not possible to form a joint probability of both the position and momentum of a quantum particle, it is not possible for the human participant to form a joint probability across decisions, e.g., whether they both trust the image and the content of the image. Not being able to form joint probabilities signals the presence of *contextuality*. A well studied example of contextuality is the curious phenomenon of entanglement: Empirical observations are collected in four measurement settings of a system of two quantum particles such as photons. Each of the four settings yields a pairwise joint probability distribution which models the observations made in that measurement setting. An entangled system is deemed "contextual" because it is not possible to combine these four pairwise joint probability distributions into a single probabilistic model such that the four pairwise empirical distributions are marginal distributions of this global model. Even though contextuality manifests within the sub-atomic realm, there is a growing body of research which is exploring whether contextuality manifests in cognition and related areas (e.g., [1, 2, 8, 9, 16, 17]). In the context of our example, contextuality arises because the image and content decision perspectives cannot be meaningfully combined into a single joint distribution.

12.4 Conclusion

In the foreseeable future, humans and autonomous systems will engage in shared decision making. Given the discrepancies between the way they arrive at decisions, whose form of rationality should be given precedence: human or machine?

One form of rationality can be termed "Bayesian rationality" in which Bayesian probability theory provides such powerful models of both human and machine cognition, that it is sometimes called "Bayesian fundamentalism" [26]. We made the point in this chapter that humans often do not adhere to Bayesian rationality, but rather to a "quantum rationality," as it is based on the the same Dirac-von Neumann axioms as quantum theory. Neither of these two rationalities should be given precedence. We argued that quantum rationality is often more suited to model human decision making, and Bayesian rationality more to model decision making by machines. So the more pertinent question is how to best align these rationalities so that shared decision making between human and machine becomes more effectual than that of each in isolation.

There are good examples where traditional (often brute force) mechanisms were unable to solve difficult problems, and which became tractable when augmented with cognitive elements. One famous example is the breaking of the Enigma code by augmenting brute force methods with simple aspects of human communication [23]. A second example is the incorporation of human reasoning in chess programs. And finally we have already seen this clearly confirmed in our own research in areas outside that of this chapter [24, 25].

Therefore, we propose to augment the rationality already available in the current systems with quantum rationality. We expect that future autonomous systems can compute decisions based on both rationalities and hence detect situations when these decisions do not align. In such cases the machine could make the human aware of the discrepancy, thus preventing a potential erosion of trust between human and machine.

References

1. S. Abramsky, Contextual semantics: from quantum mechanics to logic, databases, constraints, and complexity. Bull. EATCS **2** (113) (2015)
2. S. Abramsky, M. Sadrzadeh, Semantic unification: a sheaf theoretic approach to natural language, in *Categories and Types in Logic, Language and Physics: Essays Dedicated to Jim Lambek*, ed. by C. Casadio, B. Coecke, M. Moorgat, P. Scott (Springer, Berlin, 2014)
3. D. Aerts, Quantum structure in cognition. J. Math. Psychol. **53**(5), 314–348 (2009)
4. D. Aerts, L. Gabora, S. Sozzo, Concepts and their dynamics: a quantum-theoretic modeling of human thought. Top. Cogn. Sci. **5**(4), 737–772 (2013)
5. R. Blutner, E.M. Pothos, P.D. Bruza, A quantum probability perspective on borderline vagueness. Top. Cogn. Sci. **5**(4), 711–736 (2013)
6. R.F. Bordley, Violations of compound probability principles: toward a generalized Heisenberg uncertainty principle. Oper. Res. **46**, 923–926 (1998)
7. A. Borowitz, Putin announces historic G1 summit http://goo.gl/Hd8Jhp. Apr 2017
8. P.D. Bruza, K. Kitto, B. Ramm, L. Sitbon, A probabilistic framework for analysing the compositionality of conceptual combinations. J. Math. Psychol. **67**, 26–38 (2015)
9. P.D. Bruza, Z. Wang, J.R. Busemeyer, Quantum cognition: a new theoretical approach to psychology. Trends Cogn. Sci. **19**(7), 383–393 (2015)
10. J. Busemeyer, P.D. Bruza, *Quantum Cognition and Decision* (Cambridge University Press, Cambridge, 2012)
11. J. Busemeyer, E.M. Pothos, Can quantum probability provide a new direction for cognitive modeling? Behav. Brain Sci. **36**, 255–327 (2013)
12. J. Busemeyer, E.M. Pothos, R. Franco, J.S. Trueblood, A quantum theoretical explanation for probability judgment errors. Psychol. Rev. **118**(2), 193–218 (2011)
13. N. Chater, J.B. Tenenbaum, A. Yuille, Probabilistic models of cognition: conceptual foundations. Trends Cogn. Sci. **10**(7), 287–291 (2006)
14. E. Conte, A. Khrennikov, O. Todarello, R. De Robertis, A. Federici, J. Zbilut, On the possibility that we think in a quantum mechanical manner: an experimental verification of existing quantum interference effects in cognitive anomaly of conjunction fallacy. Chaos Complex. Lett. **4**(3), 123–136 (2011)
15. E. Conte, O. Todarello, A. Federici, F. Vitiello, M. Lopane, A. Khrennikov, J. Zbilut, Some remarks on an experiment suggesting quantum-like behaviour of cognitive entities and formulation of an abstract quantum mechanical formalism to describe cognitive entity and dynamics. Chaos, Solitons & Fractals **33**, 1076–1088 (2007)
16. E. Dzhafarov, J. Kujala, Contextuality is about identity of random variables. Physica Scripta **T163**, 014009 (2014)
17. E. Dzhafarov, R. Zhang, J. Kujala, Is there contextuality in behavioral and social systems? Philos. Trans. Royal Soc. A **374**, 20150099 (2015)
18. L. Fell, P.D. Bruza, S.K. Devitt, G. Oliver, M. Gradwell, H. Partridge, The cognitive decision space of trust: an exploratory study of image trustworthiness and the propensity to deceive, http://eprints.qut.edu.au/102009/ (2016)
19. J.E. Fisk, Conjuction fallacy, in *Cognitive Illusions: A Handbook on Fallacies and Biases in Thinking, Judgement and Memory*, ed. by R.F. Pohl (Psychology Press, Hove, UK, 2004), pp. 23–42

20. L. Gabora, D. Aerts, Contextualizing concepts using a mathematical generalization of the quantum formalism. J. Exp. Theor. Artif. Intell. **14**, 327–358 (2002)
21. L. Gabora, E. Rosch, D. Aerts, Toward an ecological theory of concepts. Ecol. Psychol. **20**, 84–116 (2008)
22. G. Gigerenzer, P.M. Todd, *Simple Heuristics That Make Us Smart* (Oxford University Press, Oxford, 1999)
23. A. Hodges, *Alan Turing: The Enigma* (Walker, New York, 2000)
24. E.C.M. Hoenkamp, P.D. Bruza, How everyday language can and will boost effective information retrieval. J. Am. Soc. Inf. Technol. **66**(8), 1546–1558 (2015)
25. E. Hoenkamp, Why information retrieval needs cognitive science: a call to arms, in *Proceedings of the 27th Annual Conference of the Cognitive Science Society*, pp. 965–970 (2005)
26. M. Jones, B.C. Love, Bayesian fundamentalism or enlightenment? on the explanatory status and theoretical contributions of Bayesian models of cognition. Behav. Brain Sci. **34**, 169–231 (2011)
27. D. Kahneman, *Thinking, Fast and Slow* (Farrar, Straus and Giroux, New York, 2011)
28. A. Khrennikov, *Ubiquitous Quantum Structure: From Psychology to Finance* (Springer, Berlin, 2010)
29. G.W. Leibniz, Specimen dynamicum, in *Philosophical Papers and Letters*, ed. by L.E. Loemker (Springer, The Netherlands, Dordrecht, 1989), pp. 435–452
30. R. Moro, On the nature of the conjunction fallacy. Synthese **171**(1), 1–24 (2009)
31. E.M. Pothos, J. Busemeyer, A quantum probability explanation for violations of 'rational' decision theory. Proc. Royal Soc. B **276**(1165), 2171–2178 (2009)
32. L.J. Savage, *The Foundations of Statistics* (Wiley, Hoboken, 1954)
33. H.A. Simon, *Models of Man: Social and Rational* (Wiley, Hoboken, 1957)
34. J.S. Trueblood, J. Busemeyer, A quantum probability explanation for order effects on inference. Cogn. Sci. **35**, 1518–1552 (2011)
35. A. Tversky, D. Kahneman, Extensional versus intuitive reasoning: the conjunctive fallacy in probability judgment. Psychol. Rev. **90**, 293–315 (1983)
36. Z. Wang, J. Busemeyer, The potential of using quantum theory to build models of cognition. Topics Cogn. Sci. **5**(4), 672–688 (2013)
37. L. Wittgenstein, *Philosophical Investigations* (Basil Blackwell, Oxford, 1953)

Chapter 13
Learning to Shape Errors with a Confusion Objective

Jason Scholz

13.1 Introduction

Automated systems typically operate on their own only if they are simple, or within a closed and managed environment. In situations when things can go wrong with serious consequences, control is typically introduced through a human-in-the-loop. Trusted Autonomous Systems (TAS) are different in that they will be required to operate in open and unmanaged environments and when human guidance may not be available. To achieve machine perception in open environments, machine learning classifiers are likely critical components in TAS. However, in order for such systems to be worthy of their trust, users and designers will want to ensure that classification errors that may result in high-consequence impact can be managed; in effect, to introduce a bias in the machine analogous to a learned human bias.

Much recent statistical machine learning [11] involves learning weights in complex, modular non-linear artificial neural networks. These weights are derived from the gradient of the objective (or loss) function. This gradient is back-propagated (Rummelhart et al. 1986) throughout the whole system. In supervised classification, the objective function typically uses some relationship between the machine's output estimated class and the training set true class. The squared error and cross entropy are popular example functions. The choice of objective function instructs the machine by reward for a correct estimate, and lack thereof for incorrect estimates. These objective functions reward all correct choices and punish all incorrect choices equally. The choice of objective function affects error distribution and in high consequence situations some types of errors may be undesirable. When errors are made, a user wants them to be of least consequence, in effect, the *lesser of evils*. There may be more at stake than simply choosing the 'best' overall performance.

Consider as illustrative, the following humourous exchange from the 20th Century Fox film 'Master and Commander: The Far Side of the World', when in the officers mess Captain Aubrey points to a plate of bread and asks Dr Maturin,

J. Scholz (✉)
Defence Science and Technology Group, Joint and Operations Analysis Division,
Box 1500 Edinburgh, SA, Australia
e-mail: jason.scholz@defence.gov.au

© The Author(s) 2018
H. A. Abbass et al. (eds.), *Foundations of Trusted Autonomy*, Studies in Systems,
Decision and Control 117, https://doi.org/10.1007/978-3-319-64816-3_13

Do you see those two weevils doctor?
I do.
Which would you choose?
(*sighs annoyed*) *Neither; there is not a scrap a difference between them. They are the same species of Curculio.*
If you had to choose. If you were forced to make a choice. If there was no other response...
(*Exasperated*) *Well then if you are going to push me...* (*the doctor studies the weevils briefly*) *...I would choose the right hand weevil; it has... significant advantage in both length and breadth.*
(*the captain thumps his fist in the table*) There, I have you! You're completely dished! Do you not know that in the service... one must always choose the lesser of two weevils.
(*the officers burst out in laughter*)

The consequences of a deep network incorrectly classifying a pedestrian as a rubbish bin may have dire consequences in a potential driver-less vehicle accident, yet incorrectly classifying a post box as a rubbish bin may be of lesser concern. Further, the nature of those consequences will be application specific and thus empirically driven. In mission and safety critical applications when the consequences of error are high, the semantic coding of errors for a machine to learn more and less important costs is an important problem. One would ideally like the designer and/or user of systems to be able to specify a profile for acceptability of error consequences for the system application at hand and is critical to trust justifications. Errors will be a reality for deployed network solutions. Despite the performance claims of Deep Networks and the rush to multi-layer engineering with GPU implementations, the vast majority of techniques remain brittle. Small changes can result in magnified errors undermining trustworthiness in the accuracy and competence of such systems. This brittleness is evident when a network trained on given data sets is exposed to independently-derived dataset. Seewald [12] illustrates an order of magnitude in error degradation when a classifier trained on MNIST and USPS datasets is tested on the independently-derived DIGITS handwritten digits dataset. Further, it appears that adversarial examples may be created that have deleterious error effects on practically all forms of network architecture, both deep and shallow [6]. Although adversarial forms appear to be rare in naturally-occurring data, they can be easily generated by an adversary in ways that are indistinguishable to humans [15], which may undermine trustworthiness in the integrity of the system.

The literature on 'cost-sensitive classification' (CSS) is concerned with class distribution imbalance, typical of situations when the training data is limited or not necessarily statistically representative, and for classifying rare and important classes (e.g. medical diagnosis of rare diseases). Despite this specific focus, the techniques may be applicable to objective functions for more general deliberate error shaping, as they are generally formulated for a cost C of confusing the actual class i with estimated class j and probability of class j given training example x,

$$L(x, i) = \sum_j P(j|x)C(i, j) \tag{13.1}$$

Approaches to CSS include reweighting (stratifying) available training data so that more costly errors will incur a larger overall cost [5], cost-sensitive boosting [13], [4] that combine multiple weak or diverse learners, or by changing the learning algorithm. Those that alter the learning algorithm appear to focus on either decision tree classifiers where cost sensitivity is achieved by pruning [10], or Support Vector Machines (SVM) that all use a hinge loss approximation of the cost-sensitive loss function [9]. A cost sensitive learning algorithm applicable as an augmentation to deep networks appeared only recently in the literature [7] and as with all the CSS has not considered the broader problem of shaping the entire error distribution, but only to make up for imbalance in the input class distribution.

The feasibility of a simple augmentation to networks to learn user-defined error profiles is examined next on the basis of a new proposed maximum likelihood objective function. The new technique is developed and explained in the context of binomial and multinomial regression. An implementation using Google's TensorFlow is then studied. A range of experiments are conducted with several independent data sets to ascertain the degree of control over error distributions where the prior distribution is unknown, on a shallow and a deep network architecture and findings are discussed.

13.2 Foundations

A derivation of maximum likelihood classifiers to match a target error profile follows. Noting, in this preliminary study, there is not yet any established theory as to how the overlaps in classification can be traded off, so the examination will necessarily be empirical.

In supervised learning, the Error Matrix, or Confusion Matrix indicates the correct and error classifications across categories from either training or test data. Each row represents the number of instances in an actual (true) class and each column represents the number of instances in an estimated class. The term confusion is used, as this representation quickly shows how often one class is confused with another. Type I errors refer to a true class X being incorrectly classified as a different class Y and is indicated in the non-diagonal values in the rows of the confusion matrix. Type II errors refer to a classified class being X when the true class was Y, and is indicated in the non-diagonal column values of the confusion matrix.

Logistic Regression for the binomial case (two class) and multinomial [3] have been well studied in the literature. However, in order for the reader to comprehend the gradient function for the pairing of the multi-class logistic function also termed 'softmax' [14] function with alternate objective functions, it is necessary to derive these from first principles. Multinomial logistic regression is also extended to the Gaussian case to gain insight and contrast the form of the gradient function with our proposed objective. The following forms a foundation for the proposed technique for multinomial softmax regression on the confusion matrix to follow in the next section.

13.2.1 Binomial Logistic Regression

Consider a network with input $\mathbf{x}_n$ for the nth data presentation, weight vector $\mathbf{w}$, and bias vector $\mathbf{b}$.

$$\mathbf{z}_n = \mathbf{w}^T \mathbf{x}_n + \mathbf{b} \tag{13.2}$$

Consider a simple binary decision classifier, with a logistic function non-linearity,

$$y_n = \sigma(z_n) = \frac{1}{1 + e^{-z_n}} \tag{13.3}$$

In training the network with N examples, the probability of the true class t given the estimated class y output for training case n is,

$$p(\mathbf{t}|\mathbf{y}) = \prod_{n=1}^{N} p(t_n|y_n) = \prod_{n=1}^{N} y_n^{t_n} (1 - y_n)^{1-t_n} \tag{13.4}$$

To maximise this probability, it is equivalent to minimise the negative log likelihood termed the loss function,

$$E = -\log p(t|y) = -\sum_{n=1}^{N} \log p(t_n|y_n) = \sum_{n=1}^{N} t_n \log y_n + (1 - t_n) \log(1 - y_n) \tag{13.5}$$

When the gradient of the loss function is zero, this corresponds to the minimum.

$$\frac{\partial E}{\partial y_n} = -\frac{t_n}{y_n} + \frac{1 - t_n}{1 - y_n} = \frac{y_n - t_n}{y_n(1 - y_n)} \tag{13.6}$$

Now consider the chain rule,

$$\frac{\partial E}{\partial \mathbf{w}} = \frac{\partial E}{\partial y_n} \frac{\partial y_n}{\partial z_n} \frac{\partial z_n}{\partial \mathbf{w}} \tag{13.7}$$

Differentiating (13.2),

$$\frac{\partial y_n}{\partial z_n} = y_n(1 - y_n) \tag{13.8}$$

Thus from (13.6),

$$\frac{\partial E}{\partial \mathbf{w}} = \frac{y_n - t_n}{y_n(1 - y_n)} y_n(1 - y_n)\mathbf{x_n} = (y_n - t_n)\mathbf{x_n} \tag{13.9}$$

The logistic function matches the negative log likelihood function to achieve a cancellation in terms and a very simple expression for the gradient and thus the minimum point, which occurs precisely when $y_n = t_n$.

13.2.2   Multinomial Logistic Regression

Consider the network as per (13.1), independently duplicated in structure for each class j,

$$z_{nj} = \mathbf{w}_j^T \mathbf{x}_n + b_j \qquad j = 1, 2, .., K \qquad (13.10)$$

Consider a multi-class decision classifier, with the generalised logistic or 'softmax' function non-linearity, for data presentation n and class j as follows,

$$y_{nj} = \sigma(z_{nj}) = \frac{e^{z_{nj}}}{\sum_{c=1}^{K} e^{z_{nc}}} \qquad j = 1, 2, .., K \qquad (13.11)$$

Softmax is thus the logistic function extended to one output for each class. This generalises the previous binary case where $K = 2$. Noting each class output is normalised with respect to all other classes, and therefore is not independent.

As before, in training the network with N examples, the probability of the true class t (a 'one-hot' vector) given the estimated output y for training case n and class j might be described with a multinomial probability mass function, normalised such that $\sum_j t_j = 1$

$$p(\mathbf{t}|\mathbf{y}) = \prod_{n=1}^{N} p(\mathbf{t_n}|\mathbf{y_n}) = \prod_{n=1}^{N}\prod_{j=1}^{K} y_{nj}^{t_{nj}} \qquad (13.12)$$

To maximise this likelihood, it is equivalent to minimise the negative log likelihood,

$$E(\mathbf{w}) = -\log p(\mathbf{t}|\mathbf{y}) = -\sum_{n=1}^{N} \log p(\mathbf{t_n}|\mathbf{y_n}) = -\sum_{n=1}^{N}\sum_{j=1}^{K} t_{nj} \log y_{nj}(\mathbf{w}_j) \quad (13.13)$$

This log-likelihood function is often termed cross entropy and can be derived from the Kullback-Liebler divergence. Where the gradient of the negative log likelihood function is zero, $\frac{\partial E}{\partial \mathbf{w}} = 0$ corresponds to the minimum. This is pursued in stages,

$$\frac{\partial E}{\partial y_{nk}} = -\frac{t_{nk}}{y_{nk}} \qquad (13.14)$$

Recalling the softmax function normalises with regard to other classes, means that the derivative of the softmax has two cases,

$$\frac{\partial y_{nk}}{\partial z_{nj}} = \begin{cases} y_{nk}(1 - y_{nk}) & j = k \\ -y_{nj}y_{nk} & j \neq k \end{cases} \tag{13.15}$$

The chain rule must account for K class paths, however, the expression may be simplified according to $j = k$ and $j \neq k$ parts only,

$$\frac{\partial E}{\partial \mathbf{w}} = \frac{\partial E}{\partial y_{nk}} \frac{\partial y_{nk}}{\partial z_{nk}} \frac{\partial z_{nk}}{\partial \mathbf{w}_k} + \sum_{j:j \neq k} \frac{\partial E}{\partial y_{nj}} \frac{\partial y_{nj}}{\partial z_{nj}} \frac{\partial z_{nj}}{\partial \mathbf{w}_j} \tag{13.16}$$

$$= -\frac{t_{nk}}{y_{nk}} y_{nk}(1 - y_{nk})\mathbf{x_n} + \sum_{j:j \neq k} \frac{t_{nj}}{y_{nj}} y_{nj} y_{nk}\mathbf{x_n} \tag{13.17}$$

$$= t_{nk}\mathbf{x_n}(y_{nk} - 1) + y_{nk}\mathbf{x_n} \sum_{j:j \neq k}^{K} t_{nj} \tag{13.18}$$

$$= \mathbf{x_n} \left(-t_{nk} + y_{nk} \sum_{j=1}^{K} t_{nj} \right) \tag{13.19}$$

Thus,

$$\frac{\partial E}{\partial \mathbf{w}} = \mathbf{x_n} (y_{nk} - t_{nk}) \tag{13.20}$$

This is the same form as for the binary case. As Eq. (13.16) makes clear from the cancellation of terms, the gradient of the softmax function *perfectly* matches the gradient of the log likelihood objective function, conspiring so as to form a significant reduction. Only one element of the training and the output is used, representing the single true (kth) condition.

Having established this foundation, alternative objective functions are considered next.

13.2.3 Multinomial Softmax Regression for Gaussian Case

Consider the case where y are Gaussian distributed. Such a distribution may not be an unreasonable approximation in the case of the limit for very large training sets. This will allow investigation of the less-than-perfect match of softmax function with this new objective function.

$$p(\mathbf{t}|\mathbf{y}) = \prod_{n=1}^{N} p(\mathbf{t_n}|\mathbf{y_n}) = \prod_{n=1}^{N} \prod_{j=1}^{K} e^{\frac{-(y_{nj}-t_{nj})^2}{2\pi\sigma^2}} \tag{13.21}$$

To maximise this likelihood, it is equivalent to minimise the negative log likelihood,

$$E(\mathbf{w}) = -\log p(\mathbf{t}|\mathbf{y}) = -\sum_{n=1}^{N} \log p(\mathbf{t_n}|\mathbf{y_n}) = \frac{1}{2\pi\sigma^2} \sum_{n=1}^{N}\sum_{j=1}^{K} \left(t_{nj} - y_{nj}(\mathbf{w}_j)\right)^2$$

(13.22)

Instead of the logarithmic relationship to the outputs y, previously, there is now a square law relationship.

As previously, the gradient of the negative log likelihood function is zero, $\frac{\partial E}{\partial \mathbf{w}} = 0$ corresponds to the minimum. Which is pursued in stages,

$$\frac{\partial E}{\partial y_{nk}} = t_{nk} - y_{nk}$$

(13.23)

Using Eqs. (13.14) and (13.15) as before,

$$\frac{\partial E}{\partial \mathbf{w}} = (t_{nk} - y_{nk})\, y_{nk}(1 - y_{nk})\mathbf{x_n} - \sum_{j:j\neq k}^{K} (t_{nj} - y_{nj})y_{nj}\, y_{nk}\mathbf{x_n}$$

(13.24)

$$= -\mathbf{x_n} y_{nk}\left((y_{nk} - t_{nk}) - \sum_{j=1}^{K}(y_{nj} - t_{nj})y_{nj}\right)$$

(13.25)

Firstly, note the sign of the gradient is reversed with respect to (13.19). Second, as the minimum is achieved only when the gradient is zero, this occurs *iff* $t_{nj} = y_{nj}$, $\forall j$. This is notably different to the previous form (13.19), which showed no dependency to any other than the kth estimates of y and training values t. Indeed, even if $t_{nk} = y_{nk}$ a gradient remainder is left over.

Notably, if a one-hot vector is used for training (which is usual for multinomial classifiers), then $t_{nj} = 0$ if $j \neq k$ so,

$$\frac{\partial E}{\partial \mathbf{w}} = -\mathbf{x_n} y_{nk}\left((y_{nk} - t_{nk})(1 - y_{nk}) - \sum_{j=1:j\neq k}^{K} y_{nj}^2\right)$$

(13.26)

This clearly shows that the incorrect estimate outputs from the right-hand term in (13.25) introduce an irreducible offset into the gradient. However, if the classifier is working well then $y_{nj}^2 << 1$ and thus this term may be small.

13.3 Multinomial Softmax Regression on Confusion

A proposal to effect control over the confusion matrix is examined. The confusion matrix summarises the overall distribution of examples from trained classes $\mathbf{t}$ versus estimated classes $\mathbf{y}$. Consider the potential to learn a user-defined target confusion matrix distribution $\mathbf{u}$. Using Bayes theorem this means maximising,

$$p(\mathbf{u}|\mathbf{t}, \mathbf{y}) = \frac{p(\mathbf{t}, \mathbf{y}|\mathbf{u})\, p(\mathbf{u})}{p(\mathbf{t}, \mathbf{y})} \tag{13.27}$$

Noting the matrix product of $\mathbf{t}$ and $\mathbf{y}$ defines the confusion matrix for some representative data batch size N, writing the likelihood as,

$$p(\mathbf{t}, \mathbf{y}|\mathbf{u}) = \prod_{i=1}^{K} \prod_{j=1}^{K} \left(\sum_{m=1}^{N} t_{im} y_{mj} \right)^{u_{ij}} \tag{13.28}$$

Unlike previous formulations this likelihood directly mixes the estimator output and training classes within the argument of the logarithm. The negative log likelihood is thus,

$$E = -\log p(\mathbf{t}, \mathbf{y}|\mathbf{u}) = -\sum_{i=1}^{K} \sum_{j=1}^{K} u_{ij} \log \left(\sum_{m=1}^{N} t_{im} y_{mj} \right) \tag{13.29}$$

In this form it resembles the multinomial regression objective, with an important distinction in the log over a training-weighted sum of estimated y values. Where the gradient of the negative log likelihood function is zero, $\frac{\partial E}{\partial \mathbf{w}} = 0$ corresponds to the minimum. Which as previously is pursued in stages,

$$\frac{\partial E}{\partial y_{nj}} = -\sum_{i=1}^{K} \left(\frac{u_{ij} t_{in}}{\sum_{m=1}^{N} t_{im} y_{mj}} \right) \tag{13.30}$$

Equations (13.14) and (13.15) similarly apply,

$$\begin{aligned}
\frac{\partial E}{\partial \mathbf{w}} = & -\sum_{i=1}^{K} \left(\frac{u_{ik} t_{in}}{\sum_{m=1}^{N} t_{im} y_{mk}} \right) y_{nk} (1 - y_{nk}) \mathbf{x_n} \\
& + \sum_{j:j \neq k}^{K} \left\{ \sum_{i=1}^{K} \left(\frac{u_{ij} t_{in}}{\sum_{m=1}^{N} t_{im} y_{mj}} \right) y_{nj} y_{nk} \mathbf{x_n} \right\}
\end{aligned} \tag{13.31}$$

$$= \mathbf{x_n} y_{nk} \sum_{i=1}^{K} \left(\frac{u_{ik} t_{in}}{\sum_{m=1}^{N} t_{im} y_{mk}} \right) (y_{nk} - 1)$$

$$+ \mathbf{x_n} y_{nk} \sum_{j:j \neq k}^{K} \sum_{i=1}^{K} \left(\frac{u_{ij} t_{in}}{\sum_{m=1}^{N} t_{im} y_{mj}} \right) y_{nj} \tag{13.32}$$

$$\frac{\partial E}{\partial \mathbf{w}} = \mathbf{x_n} y_{nk} \left(\sum_{j=1}^{K} \sum_{i=1}^{K} \left(\frac{u_{ij} t_{in} y_{nj}}{\sum_{m=1}^{N} t_{im} y_{mj}} \right) - \sum_{i=1}^{K} \frac{u_{ik} t_{in}}{\sum_{m=1}^{N} t_{im} y_{mk}} \right) \tag{13.33}$$

This does not reduce further. If $t_{in} = 0$ for all but the $i = k$ training element (one hot training vector) then from (13.31),

$$\frac{\partial E}{\partial \mathbf{w}} = \mathbf{x_n} y_{nk} \left(\sum_{j=1:j \neq k}^{K} \left(\frac{u_{kj} t_{kn} y_{nj}}{\sum_{m=1}^{N} t_{km} y_{mj}} \right) + (y_{nk} - 1) \frac{u_{kk} t_{kn}}{\sum_{m=1}^{N} t_{km} y_{mk}} \right) \tag{13.34}$$

The right hand term is zero iff $y_{nk} = 1$. This will only be the case if all $y_{nj} = 0, \forall j \neq k$. Indeed the left hand term is zero if the latter condition is true, making the left and right hand terms coupled. An alternative condition that makes the left hand term equal to zero is if $u_{kj} = 0, \forall j \neq k$ that is, all non-diagonals of the user-defined matrix are zero (i.e. u is the identity matrix). Classification performance cannot be expected to equal that of the optimum multinomial case except in this special case. Similar to the Gaussian case prior, then the gradient cannot reach a minimum. Thus *any* choice of user-defined error confusions will compromise overall performance. A question remaining is to what degree will that performance be compromised and under what conditions might that compromise be acceptable?

13.4 Implementation and Results

An implementation was chosen to make an empirical study of classifier performance. The MNIST digits database [8], was chosen due to its requiring a multinomial classifier with modest computational requirements. The error objective that a user or designer might want could take a large variety of forms, so instead of attempting some exhaustive approach, a few likely error-shaping scenarios were chosen. The plan for the empirical testing to follow is then:

- Simple Classifier Evaluation

  - 'Baseline' to establish whether the confusion objective performs as well as the standard classifier under conditions when a user does not care about the error distribution.

- 'Error trading' to examine the capacity to trade error types I and II related to specific classification classes.

• Deep Network Classifier Evaluation

- Repeat of 'error trading' used for the simple network to examine if a deeper classifier structure provides a greater capacity to trade error types I and II.
- 'Adversarial Errors' examines the potential to thwart classification decisions by producing deliberate errors.

The first candidate was a simple regression network from the TensorFlow tutorial [1]. TensorFlow is an open source software library well suited to fast multidimensional data array (tensor) processing and allows deployment of computation on multiple CPUs or GPUs. The simple network uses 784 (28 × 28 pixels) inputs and n = 10 output nodes, each output employing a softmax non-linear function (corresponding to MNIST digit categories 0–9). As is typical with this dataset, 60,000 training images and 10,000 test images were used. A batch size of m=200 was maintained throughout. Only the objective function and optimizer were modified, with the 10 × 10 user-defined error distribution tensor, **u** according to the specific applied test. Given the batch of true labels y_

```
y_ = tf.placeholder(tf.float32,[None,10])
```

Using the Adaptive Gradient optimizer,

```
train_step = tf.train.AdagradOptimizer(0.01).minimize(loss)
```

The following implements the extended square loss,

```
loss = tf.reduce_sum(tf.square(u − tf.matmul(y_,y,True,False)))
```

The following implements the extended cross-entropy loss,

```
loss = −tf.reduce_sum(u*tf.log(tf.matmul(y_,y,True,False)))
```

Noting * represents element-wise multiplication of tensors.

Seewald [12] compared machine learning performance on three independently sourced handwritten digits databases: MNIST [8]; USPS, the US Postal Service 'zip' codes reduced to individual digits[1]; and DIGITS, Seewald's own collection from school students. The value of independent data rests in the fact that the offline machine training may differ from the real world when a system is deployed. By studying the proposed error redistribution learner on independent data, its true robustness may be studied.

Table 13.1 shows the results of a baseline test to compare the simple classifier overall performance for cross entropy, square loss and the confusion objective functions as detailed in previous sections. The latter novel objective employed a 10 × 10 identity matrix as the user objective. The network was trained on the MNIST 60,000

---

[1]http://statweb.stanford.edu/~tibs/ElemStatLearn/datasets/zip.digits/.

Table 13.1 Comparison of classification performance, trained on MNIST data only

Objective/test data	Cross entropy	Square loss	Confusion objective as identity
MNIST	0.926	0.933	0.926
USPS	0.771	0.767	0.755
DIGITS	0.687	0.688	0.682

Table 13.2 Test 1. Confusion matrix for MNIST test data with a simple softmax regression network and standard cross-entropy objective function

Digit	0	1	2	3	4	5	6	7	8	9	Total	$p(i\|i)$
0	965	0	2	1	0	2	7	1	2	0	980	0.985
1	0	1106	4	5	1	2	4	2	11	0	1135	0.974
2	12	3	921	13	14	4	12	12	38	3	1032	0.892
3	2	1	22	916	0	21	4	10	25	9	1010	0.907
4	1	2	4	0	926	0	11	2	5	31	982	0.943
5	8	1	2	30	12	780	16	5	30	8	892	0.874
6	12	3	4	2	9	13	910	1	4	0	958	0.950
7	2	7	23	6	10	0	0	948	3	29	1028	0.922
8	4	3	7	15	10	21	10	12	888	4	974	0.912
9	11	5	2	11	32	14	0	13	8	913	1009	0.905
Total	1017	1131	991	999	1014	857	974	1006	1014	997	10000	N/A

image set only but tested on the MNIST 10,000 image test set, USPS 2,007 image test set and the entire DIGITS 4389 images. Table 13.1 demonstrates the expected degradation in performance for unseen data sets reported by Seewald, and that the performance of the novel objective function is comparable under baseline conditions.

The confusion matrix is shown in Table 13.2 with actual digit categories by row and estimated digit categories by column for MNIST training only on the 10,000 image test data set.

As illustrated in the row totals, not all digits are equally likely in the training data and, of course, not all were equally 'easy' for the network to recognise. Correct identification of a digit '0' was highest at $p = 0.985$ and correct identification of a '5' lowest at $p = 0.874$. The five most significant error confusions are (2,8), (9,4), (4,9), (5,8), (5,3).

13.4.1 Error Trading

In the following tests the capacity to trade errors related to a specific class (arbitrarily the digit 4) was examined. Table 13.3 summarises error trading tests with the simple network trained on MNIST data only and tested on MNIST, USPS and DIGITS

Table 13.3 Summary of five error-trading tests across three test data sets, noting the simple network was trained on MNIST training data *only*

Test objective	MNIST test data				USPS test data				DIGITS test data			
	Avg Acc	Prob (4,4)	T-I errs	T-II errs	Avg Acc	Prob (4,4)	T-I errs	T-II errs	Avg Acc	Prob (4,4)	T-I errs	T-II errs
Test 1. Standard x-entropy	0.93	0.94	56	88	0.77	0.79	42	17	0.69	0.82	79	91
Test 2. One-hot vector u(4,4) = 5	0.92	0.98	23	193	0.74	0.9	20	49	0.68	0.93	33	205
Test 3. u: r4 = 0 c4 = 0 u(4,4) = 5 flr = 0.05	0.89	0.97	34	196	0.74	0.84	33	33	0.63	0.88	53	157
Test 4. u: c4 = 0 u(4,4) = 1 flr = 0.05	0.89	0.77	229	10	0.74	0.75	50	15	0.63	0.77	101	45
Test 5. u: c4 = 0 u(4,4) = 2 flr = 0.05	0.89	0.87	126	32	0.74	0.78	44	19	0.63	0.84	72	84

Table 13.4 Test 2. Confusion matrix for MNIST test data illustrates the effect of increasing the one-hot vector for a digit '4' with respect to others

Digit	0	1	2	3	4	5	6	7	8	9	Total
0	960	0	1	2	5	2	7	1	2	0	980
1	0	1103	4	5	1	2	4	2	14	0	1135
2	10	1	913	13	25	4	13	11	38	4	1032
3	2	1	22	918	4	20	1	9	22	11	1010
4	1	1	2	0	959	0	6	2	3	8	982
5	8	2	1	30	26	773	16	5	25	6	892
6	15	3	5	2	13	19	896	0	5	0	958
7	2	7	25	5	19	0	1	943	0	26	1028
8	6	4	8	15	15	18	10	12	881	5	974
9	11	4	4	10	85	13	0	16	10	856	1009
Total	1015	1126	985	1000	1152	851	954	1001	1000	916	10000

data sets. Beginning with test 1, as the error benchmark, which is the standard cross entropy objective function, Table 13.1 shows the probability of correctly classifying a digit 4 given a digit 4 was presented, the Type I error related to the row for the digit 4 and Type II errors related to column for the digit 4, for each test data set. The results of test 1 clearly show that the nature of the errors across independent data sets is highly varied.

In test 2, the relative value of one diagonal cell in the user-defined matrix u with respect to all others was changed. This effect may be achieved by scaling one-hot vector unit values. The result of increasing the weight for digit 4 to be five times higher than any other individual digit is illustrated in Table 13.4. For the MNIST test, the average accuracy was 0.920 down slightly from 0.927. This reduced the errors across the row for the digit 4 from 56 to 23 errors, but in the column for digit 4, errors increased from 88 to 193. A reduction in type I errors is achieved at the expense of an increase in type II errors. This is shown in detail with the confusion matrix for the MNIST test data in Table 13.4.

As an analogy, consider a hunter who has a bias that improves detecting a 'lion' in long grass. This bias is better for the survival of the hunter to predation, but comes at the cost of more often perceiving a lion when in fact there was not one, which may have other consequences.

If this test is illustrative of the behaviour to be expected for a classifier, is it possible to exert more shaping control over type I and type II errors as trade for average error performance? Table 13.3 (test 3) summarises results of raising the 'floor' (flr) for diagonals in u as a compensatory allowance. A setting of $flr = 0.05$ was chosen to represent a desire to keep errors low everywhere (reasoning 0.05 is near to zero), but it is most desired to reduce errors in combinations of row '4' and column '4' where cells were set to zero. Table 13.4 illustrates the resulting confusion matrix for test. Test 3 was typical of attempts to change the same major row and column (i = j). The

Table 13.5 Test 5. Confusion matrix for MNIST test data illustrates the ability to reduce type II errors (column '4') by trading a moderate increase in type I errors (row '4')

Digit	0	1	2	3	4	5	6	7	8	9	Total
0	940	1	1	6	0	6	8	3	13	2	980
1	0	1097	1	3	0	10	5	0	19	0	1135
2	7	17	869	23	3	14	9	30	46	14	1032
3	1	5	19	902	0	24	3	17	27	12	1010
4	5	21	9	11	856	6	16	6	23	29	982
5	6	8	5	21	2	772	18	17	30	13	892
6	9	6	7	2	6	33	883	2	8	2	958
7	3	28	21	11	2	4	3	910	7	39	1028
8	12	17	12	23	4	31	9	13	836	17	974
9	17	10	3	25	15	24	7	24	17	867	1009
Total	1000	1210	947	1027	888	924	961	1022	1026	995	10000

result was in increase in both type I *and* type II errors compared with manipulating one hot vector values only. Table 13.3 shows across test data sets a relative increase in Type I errors and decrease in Type II errors compared with test 2. For the MNIST test data the most errors related to cells (9,4) and/or (4,9). Driving down errors for one of these cells had the direct effect of raising errors in the other: a kind of 'pivot' effect. The strong and direct relationship between errors appeared to be only very weakly controllable. Evidence of these pivots appear in the unshaped confusion matrix in the form of the larger and symmetrical errors in Table 13.2. Pivots were observed at (4,9):(9,4), (5,8):(8,5), (5,3):(3,5), (7,9):(9,7) for MNIST tests. Notably these pivots were different, for each of the three test data sets.

In test 4, the capacity to target a reduction in type II errors *only* for a specific classification was examined. Table 13.3 summarises the result which had a strong desired effect, at the expense of an increase in type I errors. Test 5 demonstrates a final level of 'tuning' where the type II errors were generally lower than those of test 1, with only a small increase in Type I errors. The confusion matrix for test 5 with MNIST test data is shown in Table 13.5.

Tests 4 and 5 demonstrate a level of control over errors that is not possible by one-hot vector manipulation and could not be achieved without shaping the objective function. The set of tests also demonstrates an ability to target reduction in Type I or Type II errors at the expense of the other, and was sustained across two unforeseen independent data sets.

These results led to the question 'if pivots identified in the initial confusion matrix are excluded from consideration, can significant control over errors in other arbitrary rows and columns be achieved?' For the tests attempted, the answer appears to be yes. Table 13.6 illustrates the results of an attempt to minimise errors in the column for a digit '4' (type II) and a row for a digit '7' (type I) for MNIST test data. The user defined matrix, *u* column representing a '4' contained zeros, and the row

Table 13.6 Test 6. Confusion matrix result for MNIST test data with the aim to 'minimise errors in row 7 without excessive impact on column 4 and the overall error average'

Digit	0	1	2	3	4	5	6	7	8	9	Total
0	925	0	1	5	0	6	4	25	11	3	980
1	0	1094	1	4	0	7	5	5	19	0	1135
2	6	15	855	20	3	9	12	72	34	6	1032
3	1	7	19	870	0	21	3	53	23	13	1010
4	7	21	12	7	848	7	19	10	22	29	982
5	7	8	6	16	2	740	14	63	29	7	892
6	7	7	7	1	5	35	862	27	7	0	958
7	3	8	11	7	5	0	1	979	0	14	1028
8	10	15	11	22	4	24	9	48	816	15	974
9	14	8	2	12	15	16	4	73	17	848	1009
Total	980	1183	925	964	882	865	933	1355	978	935	10000

representing a '7' contained zeros, with $u(4,4)=2$, $u(7,7)=2$, $u(i,i)=1$ if i4 and a floor $u(i,j)$ otherwise equal to 0.05. The overall test accuracy was 0.88. Notably the most significant errors at $(9,4)$ and $(7,9)$ were suppressed as desired. The ability to push these errors lower would not come without more significant cost to the overall error rate. Type II errors increased in column '8'.

In practice, users will likely want to achieve a lower misclassification rate for some arbitrary chosen cell targets relevant to their problem choice. Empirical investigation appears to confirm this is possible providing data pivot points are avoided. This is discussed later.

13.4.2 Performance Using a Deep Network and Independent Data Sources

The deep convolutional network (Convnet) chosen was from the TensorFlow tutorial.[2] The Convnet chosen uses a hierarchy of three macro-levels, each level comprises a convolutional layer, rectified linear unit layer, max pool layer, and drop out layer. At the top of all this, there is an output processing layer termed 'softmax' or normalised exponential, making 13 layers in total. This provides significantly better performance than the simple network. When trained on MNIST data, this network provided a 99.2% average classification accuracy which is near to the current state of the art of 99.77% [2].

As a comparison, the earlier successful aim to 'minimise errors in row 7 without excessive impact on column 4 and the overall error average' (test 6) is revisited.

---

[2]https://www.tensorflow.org/versions/r0.9/tutorials/mnist/pros/index.html.

Table 13.7 MNIST test errors (softmax with logits)

Digit	0	1	2	3	4	5	6	7	8	9	Total
0	751	0	32	0	5	2	6	2	2	3	803
1	1	754	52	0	72	2	13	0	7	0	901
2	0	0	834	2	1	0	3	1	4	0	845
3	1	0	50	719	2	47	0	8	4	2	833
4	0	0	7	0	793	0	0	0	3	9	812
5	5	0	11	7	2	683	3	4	14	2	731
5	10	0	0	0	85	12	679	0	3	0	789
7	1	4	117	2	5	0	0	714	4	10	857
8	8	3	27	6	9	16	5	3	709	4	790
8	0	5	24	9	31	2	0	18	26	716	831
Total	777	766	1154	745	1005	764	709	750	776	746	8192

In this case it was performed by training the Convnet on *USPS training data only* and then tested on USPS test data, MNIST test data and DIGITS test data. For the experiment 1280 USPS training images only were used. Test data used included 8192 MNIST test images, 1024 USPS test images, and 4096 DIGITS test images. The results are shown in Tables 13.7, 13.8, 13.9, 13.10, 13.11 and 13.12 inclusive. Tables 13.7, 13.9, and 13.11 show the confusion matrix achieved using the standard classifier on MNIST, USPS and DIGITS test databases with average error rates 0.897, 0.957, and 0.692, respectively. These errors are contrasted with Tables 13.8, 13.10, and 13.12 which show the achieved error performance for test 6 using the new objective function on MNIST, USPS and DIGITS test databases with average error rates of 0.784, 0.893 and 0.538 respectively. As expected, test 6 resulted in some increase in the average errors in all cases. Notably, the misclassifications for each of these databases show significantly *different* characteristics to what is now seen trained on USPS data. Pivot points for the MNIST test are different to those earlier, now with confusions (7,2) and (6,4) as most significant. Pivot points for the DIGITS test have significant confusions (7,2), (9,3), (1,4) and (1,7). Examining the objective to minimise errors in row 7, this is demonstrated in all three cases. Error confusion (7,2) were reduced from 117 to 31, and 220 to 130 in the MNIST and DIGITS tests and remained the same at 2 for the USPS test. Errors for column 4 increased from 212 to 460 and 204 to 278 for MNIST and DIGITS tests, and from 3 to 20 for the USPS test. This largely fulfilled the aim of the test, however, there are several notable 'surprises'. Table 13.8 shows significant error increases for (0,2), (0,6), and (1,6) compared with Table 13.7. Table 13.12 shows significant error increases for (0,2), (2,6) and (4,9).

Table 13.8 MNIST test errors for the objective to minimise row 7 errors (test 6)

Digit	0	1	2	3	4	5	6	7	8	9	Total
0	475	2	192	0	23	2	113	1	1	13	822
1	0	636	5	0	159	1	135	0	10	0	946
2	0	0	757	8	31	1	17	14	14	1	843
3	3	1	101	618	3	47	3	21	4	16	817
4	0	0	0	0	775	0	1	10	1	11	798
5	6	0	5	35	5	613	26	11	0	17	718
6	0	0	5	0	157	1	625	0	0	3	791
7	0	3	31	5	11	1	0	765	8	19	843
8	3	1	20	44	23	64	93	27	485	30	790
9	1	3	16	6	48	3	0	58	12	677	824
Total	488	646	1132	716	1235	733	1013	907	535	787	8192

Table 13.9 USPS test errors (softmax with logits)

Digit	0	1	2	3	4	5	6	7	8	9	Total
0	197	0	0	0	0	0	1	1	1	0	200
1	0	128	0	0	0	0	2	1	0	0	131
2	0	1	87	0	1	1	0	1	3	0	94
3	0	0	2	79	0	4	0	0	0	0	85
4	1	2	1	0	99	0	0	0	0	3	106
5	2	0	0	1	0	73	0	0	0	0	76
6	0	0	0	0	1	2	80	0	0	0	83
7	0	0	1	0	1	0	0	58	0	0	60
8	1	2	1	1	0	1	0	0	83	0	89
9	0	1	0	0	0	0	0	2	1	96	100
Total	201	134	92	81	102	81	83	63	88	99	1024

13.4.3   Adversarial Errors

Control over the occurrence of type I errors is most strikingly demonstrated by training a network to create a deliberate confusion. This may be a requirement for a cyber operations system. Table 13.13 illustrates the result to cause a deliberate confusion of a digit '7' for a '0' created by a user matrix. As expected the average error drops by 10% to 0.829. Notably this network learns never to output a '7' and instead classifies a '7' as a '0' precisely as desired. Further, the "normal" detection statistics of other categories is largely unaffected.

To produce type II errors in a specific cell of row i and column j, proved more difficult. It is necessary to maintain at least some small value for $u(j, j)$ to ensure correct classifications of i are allowed as reducing $u(j, j)$ to zero will have the

Table 13.10 USPS test errors for the objective to minimise row 7 errors (test 6)

Digit	0	1	2	3	4	5	6	7	8	9	Total
0	165	2	1	1	3	0	13	1	4	0	190
1	0	130	0	0	1	0	4	0	0	0	135
2	3	0	72	6	6	0	0	0	11	0	98
3	0	0	2	73	1	3	0	2	3	0	84
4	0	0	0	0	88	0	0	4	0	1	93
5	2	0	0	0	1	67	6	2	4	1	83
6	0	0	0	0	3	1	87	0	0	0	91
7	0	0	0	0	2	0	0	72	0	0	74
8	2	0	0	2	1	1	1	1	81	1	90
9	0	0	0	0	2	0	0	4	1	79	86
Total	172	132	75	82	108	72	111	86	104	82	1024

Table 13.11 DIGITS errors (softmax with logits)

Digit	0	1	2	3	4	5	6	7	8	9	Total
0	372	0	17	0	1	3	3	1	11	0	408
1	1	66	8	0	169	0	0	152	1	11	408
2	6	1	366	0	1	1	13	2	20	0	410
3	0	15	13	298	3	63	0	13	4	1	410
4	0	1	0	0	383	9	7	2	2	3	407
5	2	1	2	4	0	371	6	6	16	5	413
6	5	3	0	0	4	27	363	0	2	0	404
7	0	13	220	0	9	0	0	136	10	16	404
8	2	14	42	12	7	23	11	1	304	1	417
9	6	8	22	120	10	10	0	28	33	178	415
Total	394	122	690	434	587	507	403	341	403	215	4096

effect of reducing the entire column to zero, as per the example in Table 13.13, with the resulting distribution of the classification estimates redistributed without user control, into the next most similar categories. Table 13.14 illustrates a best effort test (error average 0.806) achieved with the aim of 'creating errors in cell (2,5) and minimising correct classifications in cell (5,5)'. As can be seen in Table 13.14, the desired effect is achievable, however, the consequence of the desired weakness in correctly classifying (5,5) produces more significant errors at both (5,3) and (5,8), which may be the 'next weakest' points.

Table 13.12 DIGITS errors for the objective to minimise row 7 errors (test 6)

Digit	0	1	2	3	4	5	6	7	8	9	Total
0	235	2	101	2	15	0	52	6	0	3	416
1	2	28	9	0	165	0	17	127	4	55	407
2	0	4	204	1	3	00	127	5	67	0	411
3	0	9	87	253	1	16	13	19	2	12	412
4	0	6	1	0	294	2	17	11	0	77	408
5	0	1	3	16	0	302	52	11	0	19	404
6	1	0	4	0	0	9	381	0	0	11	406
7	0	20	130	16	69	0	0	160	3	15	413
8	0	1	26	26	3	80	75	6	175	19	411
9	1	10	6	90	22	8	0	83	17	171	408
Total	239	81	571	404	572	417	734	428	268	382	4096

Table 13.13 Example confusion matrix for adversarial error creation. Most occurrences of a digit '7' are classified as the digit '0' as desired

Digit	0	1	2	3	4	5	6	7	8	9	Total
0	956	0	1	1	0	4	11	0	7	0	980
1	3	1107	4	3	1	2	5	0	10	0	1135
2	10	2	933	14	12	3	15	0	39	4	1032
3	15	1	20	917	2	22	2	0	21	10	1010
4	2	2	4	0	927	0	11	0	6	30	982
5	30	1	2	26	15	762	15	0	32	9	892
6	19	3	5	0	6	13	907	0	5	0	958
7	929	9	24	9	13	3	1	0	4	36	1028
8	23	3	8	17	8	17	11	0	883	4	974
9	25	4	0	11	30	13	1	0	8	917	1009
Total	2012	1132	1001	998	1014	839	979	0	1015	184	10000

13.5 Discussion

The hypothesis that a more expressive deep network would be significantly more capable of supporting an arbitrary redistribution of errors than a shallow network was not demonstrated in these tests. It does appear possible to trade errors under limited conditions towards arbitrarily-chosen errors which, as expected from theoretical study of the gradients, comes at some cost to the total error rate. The degree of control that can be exercised appears limited by pivots in the presented data. These pivots are in turn dependent on training data and the actual data used in practice. As demonstrated by use of three independent data sets, the 'confusion training' to meet an objective *performed best* when the training and live data were drawn from independent sources.

Table 13.14 Adversarial error result for the aim to 'create errors in cell (2,5) and minimise correct classifications in cell (5,5)'

Digit	0	1	2	3	4	5	6	7	8	9	Total
0	949	0	1	1	0	7	13	1	8	0	980
1	0	1096	3	8	0	6	5	1	16	0	1135
2	5	0	660	5	6	319	8	7	19	3	1032
3	2	1	18	893	1	43	5	10	28	9	1010
4	2	3	1	1	896	13	10	1	14	41	982
5	40	9	1	248	46	39	37	26	425	21	892
6	14	2	1	2	6	22	891	0	20	0	958
7	3	15	41	1	11	14	1	902	2	38	1028
8	13	10	5	22	14	33	9	10	842	16	974
9	8	6	9	14	42	8	0	15	18	889	1009
Total	1036	1142	740	1195	1022	504	979	973	1392	1017	10000

A critical question is 'what other loss functions might one choose, and what are the implications for that choice?' [16], Williamsion argues that a proper composite loss (objective) needs to control convexity (geometrical properties) and control statistical properties. Considering logistic loss, the deliberate *perfect* match of the logistic or softmax characteristic and the multinomial log likelihood characteristic that yields a collapse in complexity of the gradient was not afforded by our proposed choice of objective function with softmax. Yet, perhaps it remains possible to find an alternate functional composition that would allow a similar simplification?

The adversarial generation of confusions was convincingly demonstrated with the 'user specified confusion' objective function. The confusion objective function is very good at *creating* errors where they are wanted.

13.6 Conclusion

A technique has been demonstrated with the ability to learn to shape errors for both a shallow and deep network based on a novel maximum likelihood 'confusion objective' function. Results were demonstrated for some limited but useful cases in trading type I and type II errors, maintaining error objectives across independent and unforseen data sets, and an ability to create adversarial confusions. Next steps might include tests on examples with a significantly larger number of classes, and the derivation of bounds of the gradient minima for the technique to provide further insights. The technique appears sufficiently promising to warrant more thorough statistical analysis of sensitivity of total errors to specific error objectives.

Acknowledgements Thanks to: Mr. Darren Williams, Dr. Glenn Moy and Dr. Darryn Reid for reviewing earlier drafts and encouragement; Prof. Hussein Abbass for timely and critical feedback on the draft; and Dr. Alexander Seewald for kindly supplying USPS and DIGITS data sets.

References

1. M. Abadi, A. Agarwal, P. Barham, E. Brevdo, Z. Chen, C. Citro, G. Corrado, A. Davis, J. Dean, M. Devin, et al., Tensorflow: Large-scale machine learning on heterogeneous distributed systems. (2016) *arXiv preprint* arXiv:1603.04467
2. D. Ciregan, U. Meier, J. Schmidhuber, Multi-column deep neural networks for image classification. *2012 IEEE Conference on Computer Vision and Pattern Recognition (CVPR)* (IEEE, 2012), pp. 3642–3649
3. J.N. Darroch, D. Ratcliff, Generalized iterative scaling for log-linear models. *The Annals of Mathematical Statistics* (1972), pp. 1470–1480
4. P. Domingos, Metacost: A general method for making classifiers cost-sensitive. *Proceedings of the fifth ACM SIGKDD International Conference on Knowledge Discovery and Data Mining* (ACM, 1999), pp. 155–164
5. C. Elkan, The foundations of cost-sensitive learning. *International Joint Conference on Artificial Intelligence*, vol. 17 (LAWRENCE ERLBAUM ASSOCIATES LTD, 2001), pp. 973–978
6. I.J. Goodfellow, J. Shlens, C. Szegedy, Explaining and harnessing adversarial examples (2014). *arXiv preprint* arXiv:1412.6572
7. S.H. Khan, M. Bennamoun, F. Sohel, R. Togneri, Cost sensitive learning of deep feature representations from imbalanced data (2015). *arXiv preprint* arXiv:1508.03422
8. Y. LeCun, L. Bottou, Y. Bengio, P. Haffner, Gradient-based learning applied to document recognition. Proc. IEEE **86**(11), 2278–2324 (1998)
9. Y. Lee, Y. Lin, G. Wahba, Multicategory support vector machines: Theory and application to the classification of microarray data and satellite radiance data. J. Am. Stat. Assoc. **99**(465), 67–81 (2004)
10. S. Lomax, S. Vadera, A survey of cost-sensitive decision tree induction algorithms. ACM Comput. Surv. (CSUR) **45**(2), 16 (2013)
11. J. Schmidhuber, Deep learning in neural networks: An overview. Neural Netw. **61**, 85–117 (2015)
12. A.K. Seewald, On the brittleness of handwritten digit recognition models. *ISRN Machine Vision*, 2012 (2011)
13. Y. Sun, M.S. Kamel, A.K.C. Wong, Y. Wang, Cost-sensitive boosting for classification of imbalanced data. Pattern Recogn. **40**(12), 3358–3378 (2007)
14. R.S. Sutton, A.G. Barto, *Reinforcement Learning: An Introduction* (MIT press Cambridge, 1998)
15. P. Tabacof, E. Valle. Exploring the space of adversarial images (2015). *arXiv preprint* arXiv:1510.05328
16. R. Williamson. *Loss Functions*, vol. 1 (Springer, 2013), pp. 71–80

Chapter 14
Developing Robot Assistants with Communicative Cues for Safe, Fluent HRI

Justin W. Hart, Sara Sheikholeslami, Brian Gleeson, Elizabeth Croft, Karon MacLean, Frank P. Ferrie, Clément Gosselin and Denis Laurandeau

14.1 Introduction

The Collaborative Advanced Robotics and Intelligent Systems (CARIS) laboratory at the University of British Columbia has focused on the development of autonomous

---

Justin W. Hart: Work completed while affiliated with the Department of Mechanical Engineering at University of British Columbia.

J. W. Hart (✉)
Department of Computer Science, University of Texas at Austin, Austin, USA
e-mail: justinhart@utexas.edu

J. W. Hart · S. Sheikholeslami · E. Croft
Department of Mechanical Engineering, University of British Columbia,
Vancouver, Canada
e-mail: sara.sheikholeslami@alumni.ubc.ca

E. Croft
e-mail: elizabeth.croft@ubc.ca

B. Gleeson · K. MacLean
Department of Computer Science, University of British Columbia, Vancouver, Canada
e-mail: brian.gleeson@gmail.com

K. MacLean
e-mail: maclean@cs.ubc.ca

F. P. Ferrie
Department of Electrical and Computer Engineering, McGill University,
Montreal, Canada
e-mail: ferrie@cim.mcgill.ca

C. Gosselin
Department of Mechanical Engineering, Laval University, Quebec City, Canada
e-mail: clement.gosselin@gmc.ulaval.ca

D. Laurandeau
Department of Electrical Engineering, Laval University, Quebec City, Canada
e-mail: denis.laurendeau@gel.ulaval.ca

© The Author(s) 2018
H. A. Abbass et al. (eds.), *Foundations of Trusted Autonomy*, Studies in Systems,
Decision and Control 117, https://doi.org/10.1007/978-3-319-64816-3_14

robot assistants which work alongside human workers in manufacturing contexts. A robot assistant is a device that is intended to aid and support the activities of a human worker, rather than working entirely autonomously - as a welding or painting robot in a work cell at a car factory would, or entirely under human control - as a teleoperated robot would. The creation of such robot assistants presents a number of challenges to the scientists and engineers designing them. A traditional robotic arm operating in a factory work cell is physically separated from human workers by a thick glass wall secured by steel beams that are often marked with bright orange paint. This creates an environment where the robot's operation is safe, because it is away from the people who must be kept safe from it, and where it is capable of performing its tasks with minimal input because they are repetitive and not subject to variations such as those introduced by typical human behavior. Many of the technological barriers to move these robots past the physical barriers of the work cell involve establishing a clear cycle of back-and-forth communication between the robots that would work alongside humans and their human collaborators. It is important for workers collaborating with robots to know what mode each robot is in and what actions they may take.

The purpose of a robot assistant is to aid and support the actions of a human worker. Robot assistants, by necessity, are not constrained to the confines of the work cell. For a robot assistant to be effective, it is important that the worker not be overly burdened by the task of controlling it, allowing the worker to concentrate on their portion of the shared task. It is also important that the robot assistant and human workers be able to safely operate in close proximity. Contributing to efforts to accomplish these goals, the CARIS laboratory focuses on the identification and exploitation of natural communicative cues, explicit or non-explicit, which can be implemented into the interfaces of robotic systems. These are human behaviors which communicate information to other people. Explicit cues are intentional behaviors performed with the purpose of communication, such as when one performs a hand gesture [6, 8, 9, 21]. Non-explicit cues are unintentionally performed, but broadcast intentions or other important information, such as when a person looks toward where they are about to hand an object to another person [1, 17, 23]. We posit that naturalistic communicative cues allow users to quickly benefit from collaboration with robots with minimal effort and training, and that through such communicative cues, it is possible to develop interfaces that are safe, transparent, natural, and predictable. To achieve this goal, we use a three phase design process. In the first phase, studies are performed in which two people collaborate on a shared goal. Recordings of these interactions are annotated in order to identify communicative cues used between the study participants in performing their tasks. In the second, these cues are described or mathematically modeled. In the third phase, they are implemented in robotic systems and studied in the context of a human-robot interaction.

Due to the close physical proximity between human workers and robot assistants, physical communicative cues and physical human-robot interaction play an important role in this research. An example of an interaction involving naturalistic communicative cues and physical human-robot interaction is human-robot handovers. The CARIS Laboratory has invested significant effort in the development of natural, fluent human-robot handovers [2–4, 9, 10, 17]. Handover behaviors include not only visual cues [1, 2, 9, 10, 17, 23], but also important force and haptic cues [3, 4]. Other physical interaction studies, which have been carried out in conjunction with the Sensory Perception & INteraction (SPIN) Laboratory at UBC, include interactions in which participants tap and push on the robot to guide its behavior [5]. Our collaborators at Laval University developed an elegant backdrive mechanism for the robot assistant (discussed in Sect. 14.2), which allows users to interactively pose the device during interaction by pushing on the robot itself.

When people work together, their efforts become coordinated and fluently mesh with each other. Many prototype collaborative robot systems work by monitoring progress toward task completion, which can create a stop-and-go style interaction. Hoffman and Breazeal [11] propose a system in which a robot performs predictive task selection based on confidence based on estimates of the validity and risk of each task selection in order to improve interaction fluency. Moon et al. [17] demonstrate that a the performance of a brief gaze motion can improve the speed and perceived quality of robot-to-human handovers. A study by Hart et al. [10] found that study participants receiving an object via handover reach toward the location of a handover prior to the completion of the handover motion, indicating the importance of short-term predictions in human handover behavior. In current work, we are investigating systems which make short-term motion cue based predictions of human behavior in order to fluently and predictively act on these cues, rather than acting on completed motion gestures or task state. The intention of this work is to not only improve the efficiency with which tasks are carried out, but also the fluency of the interaction.

In order to assure the relevance of this work to manufacturing, studies at the CARIS Laboratory are grounded in real-world scenarios. Partnering with manufacturing domain experts through industrial partners provides the us and our collaborators with important insights into the applicability of our research to actual manufacturing scenarios, practices and operations. Recently the CARIS Laboratory completed a three year project that was carried out with academic partners at the Sensory Perception & INteraction (SPIN) Laboratory at UBC, the Artificial Perception Laboratory (APL) at McGill University, and the Computer Vision and Digital Systems Laboratory and Robotics Laboratory at Laval University. Importantly, General Motors of Canada served as our industrial partner in this endeavor. This project, called the Collaborative Human-Focused Assistive Robotics for Manufacturing, or CHARM project, investigated the development of robot assistants in supporting workers performing assembly tasks in an automotive plant. Partnering with General Motors provided us with manufacturing domain experts who could guide our vision based on industry best practices and provide insights based on real-world experience in manufacturing.

This chapter presents an overview of work in the CARIS Laboratory towards the construction of robot assistants. Section 14.2 provides an overview of the CHARM project highlighting the interdisciplinary nature of work in autonomous human-robot interaction and describe the robot-assistant developed for this project. Section 14.3 describes the methodology by which we study communicative cues, from observing them in human-human interactions to description and modeling to experiments in human-robot interaction. Section 14.4 presents experimental findings that were made using this methodology in projects related to CHARM conducted at UBC. Section 14.5 describes current and future directions in which we are taking this research and conclude.

14.2 CHARM - Collaborative Human-Focused Assistive Robotics for Manufacturing

The construction of robot assistants is a large undertaking involving work in several disciplines from the design of the robot itself, to sensing and perception algorithms, to the human-robot interaction design describing how the robot should behave when performing its task. To complete the construction of a state-of-the-art working system, we formed a group of laboratories specializing in these various disciplines and structured work on the robot assistant in such a way that it allowed each lab to independently pursue projects that highlight each group's expertise, while working towards the common goal of constructing an integrated robot assistant system. Insurance of progress towards a common goal was established through several avenues. At the start of the project, the plan of work was divided into a set of three research streams and three research thrusts, with teams of investigators assigned to areas which highlighted the strengths of their laboratories. Doing so allowed a large degree of freedom in the selection and pursuit of individual research projects, while carefully describing the relationships of these streams and thrusts to each other assured that the overarching research efforts of the group remained unified. Coordination of research efforts was maintained through regular teleconferences and inter-site visits by research personnel. These meetings culminated in annual integration exercises referred to as Plugfests, in which each team's developments over the past year were integrated onto a common robot assistant platform. For each Plugfest, the team agreed upon a shared collaborative assembly task to be performed by a dyad comprising the robot assistant and a human collaborator that would be implemented as a unified team spanning the research teams from each institution. This shared task would highlight the research of each group over the past year, and would be used in order to evaluate overall progress towards the shared goal of developing a robot assistant for use in a manufacturing operation. Extensive evaluation and discussion with manufacturing domain experts from General Motors provided the academic partners with the knowledge and expertise to understand and direct the impact and applicability of the work in their laboratories work in real-world usage.

14.2.1 The Robot Assistant, Its Task, and Its Components

Operating on a common platform and regular integration enabled us to make concrete progress toward our goal by constructing a real robot assistant that works on a real-world problem. The development of autonomous robot assistants will require advancements in more than one contributing area. Without better sensing, robots will be unable to detect, track, and model all of the objects required. Without better control, they will not be able to physically interact with people in ways that users will expect them to. Without advances in human-robot interaction, they will neither understand nor be understood by their human collaborators. It was important for partners on the CHARM project to be chosen from a diverse fields representative of the challenges of the project, given the autonomy to pursue the research necessary for advancement, and be coordinated enough to remain focused on their shared goal. Keeping each other apprised of our progress through monthly teleconferences and regular email exchanges fostered an environment of collaboration between each subteam's individual efforts. As each Plugfest approached, the group would begin to focus on ways in which software could be integrated in onto the common platform. Yearly integration into the shared robot-assistant provided a real-world system to act as the testbed for these integration efforts.

14.2.1.1 Car Door Assembly

Car door assembly was chosen as the shared task for the human operator and robot assistant to collaborate on because it is presently one of the most complex and labor-intensive operations in vehicle assembly. The development of technology in this area could potentially both improve the effectiveness of workers in assembly and relieve them of portions of the assembly task that are ready to be automated. Through a series of off-site meetings, teleconferences, and site visits to vehicle manufacturing facilities, two objectives for car door assembly were identified:

- To reduce error rates and improve manufacturing quality
- To maintain or improve worker safety.

A plan was formulated to improve manufacturing quality by interactively passing parts from the robot to the worker. The robot assistant would present parts to the worker assembling a car door in the sequence in which they should be inserted, and tools in the sequence corresponding to the parts of the assembly presently worked on. This would ensure that the worker attaches the correct parts to the car door at the correct times, even as different models come down the line or as the assembly process changes. An improvement in quality would be accomplished through this process by reducing worker error rates, preventing the need to redo work that was performed incorrectly and reducing the error rate in finished products. By the end of the CHARM project, parts were presented to a worker in the test environment in an interactive fashion using an elegant handover controller which was developed by the CARIS Laboratory, with scenarios that could be either programmed into a State Machine Controller (SMC) or reasoned about automatically using a planner developed by CARIS for CHARM.

Worker safety was identified as an important concern not because present practices are unsafe, but because of two factors. The first is that safety is a priority in the both the

Fig. 14.1 The robot-assistant supporting the activities of a worker performing a simulated car door assembly task

laboratory and manufacturing environments. None of the stakeholders in this project would want to be part of a project that put the safety of study participants, factory workers, or themselves at risk. The second is that the concept of having workers working directly alongside robots is still an evolving one, in which safety standards are currently emerging. Progress on this is proceeding at a pace that observes both an appropriate level of caution and an acknowledgement that we must embrace physical human-robot interaction in order to achieve our ultimate goal of having robots which work directly alongside humans.

14.2.1.2   Robot Assistant Hardware

The robot assistant, also referred to as the Intelligent Assist Device, Fig. 14.1, in its final form consists of a robotic arm (Kuka LWR-4) with a dextrous gripper (Robotiq gripper) mounted to an overhead gantry. These devices operate on a single controller, and redundancy resolution is implemented for the entire IAD system. The IAD can be treated as a single, integrated device, rather than as its individual components. The system implements variable stiffness in all actuators, providing compliance along the robot's entire kinematic chain. This compliance contributes to worker safety by allowing the system to softly collide and interact with obstacles, and serves as an interface to the system, allowing the entire unified device to be backdriven by pushes and shoves against the robot itself.

Sensing is performed using Microsoft Kinect and PrimeSense RGBD cameras. The Kinect cameras surround the work cell in which the human and robot assistant collaborate on their shared assembly task. Their point clouds are merged using custom software, providing a global view of the scene. The PrimeSense camera is physically smaller, and focuses on a narrower field of view. It is mounted to the Kuka LWR robotic arm, allowing the system to focus on objects to be manipulated, and providing a directional sensor which can be used for tasks such as object identification and manipulation. One benefit of such as system is that it is able to perform non-contact sensing for uninstrumented human workers and parts. No markers are required to be placed on any worker in the work cell or any part that is to be manipulated using current techniques.

14.2.1.3   Robot Assistant Software

The system presented at Plugfest III represents the most complete version of the robot assistant software developed for the CHARM project. It features many important components which echo familiar components in modern robotics design. The entire system operates over a high speed computer network with nodes responsible for sensing, situational awareness, control, and planning. Nodes operating on the robot assistant communicate via a communications protocol called DirectLink. One feature of DirectLink is that it is designed to present the most recent data provided by sensors, rather than the entire log of historical data, in order to provide fast responses by the robot system and avoid latency which may be introduced by processing backlogged data. The entire history of data communicated to the system can be retrieved from the Situational Awareness Database (SADB) [22], which contains both raw sensor data and perceptual data which has been processed via systems such as the device's computer vision system. It also contains information such as commands issued to the robot, either via its State Machine Controller (SMC) or planner. The SADB can be quickly accessed via a NoSQL interface, and incurs very low latencies due to mirroring across nodes. The ability to query representations in this fashion is useful for software such as planners, because this prevents them from needing to maintain this representation internally.

Worker poses and world state are measured using the aforementioned group of Kinect and PrimeSense cameras. The system is capable of merging the multiple perspectives of the Kinect cameras into a unified point cloud which is used to reliably track the worker's motion in the workspace using a skeleton track representation [20]. It is also able to track multiple objects in the work space in this fashion. As such, progress on the shared task can be measured through actions such as mounting components onto the car door. These data are stored into the SADB, where they can be processed by either the SMC or the planner. The planner developed for CHARM uses a standard representation in Planning Domain Definition Language (PDDL) [14], which is augmented with measured times for the completion of various tasks. It can choose courses of action based on this representation in order to adapt to problems which may arise, such as a worker discarding a faulty part. Additionally, it is able to use its recorded timing data in order to plan optimal timings for task execution and to begin robot motion trajectories [9, 10]. As will be discussed later, this contributes to our present work on interaction fluency.

14.2.2   CHARM Streams and Thrusts

Dividing CHARM into a set of complimentary streams and thrusts provided a framework within which participating investigators and laboratories could bring their best work to the project, focus on making the biggest contribution that they could to the advancement of the projects goals, and ultimately be assured that their work fit into the project's scope and made a meaningful contribution. CHARM was structured around three interconnecting research streams that link three research thrusts representing the main research directions of the project. These are shown in Fig. 14.2. The three streams that connected the development of robot assistants across the three research thrusts of CHARM are:

Stream 1 Define the Human-Robot Interaction with the robotic assistant.
Stream 2 Develop relevant Situational Awareness and Data Representation.
Stream 3 Coordinate developments through an Integration Framework.

The ordering of the streams in Fig. 14.2, represents the focus on the human as central figure in the interaction with the robot assistant, and the understanding and management of situational awareness and data representation as key to supporting this interaction. An effective and strong integration framework for communication,

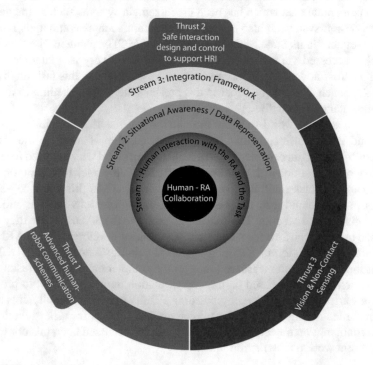

Fig. 14.2 Streams and thrusts of the CHARM project

control and perception is crucial to the successful development of a robot assistant that can effectively support a human worker at their task.

While providing support to the research thrusts, Stream 1 and Stream 2 also represent research activities that inform and are implemented within the three research thrusts, while Stream 3 is a research support activity that ties the research thrust developments together into a working prototype system.

The three research thrusts that drive the structure of Fig. 14.2 are:

Thrust 1 Advanced human-robot communication and cooperation schemes.
Thrust 2 Safe interaction design and control to support HRI.
Thrust 3 Vision and Non-Contact Sensing.

To each stream and thrust, domain experts were assigned, providing a structured exchange of knowledge and a focal point for specific needs of the project. Organizing the project in this fashion assured that responsible parties could be reached to discuss every relevant aspect of the project and that individual investigators knew who to contact, while also assuring that these investigators had the autonomy to pursue state-of-the-art work and bring it to bear on our shared task. Each stream and thrust made unique and necessary contributions to CHARM.

14.2.2.1 Stream 1

Stream 1 designed the interaction between the robot assistant and the human operator. Key systems were developed to allow human operators to interact with the robot through taps and pushes on the robot itself. Systems were developed in response to studies carried out under Thrust 1, enabling the robot to perform tasks such as elegant robot-to-human handovers. Stream 1 also developed a planner [9] that allowed the robot assistant to interactively replan for Plugfest III, as part of efforts toward using timing for fluent HRI.

14.2.2.2 Thrust 1

Thrust 1 performed studies in human-robot interaction in order to develop novel communicative cues for the robot assistant system. This thrust followed the paradigm of performing human-human studies in order to identify communicative cues, modeling and describing these cues, and then deploying them on robotic systems in order to study them in the context of Human-Robot Interaction. This process is described in greater detail in Sect. 14.3. Studies performed using this process can be found in Sect. 14.4. Contributions to the robot assistant include an elegant handover controller for robot-to-human handovers, studies in gestures, and studies in collaborative lifting.

14.2.2.3 Stream 2

Stream 2 developed the Situational Awareness Database (SADB) [22] and the basic world-state and sensor data representations used in the robot assistant system. Techniques for the modeling of and storage and retrieval of data are necessary in order to enable the system to autonomously reason about its environment and the shared task that it participates in.

14.2.2.4 Thrust 2

Thrust 2 developed the robot assistant Intelligent Assist Device (IAD) hardware and the control algorithms that drive this system. The software which integrates the gantry, robot arm, and robot gripper under a single controller was developed under this thrust, as well as capabilities such as the compliant control of the kinematic chain and backdrive capabilities [7, 12, 13].

14.2.2.5 Stream 3

Stream 3's primary focus was to integrate components from the various streams and thrusts of CHARM. As such, inter-group communication and Plugfests were of a primary concern. Stream 3 contributed the basic communication protocol, DirectLink, which is used for communications between compute nodes in the robot assistant architecture, and the State Machine Controller (SMC), which could be used for high-level control of the robot-assistant.

14.2.2.6 Thrust 3

Thrust 3's contribution was in the form of vision and non-contact sensing. For earlier revisions of the system, this involved tracking objects and actors in the interaction through the use of a Vicon motion tracking system. Later, this was replaced by a network of Microsoft Kinect RGBD cameras, and a PrimeSense camera mounted to the IAD's robotic arm. Software developed under Thrust 3 allowed the system to perform recognition and tracking, as well as to track the motion and poses of human actors in the interaction [20].

14.2.3 Plugfest

Plugfests served as a focal point for the effort of integrating components developed over each year and evaluating overall progress towards collaborative robot assistant technology. Each Plugfest was preceded by about two months of planning and preparation, with inter-site visits between investigators in each stream and thrust in order to assure that their systems properly integrated. The progress reports at the end of each Plugfest provide a picture as to how the robot assistant system matured over time.

14.2.3.1   Plugfest I

At the end of the first year, the group was still in a phase of defining requirements and specifications. Each team presented initial capabilities that represented a current state of the art in their respective areas. Reports were presented describing the findings of initial studies regarding capabilities that the labs wanted to contribute to the system, as well as findings from site visits to General Motors automotive manufacturing plants. As hardware was still being acquired, the robot was not in its final form.

At this juncture the team was able to produce an integrated system which gave a vision of what was to come. A worker instrumented with markers for a Vicon motion tracker interacted with the robot which was under the control of the State Machine Controller. The robot did not yet have its arm on it, and presented parts to the worker on a tray. Data were stored and queried from a preliminary version of the Situational Awareness Database.

14.2.3.2   Plugfest II

For Plugfest II, the Vicon system was removed in favor of a Kinect-based markerless tracking system. The system was able to register the point clouds from the independent Kinect sensors into a single point cloud and perform person tracking, where the representation of the worker was as a blob of points. A gesture-based system was developed in which the worker would make requests to the robot and the robot could communicate back to the worker through gestures. At this time, the robot arm was integrated into the system and a handover controller was used to hand objects to the worker. Improvements were made to the State Machine Controller and DirectLink protocols improving the overall responsiveness of the system.

14.2.3.3   Plugfest III

For Plugfest III human tracking had been updated to track a full skeletal model and the SADB was capable of high-performance transactions and replication across nodes. The gantry, gripper, and arm of the robot assistant were integrated into a single controller allowing compliant control and backdrive across the entire system. Gesture-based control had largely been replaced by a system which monitored the work state of the system, removing the need for some of the gesture-based commands, which had been found to slow down the worker during the interaction. An interactive planner had been added which allowed the system to reason about the scenario using PDDL [14].

CHARM provided its stakeholders with a project that enabled us to develop an integrated robot assistant system which brought the best of current technology to bear on the problem. By building a real, integrated system, we were able to see how our contributions impacted a real-world application and how these systems interacted with each other.

14.3   Identifying, Modeling, and Implementing Naturalistic Communicative Cues

The CARIS Laboratory uses a three phased method to identify, model, and implement naturalistic communicative cues in robotic systems. These phases comprise the following steps:

Phase 1 Human-Human Studies
Phase 2 Behavioral Description
Phase 3 Human-Robot Interaction Studies

To aid in describing this process, this section will use the example of recent work in hand gestures carried out in the CARIS Laboratory [21]. In this work, first, a human-human study of a non-verbal interaction between dyads of human participants performing an assembly task is carried out in order to witness and identify the hand gestures that they use in order to communicate with each other. This is Phase 1 from the three phase method. From the data collected during this study, a set of communicative hand-gestures is identified by the researchers from annotated data. This set of gestures is validated using an online study of video-recorded example gestures, ascertaining whether study participants recognize the same gestures and intentions as those identified by the experimenters. This is Phase 2, in which an accurate description of the human behavior is identified. In Phase 3, these gestures are then programmed into a Barrett Whole-Arm Manipulator (WAM) for use in a human-robot study. In a step mirroring Phase 2, the robot gestures are studied in an online study, with results reported for recognition and understanding by participants.

14.3.1   Phase 1: Human-Human Studies

The purpose of the Human-Human study phase is to elicit behaviors on the part of humans in a collaboration so that they can be characterized and understood, and then replicated on a robotic platform.

For this work, a study of non-verbal interaction between human dyads performing a car door assembly task was performed. The door is instrumented in seven locations with Velcro™ strips where parts can be mounted. Correspondingly, six parts that are to be mounted onto the car door are instrumented with corresponding Velcro™ strips. Study participants were provided with a picture of a completed assembly of these parts mounted onto the car door, and asked to non-verbally communicate the proper placement and orientation of these parts on the door to a confederate through the use of hand gestures. After this assembly was completed, a second picture was presented to the participants, changing the location and orientation of four of the parts on the door. At this stage, participants were asked to direct the confederate to modify the arrangement of the parts on the door as indicated in the new picture. Items were placed on a table between the study participant and confederate in order to provide easy access to both the door and the items, as in Fig. 14.3. The participants were required to perform this task in accordance with a set of provided rules, which assure that relevant communication could be performed only via hand gestures.

- Only use one hand to direct the worker.
- Only make one gesture and hold only one part at a time.
- You must wait for the worker to complete the task before making your next gesture.
- You must remain in your home position at all times.

A group of 17 participants (female: 7, male 10) between 19 and 36 years of age participated in this study.

Fig. 14.3 Participants in a Phase 1 human-human study of non-verbal interaction

14.3.2 Phase 2: Behavioral Description

In the second phase, behaviors exhibited during the human-human study are characterized. In this study, gestures used by the participants were identified from annotated video data. Gestures were selected from the annotated data according to the following criteria:

- They should be understandable without trained knowledge of the gesture.
- They should be critical to task completion.
- They should be commonly used among all participants.

Based on these criteria, the experimenters identified directional gestures as a category of interest for further study, and narrowed the gestures into four categories. These were "Up," "Down," "Left," and "Right." They also identified that each of these gestures could be performed with an Open-Hand or Finger-Pointing hand pose. These gestures appear as in Fig. 14.4.

Video clips of these gestures were used in an online study of 120 participants in which participants were asked three questions. The first asks, "What do you think the worker should do with this part?" where participants were instructed to answer "I don't know" if they did not understand the gesture. In the second they were asked to rate, "How easy was it for you to understand the meaning of this gesture (on a scale from 1 (very difficult) to 7 (very easy))?" In the third they were asked, "How certain are you of your answer to question 1 (on a scale from 1 (very uncertain to 7 (very certain))?" At the end of this phase, survey responses to the latter two questions were found to have a high degree of internal consistency (Cronbach $\alpha = 0.891$).

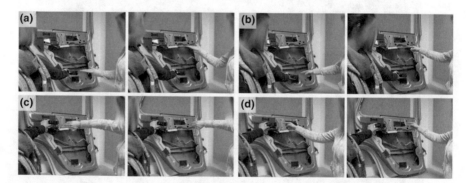

Fig. 14.4 Directional Gestures and frequently observed accompanying hand poses identified from annotated data - "Up" and "Down" gesture with Open-Hand (**a**), and Finger-Pointing (**b**) poses, and "Left" and "Right" gesture with Open-Hand (**c**), and Finger-Pointing (**d**) poses

14.3.3 Phase 3: Human-Robot Interaction Studies

The purpose of Phase 3 is to attempt to replicate the identified and described communicative cues in a human-robot interaction. To do this, the study in Phase 2 was replicated with a robotic arm, a 7 Degree of Freedom (DoF) Barrett Whole-Arm Manipulator equipped with a 3-fingered BarrettHand. Gestures were programmed into the arm and presented as video clips in an online study of 100 participants. In the robotic arm case, each gesture was presented using one of three hand poses: one with an Open-Hand (OH), one with a Finger-Pointing hand pose (FP), and one with a Closed-Hand (CH), as in Fig. 14.5.

Sheikholeslami et al. [21] report recognition rates for these gestures, comparing against the human and robot conditions. Results are shown in Fig. 14.6. Their results demonstrate a similar degree of recognition and understanding of these gestures in both the human and robot conditions, and show whether they are more easily-understood for various poses of the robot's manipulator.

The purpose of the this methodology is to directly identify, analyze, and implement human communicative behaviors on robotic systems in order to create human-robot interactions which are naturalistic and intuitive. The steps of performing human-human studies and characterizing the behaviors of the participants provide the data required to reproduce these behaviors, while human-robot interaction studies validate the effectiveness of the reproduced communicative cues. In the CARIS Laboratory, a central goal is to create human-robot interactions in which robots are able to interact with humans in a collaborative fashion, rather than in a manner that is more akin to direct control through a keyboard or teach-pendant interface. Exploiting naturalistic communicative behaviors is a key avenue by which we are attempting to achieve this goal.

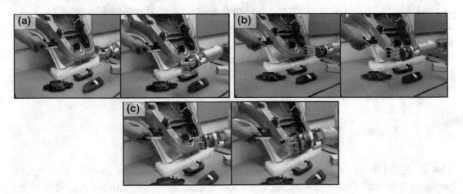

Fig. 14.5 "Left" and "Right" gestures implemented on the Barrett WAM with Closed-Hand (**a**), Open-Hand (**b**), and Finger-Pointing (**c**) hand poses implemented on the BarrettHand

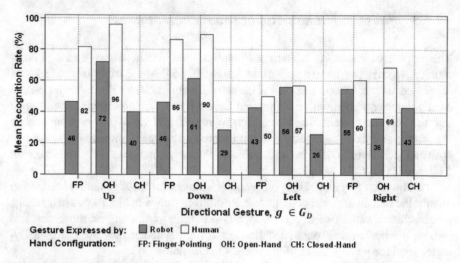

Fig. 14.6 Comparison of recognition rates for hand gestures by gesture and hand configuration between robot and human cases. OH - Open-Hand, FP - Finger-Pointing, CH - Closed-Hand

14.4 Communicative Cue Studies

The CARIS Laboratory has done extensive research exploring the use of communicative cues for human-robot interaction. Many of these cues are naturalistic cues; cues that emulate natural human communicative behaviors such as can be found through human-human studies and witnessed in everyday human interactions. Some of these cues are non-naturalistic, as in the case of tapping and pushing on the robot to guide its motion [5]. These cues can also be divided into explicit and non-explicit cues. An example of explicit cues would be hand gestures, as described in Sect. 14.3. An example of a non-explicit cue would be when a person inadvertently looks towards the location where they intend to hand an object over to someone else [1, 17, 23]; or the sensation of feeling another person managing the weight of an object that is handed over, allowing it to be released [3, 4].

14.4.1 Human-Robot Handovers

The CARIS lab has extensively studied the process of handing an object from one party to another. This is an example of an interaction that is mostly mediated by naturalistic, non-explicit cues. Important cues occurring during handovers that have been explored by CARIS include the forces acting on the object being handed over [3, 4], gaze behaviors during the handover interaction [17], motion trajectories [9, 10], and kinematic configurations [2].

To explore forces acting on an object during handovers, Chan et al. [3] constructed a baton that is instrumented with force sensing resistors (FSRs), an ATI force/torque sensor, and inertial sensors, as can be seen in Fig. 14.7. Nine pairs of participants were recruited to perform 60 handovers each in 6 different configurations. The investigators found that distinct roles for the giver and receiver of the object emerge in terms of the forces acting on the baton. The giver assumes responsible for the safety of the object by assuring that it does not fall, whereas the receiver assumes responsible for the efficiency of the handover by taking it from the giver. This can be measured as a characteristic of grip and load forces over time on behalf of the giver and receiver, with the handover concluding when the giver experiences negative load force as the receiver slightly pulls it out of their hand. This has direct implications for the design of a controller for robot-to-human handovers. Chan et al. [4] implemented such a controller on the Willow Garage PR2 robot, which mimics human behavior by regulating grip forces acting on the transferred object.

Moon et al. [17] followed up on this study by adding a gaze cue to the PR2's handover software. In a three condition, intra-participant design, 102 study participants were handed water bottles and asked which of three handovers they preferred. The robot indicated its gaze direction by tilting its head to either look down and away

Fig. 14.7 Baton instrumented for measuring forces acting on an object during handover

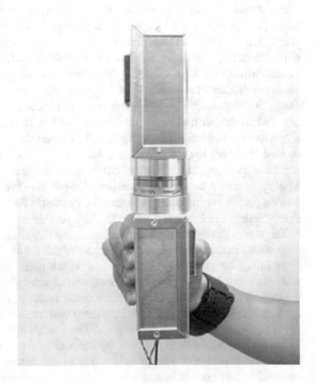

from the participant (No Gaze), look towards the water bottle (Shared Attention), or look to the water bottle and then up at the participant (Turn-Taking). The study found that participants reach for the water bottle significantly earlier in Shared Attention ($M = 1.91$s, $SD = 0.52$) condition over the No Gaze condition ($M = 2.54$s, $SD = 0.79$)($p < 0.005$). This is measured with respect to the time when the robot grasps the water bottle to be handed over and starts its trajectory toward the handover location. No significant difference was found between the Shared Attention and Turn Taking ($M = 2.26$, $SD = 0.79$) conditions or between the Turn Taking and No Gaze conditions. Their results also suggest that participants may prefer conditions in which the robot makes eye contact.

Additional work in conjunction with University of Tokyo went on to study how the orientation of an object interacts with how it is handed over [2]. Current work in CARIS investigates motion cues that can be exploited to detect the timing and location of handovers, for fluent human-to-robot handover behaviors [10].

14.4.2 Hesitation

Another example of a non-explicit communicative cue is motor behavior when hesitating. Moon et al. [15, 16] recorded human motion trajectories in an experiment where dyads of study participants reach for a target placed between them on a table when prompted by tones in sets of headphones. In the study, randomly-timed tones are played separately in each pair of headphones, such that only sometimes are both participants required to reach for the target at the same time. This causes participants to sometimes hesitate in accessing the shared resource placed between them, ceding access to the other participant. The study separates recorded arm motions into three categories: successful (S) - when the participant accesses the shared resource, retract (R) - when the participant retracts their hand from a trajectory directed toward the target as an act of hesitation, and pause (P) - when the participant pauses along their motion trajectory toward the target in hesitation. Accelerometer data were recorded for these motions, and a motion profile was derived from these recorded human hesitations called the Acceleration-Based Hesitation Profile (AHP). A retract-based motion profile was used to plan motion trajectories for a robot arm during a similar shared task with a human collaborator. This motion was video recorded for use in an online study. Results of the study demonstrate that study participants recognize the reproduced hesitation behavior on the part of the robot.

A follow-up study was conducted in which participants perform a shared task with the robot [16]. In this experiment, participants sit across from the robot with the task of taking marbles from a bin located in the center of the workspace shared with the robot one at a time, matching them with shapes from another bin according to a set of exemplar marble-shape pairs. The robot's task is to inspect the marble bin by moving back and forth between the bin and its starting position, see Fig. 14.8. A total of 31 participants took part in a within-participant study in which they were exposed to three conditions; Blind Response - in which the robot continues along

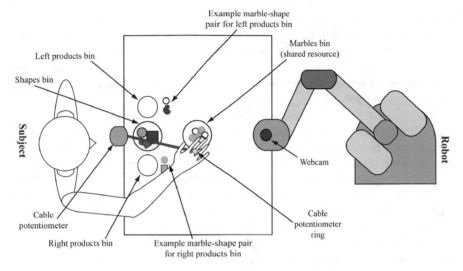

Fig. 14.8 Diagram illustrating the experimental setup of a study in which human study participants interact with a robotic arm which performs hesitation behaviors during an interaction involving access to a shared resource

its trajectory, Robotic Avoidance - in which the robot arrests its motion to wait for the participant to complete their motion, and Hesitation Response - in which the robot responds with an AHP-based trajectory. Results do not show improvements in task completion or perceptions of the robot, but do demonstrate that participants recognize the hesitation behavior.

14.4.3 Tap and Push

Our group and our collaborators are also interested in communicative cues which are non-naturalistic, but nonetheless may be highly intuitive for human collaborators, or which are based on common human interactions but not on specific communicative cues. One example of this is a study in tap-and-push style interactions in which a human study participant taps and pushes on a robot. This study was conducted in conjunction with the Sensory Perception & INteraction (SPIN) Laboratory at UBC, by Gleeson et al. [5].

In order to study tap-and-push style interactions with a robot arm, Gleeson et al. [5] performed a series of studies in which human workers and robot assistants collaborate on an assembly task. These studies compare a set of commands based on tapping and pushing on a device to commands via a keyboard interface. In the first study, participants interact with a Phantom Omni desktop haptic device in a scripted collaborative fastener insertion task which simulates the placement and tightening of four bolts. Participants pick up bolts from open boxes, place a bolt in one of four locations on a board, and then command the Omni to touch the bolt, simulating a tightening operation. See Fig. 14.9a. After performing its tightening operation, the Omni automatically moves to the next position. Gleeson et al. [5] found that on this task keyboard commands slightly outperform direct physical commands in the forms of taps and pushes on the robot in quantitative task performance metrics and qualitative user preference.

The second study comprises two more complex tasks which were performed as interactions with a Barrett Whole-Arm Manipulator (WAM) robotic arm. The first task is a bolt insertion task similar to that in the first study, but in which bolts are inserted in a random order that is presented to the user on notecards, and in which the robot does not automatically advance to the next bolt. See Fig. 14.9b. In the second task, participants interactively position the arm using discrete tap-and-push cues to position it over a series of cups, and then continuously guide the robot to the bottom of the cup, Fig. 14.9c. Gleeson et al. [5] found that in these more complex and less scripted tasks participants are able to more quickly complete the collaborative tasks, and that they prefer the physical interaction over the keyboard interface.

14.5 Current and Future Work

Current work in the CARIS Laboratory continues our study of communicative cues. Part of Matthew Pan's current work involves collaboration between two parties lifting an object. This expands on work by Parker and Croft [18, 19] on the development of controllers that enable robots to elegantly respond to cues based on the motion of an object that is in the process of being lifted. Another aspect of Pan's current work is the automatic detection of the intention of a person to handover an object, thus enabling a robot to identify this intention and respond by grasping the object.

Hart, Sheikholeslami, and Croft are currently working on approaches to extrapolate human motion trajectories based on hand and skeletal tracks [10]. Such an extrapolation would allow the prediction of the timing and location of the endpoint of the motion, and of what the person is attempting to do. In the context of a reach, this can inform a robot of what the person is reaching toward. Hoffman and Breazeal [11] noted that collaborative robotic systems that are based on observations of the current state of the shared task experience a stop-and-go style interaction. This is in part due to the need to respond to the current world state, rather than what is about to happen. They describe the ability to smoothly mesh the actions of collaborators as interaction fluency. By making guesses at a collaborator's intentions and extrapolations of their motions, a robot can act on these predictions. Precise predictions of the timing and location and choice of actions on a shared task can enable a robot to preemptively act on these predictions. For instance, if a person is reaching to a fastener, the robot can reach to the tool that attaches it. We have also explored optimal timing of a robot's behavior based on prior task performance on the part of both the robot and the worker.

One application that we are currently pursuing is combining motion extrapolation techniques with Matthew Pan's work on handover detection, allowing us to predict

(a) Phantom Omni haptic device, interacted with via taps and pushes,
or via keyboard in a bolt placement task.

(b) Barrett Whole-Arm Manipulator, interacted with via taps and pushes,
or via key board in a bolt placement task.

(c) Barrett Whole-Arm Manipulator, interacted with via taps and pushes,
or via keyboard in a guided manipulation task.

Fig. 14.9 Experimental setups for studies in tap and push interactions with robotic devices

the timing and location of a handover, in addition to preemptively predicting the intention based on a predicted motion trajectory. In a study of handover motions [10], we observed that the receiver of an object being handed over begins their motion to the ultimate location of the handover prior to the arm being fully-extended to this location. We also observed that we could determine the path that the arm would follow from only a few frames of motion tracking data (as long as the track is stable). In the context of a handover motion, this can inform the timing and location at which person intends to hand over the object, thus enabling the robot to begin its trajectory in a manner similar to that of a person reaching out to accept the object. Studies on this are in progress.

The robot-assistant project itself is also exploring new directions. We are currently in the process of designing interactions around the use of advanced composite materials such as carbon fiber reinforced plastics and the construction of large components. Techniques such as gesturing to the robot will enable non-verbal communication at the longer distances required to collaborate on large composite components, whereas other naturalistic communicative cues may be combined with advanced projection mapping and augmented reality technologies to enable the robot to communicate important information back to the user. Sensing, situational awareness, and proxemics will play key roles in designing these interactions.

A closed loop of communication between a human worker and its robot collaborator is key for progress in collaborative human-robot interaction. With humans and robots working in close proximity, physical HRI techniques will become key in enabling a robot to live up to worker expectations and interpret worker intentions. For worker safety and productivity, it is crucial that robots and human collaborators are able to transparently communicate with each other and understand each other's actions and intentions. For robot assistants to act as collaborators, rather than tools directly under the control of human operators, they must behave in predictable manners and in ways that operators are able to intuitively control. Their development requires contributions from multiple disciplines in computer science, mechanical engineering, design, sensing, and control. The study of communicative cues provides a route to establishing a closed loop of transparent communication between a human worker and its robot collaborator, while behavioral predictions provide us with a route to performing this communication fluently.

Acknowledgements We would like to thank our collaborators on CHARM for their incredible contributions to this project. In addition to the authors of this chapter, CHARM collaborators include Ergun Calisgan, Jacques-Michel Haché, Dominic Beaulieu, AJung Moon, Marc-Antoine Lacasse, Olivier St-Martin Cormier, Andrew Phan, Brian Gleeson, Denis Ouellet, Boris Mayer St-Onge, Thierry Moszkowicz, Oleg Boulanov, Roland Y. Menassa, Stephen Hart, Muhammad Abdallah, Jim Wells, Justin Gammage and Robert Tilove. CHARM was made possible thanks to the generous support of General Motors of Canada and the Natural Sciences and Engineering Research Council of Canada under contract 140570247. Studies carried out at UBC that are reported in this chapter were approved by the University Behavioral Research Ethics Board under approval number H10-00503-A020.

References

1. H. Admoni, A. Dragan, S.S. Srinivasa, B. Scassellati, Deliberate delays during robot-to-human handovers improve compliance with gaze communication. *Proceedings of the Ninth Annual ACM/IEEE International Conference on Human-robot Interaction (HRI '14)* (Bielefeld, Germany, 2014), pp. 49–56
2. W.P. Chan, M.K.X.J. Pan, E.A. Croft, M. Inaba, Characterization of handover orientations used by humans for efficient robot to human handovers. *Proceedings of the 2015 IEEE/RSJ International Conference on Intelligent Robots and Systems* (Hamburg, Germany, September 2015), pp. 1–6
3. W.P. Chan, C.A.C. Parker, H.F.M. Van der Loos, E.A. Croft, Grip forces and load forces in handovers: Implications for designing human-robot handover controllers. *Proceedings of the Seventh Annual ACM/IEEE International Conference on Human-Robot Interaction (HRI '12)* (Boston, MA, USA, 2012), pp. 9–16
4. W.P. Chan, C.A.C. Parker, H.F.M. Van der Loos, E.A. Croft, A human-inspired object handover controller. Int. J. Robot. Res. 32(8), 971–983 (2013)
5. B. Gleeson, K. Currie, K. MacLean, E. Croft, Tap and push: Assessing the value of direct physical control in human-robot collaborative tasks. J. Hum. Robot Interact. 4(1), 95–113 (2015)
6. B. Gleeson, K. MacLean, A. Haddadi, E.A. Croft, J.A. Alcazar, Gestures for industry intuitive human-robot communication from human observation. *Proceedings of the Eigth Annual ACM/IEEE International Conference on Human-Robot Interaction (HRI '13)* (Tokyo, Japan, March 2013), pp. 349–356
7. C. Gosselin, T. Laliberté, B. Mayer-St-Onge, S. Foucault, A. Lecours, V. Duchaine, N. Paradis, D. Gao, R. Menassa, A friendly beast of burden: A human-assistive robot for handling large payloads. IEEE Robot. Autom. Mag. 20(4), 139–147 (2013)
8. A. Haddadi, E.A. Croft, B. Gleeson, K. MacLean, J. Alcazar, Analysis of task-based gestures in human-robot interaction. *Proceedings of the IEEE International Conference on Robotics and Automation (ICRA '13)* (Karlsruhe, Germany, May 2013), pp. 2146–2152
9. J.W. Hart, B.T. Gleeson, M.K. X.J. Pan, A. Moon, K. MacLean, E.A. Croft, Gesture, gaze, touch, and hesitation: Timing cues for collaborative work. *Proceedings of the HRI Workshop on Timing in Human-Robot Interaction (HRI '14)* (Bielefeld, Germany, March 2014)
10. J.W. Hart, S. Sheikholeslami, M.K.X.J. Pan, W.P. Chan, E.A. Croft, Predictions of human task performance and handover trajectories for human-robot interaction. *Proceedings of the HRI Workshop on Human-Robot Teaming (HRI '15)* (Portland, OR, USA, March 2015)
11. G. Hoffman, C. Breazeal, Cost-based anticipatory action selection for human-robot fluency. IEEE Trans. Robot. 23(5), 952–961 (2007)
12. A. Lecours, B. Mayer-St-Onge, C. Gosselin, Variable admittance control of a four-degree-of-freedom intelligent assist device. *Proceedings of the IEEE International Conference on Robotics and Automation (ICRA)* (St. Paul, MN, USA, May 2012), pp. 3903–3908
13. A. Lecours, C. Gosselin, Computed-torque control of a four-degree-of-freedom admittance controlled intelligent assist device. *Experimental Robotics: The 13th International Symposium on Experimental Robotics*, ed. by P. Jaydev Desai, G. Dudek, O. Khatib, V. Kumar (Springer, Heidelberg, Germany, 2013), pp. 635–649
14. D. Mcdermott, M. Ghallab, A. Howe, C. Knoblock, A. Ram, M. Veloso, D. Weld, D. Wilkins, Pddl - the planning domain definition language. Technical Report TR-98-003, Yale Center for Computational Vision and Control, 1998
15. A.J. Moon, C.A.C. Parker, E.A. Croft, H.F.M. Van der Loos, Did you see it hesitate? Empirically grounded design of hesitation trajectories for collaborative robots. *Proceedings of the IEEE/RSJ International Conference on Intelligent Robots and Systems (IROS '11)* (San Francisco, CA, USA, September 2011), pp. 1994–1999
16. A.J. Moon, C.A.C. Parker, E.A. Croft, H.F.M. Van der Loos, Design and impact of hesitation gestures during human-robot resource conflicts. J. Hum.-Robot Interact. 2(3), 18–40 (2013)

17. A.J. Moon, D.M. Troniak, B. Gleeson, M.K.X.J. Pan, M. Zheng, B.A. Blumer, K. MacLean, E.A. Croft, Meet me where I'm gazing: How shared attention gaze affects human-robot handover timing. *Proceedings of the Ninth ACM/IEEE International Conference on Human-robot Interaction(HRI '14)* (New York, NY, USA, March 2014), pp. 334–341
18. C.A.C. Parker, E.A. Croft, Design & personalization of a cooperative carrying robot controller. *Proceedings of the IEEE International Conference on Robotics and Automation (ICRA '12)* (St. Paul, MN, USA, May 2012), pp. 3916–3921
19. C.A.C. Parker, E.A. Croft, Experimental investigation of human-robot cooperative carrying. *Proceedings of the IEEE/RSJ International Conference on Intelligent Robots and Systems ((IROS '11))* (San Francisco, CA, USA, September 2011), pp. 3361–3366
20. A. Phan, F.P. Ferrie, Towards 3d human posture estimation using multiple kinects despite self-contacts. *Proceedings of the 14th IAPR International Conference on Machine Vision Applications (MVA)* (Tokyo, Japan, May 2015), pp. 567–571
21. S. Sheikholeslami, A.J. Moon, E.A. Croft, Exploring the effect of robot hand configurations in directional gestures for human-robot interaction. *IEEE/RSJ International Conference on Intelligent Robots and Systems (IROS '15)* (Hamburg, Germany, September–October 2015), pp. 3594–3599
22. O. St-Martin Cormier, A. Phan, F.P. Ferrie, Situational awareness for manufacturing applications. *12th Conference on Computer and Robot Vision* (Halifax, Nova Scotia, Canada, 2015), pp. 320–327
23. K. Strabala, M.K. Lee, A. Dragan, J. Forlizzi, S. Srinivasa, M. Cakmak, V. Micelli. Towards seamless human-robot handovers. J. Hum.-Robot Inter. (2013)

Part III
Trusted Autonomy

Chapter 15
Intrinsic Motivation for Truly Autonomous Agents

Ron Sun

15.1 Introduction

In order to deal with complexity, uncertainty, and unpredictability, which are inevitable in many real-world tasks and environments, agents need to be intrinsically motivated. Intrinsically motivated agents are those that have human-like (or animal-like) internal motivational processes, with internally generated, self-determined needs and preferences, which may or may not be influenced externally. It is the ability and the inclination of an agent (e.g., a human or a robot) to act autonomously, at its own discretion [6, 48]. For true autonomy necessary for dealing with highly complex, uncertain, or unpredictable environments, intrinsic motivation would be a highly desirable, or even necessary, part of being autonomous agents functioning in such environments. In highly complex, uncertain, or unpredictable environments, specific motivations and preferences cannot be easily pre-specified for a system from the outside, and thus intrinsic motivation is important for the sake of autonomy and for coping with such environments [18, 59].

In past work on intelligent agents, including past work on learning, planning, and problem solving for such agents, the need for intrinsic motivation has been down-played (although not completely ignored; more on this later). Thus, by now, the shortcomings of existent autonomous agent models and systems are quite evident, for example, with regard to their acceptance and their deployment in complex, uncertain, or unpredictable environments. Clearly, we need to seriously rethink some of these old approaches based on old (and often outdated) assumptions and methodologies, and move forward to the development of new, different, and better approaches, models, and theories, especially those that involve human-like intrinsic motivation.

Having intrinsic motivation is also important to achieving trust of autonomous agents and systems (such as autonomous robots) by humans (and by other autonomous agents and systems). In fundamentally unpredictable environments, a key aspect that one can be certain of is stable internal needs and preferences–that is, intrinsic moti-

R. Sun (✉)
Cognitive Sciences Department, Rensselaer Polytechnic Institute, Troy, NY 12180, USA
e-mail: dr.ron.sun@gmail.com

© The Author(s) 2018
H. A. Abbass et al. (eds.), *Foundations of Trusted Autonomy*, Studies in Systems,
Decision and Control 117, https://doi.org/10.1007/978-3-319-64816-3_15

273

vation. Thus, in order to have trust and confidence in someone else, one has to have an understanding of what motivates the other [47, 48].

We may term human-like intrinsic motivation and autonomous choice of action (in accordance with intrinsic motivation) "free will". Self-determined intrinsic motivation (or "free will") in humans includes not only power, achievement, and other individualistic tendencies, but also adherence to social norms, affiliation with other individuals, and other tendencies related to social interactions and interdependencies ([48]; see details later). These motives are the results of evolution over a long period of human prehistory in the context of the struggles to survive within social groups. Real trust is trust among such "free willed" individuals. Limited, simpler forms of "trust" that one typically places on currently available machines such as self-driving automobiles or robotic vacuum cleaners (as they stand currently) cannot be construed as real trust (or full trust) and, I believe, is far from sufficient for the future. See, for example, Lee and See [23] or Abbass et al. [1] for characterization of such limited forms of trust. The question is: How do we move beyond that?

To achieve real trust, I believe that we need to delve into natural human tendencies to trust other individuals with intrinsic motivations that are similar to ours and similarly "free willed". Humans do have such tendencies, necessitated by their collective need for survival, evolved during their collective struggles to survive for tens of thousands of years. Such trust may start from predictability of behavior, as a result of similarly endowed (innate or acquired) motives. Understanding others' motivation leads to predictability of their behavior, which in turn leads to more complex and deeper forms of trust (e.g., involving affective or emotional processes). Only in this way, through understanding and exploiting such natural human tendencies, may we achieve truly autonomous agents, robots, and machines that may be given our real and full trust and that may also achieve real mutual trust amongst themselves.

Taking all of these issues into consideration, it is evident that we need to develop a deeper perspective on future autonomous systems, which should include intrinsic motivation in particular.

In the remainder of this chapter, first, background of some past work on human motivation is reviewed, as well as past work on computational cognitive architectures in relation to motivation. Then, a particular cognitive architecture (namely, the Clarion cognitive architecture) that is integrative and comprehensive and includes a more complete motivational subsystem is detailed, especially the interaction between its motivation and cognition [50]. Some examples of simulations using this cognitive architecture are described, which show briefly how this cognitive architecture integrates cognition and motivation and enables agents to function autonomously and appropriately. Some concluding remarks end this chapter.

15.2 Background

15.2.1 Previous Work on Intrinsic Human Motivation

It has been argued that, currently, many kinds of intelligent artifacts–autonomous agents, systems, and robots–are not truly autonomous, capable of dealing with complex, uncertain, and unpredictable environments independently, that is, truly autonomously. For one thing, they do not seem to possess independent, intrinsic motivations, needs, and preferences by themselves [48]. Notably, in highly complex, uncertain, or unpredictable environments, specific motivations and preferences cannot be easily pre-programmed; intrinsic motivation is therefore important to achieving autonomy and consequently crucial to successful coping in such environments [18, 59]. We need to make an effort to redress this current state of affairs.

Furthermore, we also need to address social interaction of and with autonomous agents, systems, and robots. For example, in social transactions, how and when one can place trust in such a system is a major issue, as mentioned earlier [24, 36, 39]. For humans, to truly trust and have confidence in someone else, one has to have an understanding of what motivates the other [47]. Having stable intrinsic motivation, as humans usually do, helps in this regard. For another example, social impasses may often result from incompatible motivations of multiple people (or agents); understanding each other's motivations may go a long way in helping to resolve such impasses [47].

Work on intrinsic human motivations has had a long history. Some particularly relevant work will be briefly discussed here [29, 34], in relation to our own theory of human motivation as embodied in the Clarion cognitive architecture mentioned earlier [46, 48, 50]. Understanding and replicating the human motivational subsystem can be highly beneficial to building autonomous intelligent agents and systems, because of its power, flexibility, and adaptability [48].

First of all, very early on, Murray [29] proposed a pertinent set of basic needs (i.e., primary drives in our terminology, as used in Clarion). Murray's proposal [29] included the need for conservance, the need for order, the need for retention, the need for acquisition, the need for inviolacy, and so on (note that these needs are included as or covered by primary drives in Clarion, as will be detailed later). Some other needs identified by Murray, such as contrarience, aggression, abasement, rejection, succorance, exposition, construction, and play, may not be fundamental needs (or primary drives) in our view—they are likely the results of more fundamental needs (i.e., primary drives) or their combinations. Murray's proposal also included some low-level (physiological, or viscerogenic in Murray's term) needs (which may be attributed to some combinations of low-level primary drives in Clarion).

More recently, Reiss [34] proposed another set of basic needs (i.e., primary drives), which was highly similar to Murray's, but with some differences. For example, as proposed by Reiss [34], there are the need for saving, the need for order, the need for family, the need for vengeance, the need for "idealism", the need for status, the need for acceptance, as well as the need for eating, the need for tranquility, the need for physical exercises, the need for romance, and so on. (Again, these needs are included as or covered by primary drives in Clarion as will be detailed later.)

As alluded to above, in Sun [48, 50], a detailed model of human motivation (as embodied in Clarion) was presented. The Clarion cognitive architecture incorporates multiple, interacting subsystems. In particular, within the motivational subsystem, there are implicit drives and explicit goals (with goals being primarily determined based on drives). While some drives, denoting essential needs and desires, are primary and built-in, some other drives may be acquired and secondary. The primary drives include: Affiliation & Belongingness, Dominance & Power, Recognition & Achievement, Autonomy, Deference, Similance, Fairness, Honor, Nurturance, Conservation, Curiosity, as well as some low-level primary drives [48]. With its built-in mechanisms and processes, especially the motivational mechanisms, Clarion is able to capture, account for, and explain many psychological data and phenomena related to human motivation. There have been various efforts at verifying those drives through experiments and data analysis [34, 48]. See further discussions of Clarion below, and also see Sun [46, 50].

Relatedly, Schwartz's [40] 10 universal values, although addressing a different aspect of human behavior (i.e., human "values"), bear some resemblance to the essential needs (i.e., primary drives) identified above [48]. Moreover, each of these values can be derived from some primary drive or some combination of these primary drives [48].

McDougall [27] proposed a framework that was concerned with "instincts". Instincts, in our framework, refer to (more or less) evolutionarily hard-wired (i.e., innate) behavior patterns or routines that can be relatively easily triggered by pertinent stimuli in pertinent situations. As discussed earlier, basic needs (or primary drives as termed in Clarion) are essential driving forces of behaviors. Instincts are different from basic needs, because one does not have to follow instincts when there is no pertinent stimulus, and even when pertinent stimuli are present, one may be able to refrain from following instincts (at least more easily than from basic needs or primary drives). In other words, they are pre-set routines: while they are relatively easily triggered, they are not inevitable. McDougall listed the following instincts: imitation, emulation or rivalry, pugnacity/anger/resentment, sympathy, hunting, fear, appropriation/acquisitiveness, constructiveness, play, curiosity, sociability and shyness, secretiveness, cleanliness, modesty and shame, love, jealousy, parental love, …, and so on (see also [19]). As evident from the list above, many of these instincts are results of primary drives or basic needs (such as "curiosity" and "parental love"), or are derived, by some means, from primary drives or basic needs (such as "play" and "constructiveness"). Some other instincts are not because they do not represent basic needs (e.g., "hunting" or "jealousy"). (See more discussions of primary drives within Clarion later.)

There have also been some less psychologically validated models of motivation. Such models include Doerner's model and Sloman's model. In Sloman's motivational model [67], goals ("motives") are generated from a suite of modules ("generactivators"), each of which expresses a single "concern" (such as caring for dependents or removing damaged dependents). Each of these modules may search through a database of beliefs; if it finds a match, a declarative representation of a goal (a "motive") is generated. On that basis, the resource management system takes goal representations and generates intentions for action. Although the model bears some resemblance to Clarion, the model has not been used to capture or explain psychological data in any detail. In addition, computationally speaking, searching through databases is cumbersome and may not be cognitively realistic.

Doerner [15] (see also Bach [5]) described the PSI theory, which included internal deficits, displeasure signals (due to deficits), negative reinforcement (from displeasure signals), urges, goals, action learning through random exploration (based on reinforcement), and so on. At an abstract level, the model is similar to Clarion to some extent [48, 50], but it appears less psychologically grounded or validated. In addition, its computational mechanisms appear less well developed algorithmically.

15.2.2 Previous Work on Cognitive Architectures

It has been suggested [30] that cognitive theories (including computational cognitive models) should be developed that satisfy multiple criteria, in order to avoid theoretical myopia. There have been steady developments of generic computational cognitive models, that is, cognitive architectures, for the past three decades since that seminal suggestion.

Early cognitive architectures often took the form of production systems and were (more or less) concerned with various psychological phenomena [20]. However, other forms of cognitive architectures have also been developed over the years — they may be in the form of a connectionist model, a constraint satisfaction network, a hybrid system of different models, and so on. Some of them may be more concerned with applications to building artificial systems than capturing and explaining empirical psychological phenomena.

Computational cognitive architectures provide the best hope for integrated systems that incorporate not just cognitive capabilities, but also motivation, emotion, personality, and many other capacities and capabilities needed for an autonomous agent. In particular, computational cognitive architectures based firmly on psychological data and findings and thus well-grounded empirically can be especially illuminating—they provide a glimpse into how human minds work, for example, in terms of the interaction between cognition and motivation, as well as their interaction with the environments (simple or complex). The human mind provides the best example of a truly autonomous intelligent system, and thus can lead to better understanding of intelligence and autonomy. Such cognitive architectures, like humans on which they are based, are capable of being truly autonomous, because they include a wide range of cognitive, motivational, and other capabilities and these capabilities function together to cope with different tasks and environments. Let us look into three examples, in chronological order.

Soar, the first proposed cognitive architecture, has been developed over the past thirty years, based essentially on a production system model. It has mostly been used for the purpose of building application systems [21, 30, 35]. In Soar, based on the framework of a state space and operators for searching the state space, decisions are made by different productions proposing different operators, when there is a goal on a goal stack. When a sequence of productions leads to achieving a goal, chunk-ing occurs, which creates a single production that summarizes the process (using explanation-based learning). However, it lacks sophisticated motivational structures and processes. In addition, a large amount of initial (a priori) knowledge about states and operators is required for Soar to work.

Another series of cognitive architectures were also proposed fairly early on: in par-ticular, ACT* and ACT-R [3]. ACT* is made up of declarative knowledge (captured in a semantic network) and procedural knowledge (captured in a production system). Procedural knowledge (in productions) is acquired through "proceduralization" of declarative knowledge, modified through use by generalization and discrimination (i.e., specialization), and have strengths associated with them (which are used for firing). ACT-R is a descendant of ACT*, in which procedural learning is limited to production formation through mimicking and production firing is based on log odds of success. There have been some later additions to ACT-R, including visual and motor modules, but there have not been any sufficiently complex motivational structures.

Clarion has been a comprehensive cognitive architecture [45, 46, 50]. The Clarion cognitive architecture, as mentioned earlier, consists of multiple, interacting subsys-tems. It is also distinguished from other existing cognitive architectures by its focus on the separation and the interaction of implicit and explicit knowledge and processes (in these different subsystems, respectively). More importantly, in relation to motiva-tional issues, compared with other cognitive architectures, Clarion is distinguished by the fact that it contains built-in motivational constructs and built-in metacogni-tive constructs. These features are not commonly found in other existing cognitive architectures. Nevertheless, these features are crucial to the cognitive architecture, as they capture important elements in the interaction between an agent and its phys-ical and social world [50]. With these mechanisms, especially the motivational and metacognitive mechanisms, Clarion attempts to explain their functioning in concrete computational terms.

15.3   A Cognitive Architecture with Intrinsic Motivation

15.3.1   Overview of Clarion

Clarion provides structural and algorithmic specifications of a wide range of generic psychological processes. In particular, Clarion accounts for basic human motives, which provide the underlying basis for behavior. This emphasis on human motivation facilitates the integration of general cognitive capacities with considerations of motivation (as well as personality, emotion, culture, sociality, and so on) in a comprehensive and unified theory/model.

Only a sketch of Clarion can be presented below; the vast majority of technical details are omitted due to the page limit. See Fig. 15.1 for the overall structure of Clarion.

As shown by the figure, Clarion consists of a number of subsystems: the action-centered subsystem (denoted as the ACS), the non-action-centered subsystem (denoted as the NACS), the motivational subsystem (the MS), and the metacognitive subsystem (the MCS). The role of the action-centered subsystem is to control actions (regardless of whether they are for external physical movements or for internal mental operations), utilizing and maintaining procedural knowledge. The role of the non-action-centered subsystem is to maintain and utilize declarative knowledge. The role of the motivational subsystem is to provide underlying motivations for perception, action, and cognition (in terms of providing impetus and feedback). The role of the metacognitive subsystem is to monitor, direct, and modify the operations of the other subsystems dynamically

Each of these interacting subsystems consists of two "levels" of representations (i.e., a dual-representational structure, as theoretically posited in [45]). Generally speaking, in each subsystem, the "top level" encodes explicit knowledge[1]

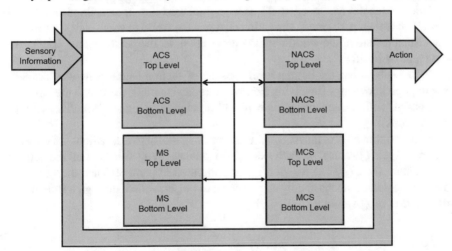

Fig. 15.1 The Clarion cognitive architecture. The subsystems of Clarion are shown. The major information flows are shown with arrows. ACS stands for the action-centered subsystem. NACS stands for the non-action-centered subsystem. MS stands for the motivational subsystem. MCS stands for the metacognitive subsystem

---

[1]Roughly speaking, explicit knowledge is directly consciously accessible (i.e., conscious or potentially conscious), while implicit knowledge is consciously inaccessible directly. Explicit processes involve explicit knowledge, while implicit processes involve implicit knowledge. The distinction has been based on voluminous empirical findings in many domains, but involves some nuances and some controversies. See [45, 50] for details.

(using symbolic/localist representations) and the "bottom level" encodes implicit knowledge (using distributed representations [32, 37]). The two levels interact, for example, by cooperating in action decision making, through integration of the action recommendations from the two levels of the ACS respectively, as well as by cooperating in learning through a "bottom-up" and a "top-down" learning process [45, 55].

Existing theories tend to confuse implicit and explicit processes; hence the "perplexing complexity" [43]. In contrast, Clarion generally separates and integrates implicit and explicit processes in each of its subsystems. With such a framework, Clarion can provide better explanations of empirical findings in a wide range of domains (for details, see [17, 45, 55]).

15.3.2 The Action-Centered Subsystem

The ACS captures the process of human action decision making as follows: Observing the current (observable) state of the world (including one's own motivational state), the two levels within the ACS (implicit or explicit) make their separate action decisions in accordance with their respective procedural knowledge (implicit or explicit), and their outcomes are "integrated". Thus, a final selection of an action is made and the action is then performed. The action changes the world in some way. Comparing the changed state of the world with the previous state, the person learns. The cycle then repeats itself.

In this subsystem, the bottom level consists of "action neural networks" encoding implicit knowledge (involving distributed representations [37]), and the top level consists of "action rules" encoding explicit knowledge (using symbolic/localist representations).

At the bottom level of the ACS, using an action neural network, actions are selected based on their Q values. At each step, given state x, the Q values of all the actions in that state (i.e., $Q(x, a)$ for all a's) are computed in parallel. Then the Q values are used to decide stochastically on an action to be performed, through a Boltzmann distribution of Q values:

$$p(a|x) = \frac{e^{\frac{Q(x,a)}{\tau}}}{\sum_i e^{\frac{Q(x,a_i)}{\tau}}}$$

where $p(a|x)$ is the probability of selecting action a, τ (temperature) controls the degree of randomness of action decision making, and i ranges over all possible actions. (This is known as Luce's choice axiom [61].)

For capturing learning of implicit knowledge at the bottom level (i.e., the Q values), the Q-learning algorithm [61], a reinforcement learning algorithm, may be applied. With this algorithm, Q values are gradually tuned through successive updating of a neural network, which enables reactive sequential behavior to emerge through trial-and-error interaction with the world (for details, see [45, 61]).

For capturing learning of explicit knowledge at the top level (i.e., action rules), a variety of algorithms may be applied, including the Rule-Extraction-Refinement (RER) algorithm [11] for a "bottom-up" learning process that relies on implicit knowledge from the bottom level to learn explicit knowledge at the top level [45]. In the reverse direction, "top-down" learning can also occur.

For stochastic selection of the outcomes of the two levels, at each step, each level (or a component within) is selected with a certain probability. There exists some psychological evidence for such intermittent use of rules [45]. The selection probabilities may be variable, determined by the metacognitive subsystem (by its processing mode module; more later; [50]).

15.3.3   The Non-Action-Centered Subsystem

The NACS is for dealing with declarative knowledge (which is not action-centered). It stores such knowledge in a dual representational form (the same as in the ACS): that is, in the form of explicit "associative rules" (at the top level), and in the form of implicit "associative memory networks" (at the bottom level). Its operation is under the control of the ACS and in the service of the ACS.

First, at the bottom level of the NACS, associative memory networks encode implicit declarative knowledge. Associations are formed by mapping an input pattern to an output pattern (e.g., using Backpropagation networks or Hopfield networks [37]).

Second, at the top level of the NACS, explicit declarative knowledge is stored. As in the ACS, each "chunk" node (denoting a concept) at the top level is linked to its corresponding microfeature nodes present at the bottom level. Additionally, in the top level, links between chunk nodes encode explicit associative rules. Explicit associative rules may be learned in a variety of ways [50].

As in the ACS, top-down or bottom-up learning may take place in the NACS, either to extract explicit knowledge at the top level from the implicit knowledge at the bottom level, or to assimilate the explicit knowledge of the top level into the implicit knowledge at the bottom level.

With the interaction of the two levels, the NACS carries out rule-based, similarity-based, and constraint-satisfaction-based reasoning (details can be found in [17, 50]). Their interaction enables the NACS to capture much of human reasoning [50].

15.3.4   The Motivational Subsystem

The MS is a critical part of the cognitive architecture. It is concerned with why an individual does what he/she does. The importance of the MS to the ACS lies in

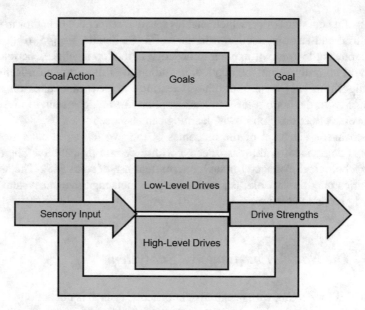

Fig. 15.2 The basic structure of the motivational subsystem

the fact that it provides the context in which goals and reinforcements of the ACS are determined. It thereby influences the working of the ACS (and by extension, the working of the NACS).

A dual motivational representation is in place in the MS. The explicit goals at the top level of the MS (such as "find food"), which are essential to the working of the ACS, may be generated based on implicit drives at the bottom level of the MS (e.g., "hunger"). See Fig. 15.2. For justifications, see [48].

At the bottom level of the MS, primary drives are those motives essential to an individual and most likely built-in (hard-wired) to a significant extent to begin with (i.e., they are "intrinsic"). Low-level primary drives (concerning mostly physiological needs) include: food, water, reproduction, and so on. Beyond low-level primary drives, there are also high-level primary drives: for example, achievement and recognition, affiliation and belongingness, dominance and power, fairness, autonomy, and so on (see [29, 34, 48, 58, 62]).[2] These primary drives have been justified in prior writings (as cited above).[3] See Table 15.1 for their specifications. On the basis of primary drives, secondary (derived) drives may be acquired.

---

[2]Note that a generalized notion of "drive" is adopted in Clarion. As discussed in [48], it is a generalized notion that transcends controversies surrounding the stricter notions of drive [18].

[3]Briefly, this set of hypothesized primary drives bears close relationships to Murray's needs [29], Reiss's motives [34], Schwartz's universal values [40], and so on. The prior justifications of these frameworks may be applied, to a significant extent, to this set of drives as well (see [25, 29, 34, 48]).

Table 15.1 Descriptions of the primary drives

Drives	Specifications
Food	The drive to consume nourishment
Water	The drive to consume liquid
Sleep	The drive to rest
Reproduction	The drive to mate
Avoiding Danger	The drive to avoid situations that have the potential to be harmful
Avoiding Unpleasant Stimuli	The drive to avoid situations that are physically (or emotionally) uncomfortable or negative in nature
Affiliation & belongingness	The drive to associate with other individuals and to be part of social groups
Dominance & power	The drive to have power over other individuals
Recognition & achievement	The drive to excel and be viewed as competent
Autonomy	The drive to resist control or influence by others
Deference	The drive to willingly follow or serve a person of a higher status
Similance	The drive to identify with other individuals, to imitate others, and to go along with their actions
Fairness	The drive to ensure that one treats others fairly and is treated fairly by others
Honor	The drive to follow social norms and codes and to avoid blames
Nurturance	The drive to care for, or attend to the needs of, others who are in need
Conservation	The drive to conserve, to preserve, to organize, or to structure (e.g., one's environment)
Curiosity	The drive to explore, to discover, and to gain new knowledge

Table 15.2 Approach versus avoidance primary drives

Approach drives	Avoidance drives	Both
Food	Sleep	Affiliation & belongingness
Water	Avoiding danger	Similance
Reproduction	Avoiding Unpleasant Stimuli	Deference
Nurturance	Honor	Autonomy
Curiosity	Conservation	Fairness
Dominance & Power		
Recognition & Achievement		

Some of these primary drives are approach-oriented, while others are avoidance-oriented. This distinction has been argued by many (e.g., [12, 16, 43]). The approach system is sensitive to cues signaling rewards, and results in active approach. The avoidance system is sensitive to cues of punishment, and results in avoidance, characterized by anxiety or fear. See Table 15.2 for this division of drives.

The processing of these drives within the bottom level of the MS involves a number of modules [50]. In particular, the core drive module determines drive strengths (using neural networks) based roughly on:

$$ds_d = gain_d \times stimulus_d \times deficit_d + baseline_d$$

where ds_d is the strength of drive d, $gain_d$ is the gain for drive d, $stimulus_d$ is a value representing how pertinent the current situation is to drive d, $deficit_d$ indicates the perceived deficit in relation to drive d (which represents an individual's internal inclination toward activating drive d), and $baseline_d$ is the baseline strength of drive d. The justifications for this may be found in the literature [50, 58, 60].

Motivational adaptation (learning) is also possible and has been tackled [50]. In addition, new drives (termed "derived drives") may be acquired. They may be gradually acquired through some kind of "conditioning", or may be externally set through externally provided instructions, on the basis of primary drives.

15.3.5 The Metacognitive Subsystem

Metacognition refers to active monitoring and consequent regulation and orchestration of one's own psychological processes [26, 33]. In Clarion, the MCS is closely tied to the MS. The MCS monitors, controls, and regulates other processes [42]. Control and regulation may be in the forms of setting goals (which are then used by the ACS) on the basis of drives, generating reinforcement signals for the ACS for learning (on the basis of drives and goals), interrupting and changing ongoing processes in the ACS and the NACS, setting essential parameters of the ACS and the NACS, and so on.

Structurally, this MCS may be divided into a number of functional modules, including:

• the goal module,
• the reinforcement module,
• the processing mode module.
• the input filtering module,
• the output filtering module,
• the parameter setting module (for setting learning rates, temperatures, etc.),

and so on. See Fig. 15.3.

For instance, the goal module selects goals to pursue (by the ACS). In order to select a new goal, it first determines goal strengths, based on information from the MS (e.g., the drive strengths) and the current sensory input. Then, a new goal is stochastically selected on the basis of the goal strengths (e.g., using a Boltzmann distribution). For arguments in support of goal setting on the basis of implicit motives (i.e., drives), see, for example, Tolman [59] and Deci [13]. In the simplest case, the following calculation is performed:

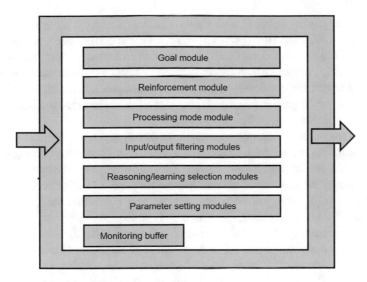

Fig. 15.3 The main modules within the metacognitive subsystem

$$gs_g = \sum_d relevance_{d,s \to g} \times ds_d$$

where gs_g is the strength of goal g, $relevance_{d,s \to g}$ is a measure of how relevant drive d is to goal g with regard to current situation s (which represents the support that drive d provides to goal g), and ds_d is the strength of drive d (from the MS). Once calculated, the goal strengths are turned into a Boltzmann distribution (as discussed earlier) and the new goal is chosen stochastically from that distribution.

For another instance, the processing mode module determines the probability of each component (a level or a component within) for the integration of outcomes from the two levels of the ACS (see the discussion of the ACS earlier). These probabilities may be determined through the notion of "probability matching": the probability of selecting a component is determined based on the relative success ratio of that component (see [45, 50] for details). However, these probabilities may be modulated multiplicatively by another parameter: the strength of avoidance-oriented primary drives (which corresponds to "anxiety" [65, 66]; see more details below).

15.4 Some Examples of Simulations

Clarion, as a framework, has been justified and validated extensively on the basis of psychological data and their simulations; see, for example, Sun [45, 50] for summaries of such justifications and validations. In particular, Clarion has been successful in simulating, accounting for, and explaining a wide variety of psychological data. For example, a number of well-known skill learning tasks have been captured, simulated, and explained using Clarion that span the spectrum ranging from simple reactive skills to complex cognitive skills. Simulations have also been done with reasoning tasks, metacognitive tasks, and motivational tasks, as well as social interaction tasks, all of which are important to autonomous intelligent agents. While accounting for various psychological data, Clarion provides explanations that shed new light on relevant phenomena, especially on the basis of motivation.

Let us look into an example of social simulation involving agents with intrinsic motivation (as originally described in [51]). A significant shortcoming of many computational social simulations is that they assume very rudimentary cognition/psychology on the part of agents. Although agents are often characterized as being "cognitive", there has been relatively scarce effort that carefully captures human psychology in detailed, process-based, and quantitative ways. Models of agents have frequently been custom-tailored to the task at hand, often amounting to a set of highly domain-specific rules. Although such an approach may be adequate for achieving some limited objectives of some specific simulations, it falls short intellectually [53, 54]: It not only limits the realism of social simulations, but also precludes the possibility of fully tackling the question of the micro-macro link [2, 38, 44] in terms of psychological-social interaction, for example, how the social emerges from the psychological [44, 47, 51].

Thus, let us first look into an existing social simulation as an illustration. In the work of Cecconi and Parisi [10], social groups (tribes) were simulated. In these groups, to survive and to reproduce, an agent must possess resources. A group in which each agent uses only its own resources is said to adopt an individual survival strategy. However, in some other groups, resources may be transferred among agents—such a group is said to adopt a social survival strategy. For instance, the "central store" (CS) is a mechanism to which all the individuals in a group transfer (part of) their resources. The resources collected by the CS can be redistributed to the members of the group in some fashion [10].

In Cecconi and Parisi [10], a number of simulations were conducted comparing groups adopting individual strategies with groups adopting CS strategies. Agents survived and reproduced differentially based on the quantity of food that they were able to consume. This task has the potential for exploring a wide range of issues, ranging from individual behaviors to social institutions, and from individual learning to evolution.

However, in that work, there was very little in the way of psychological processes. Such work needs a better understanding, and better models, of the individual mind, because only on the basis of that understanding, better understanding of aggregate processes can be developed. Accurate, detailed cognitive/psychological models may provide better grounding for understanding multi-agent social phenomena, by incorporating realistic constraints, capabilities, and tendencies of individual agents. This point was argued in Sun [44]. In Axelrod [4], it was shown that even adding a cognitive factor as simple as memory of past several events into an agent model can completely alter the dynamics of social interaction (e.g., in the iterated prisoner's dilemma).

Thus, we conducted simulations based on the Clarion cognitive architecture. In our simulation, the world was made up of a 200×200 grid. Each of these 40,000 locations might contain (at most) one food item. At the beginning and every 40 cycles, the grid was replenished: Randomly selected locations were restocked with food items (until the grid had 2400 food items). A more benign condition, in which 3600 locations contained one food item each, and a harsher condition, in which 1200 locations contained one food item each, were also tested. A food item contained 50 energy units.

Each agent began with 60 units of energy, and consumed one unit of energy per cycle. Each agent lived for a maximum of 350 cycles, but might die early due to lack of food. There were initially 120 agents to begin with, and the number of agents fluctuated due to birth and death, within the bound of a maximum of 120 agents.

At each moment, each agent was in a particular location on the grid. It faced a certain direction (north, south, east, or west). Each received inputs regarding the location of the nearest food. Each agent could generate an action output: either (1) turn 90 degrees right, (2) turn 90 degrees left, (3) move forward, (4) pick up food and contribute a portion, (5) pick up food and keep all of it, or (6) reproduce.

As in the previous simulations, procreation was asexual; procreation occurred if an agent had reached 120 energy units or more, and there were fewer than the maximum number of agents in the world. The parent handed out 60 energy units to the child upon its birth. The child inherited its parent's internal makeup, although when a child was spawned, there was a 10% chance of minor mutation.

When a central store was involved, an agent was required to contribute 20 energy units to the central store when it picked up a food item (50 energy units). At each cycle, agents with 10 or less energy might receive 5 energy units each from the central store; up to a maximum of 10% of the agent population might get energy from the central store at each cycle. Each agent, when picking up a piece of food, decided whether to contribute to the central store or not. There were three variations on cheater detection and punishment, ranging from full detection and full punishment to no detection and no punishment.

Each agent had three intrinsic drives: food, reproduction, and honor. They competed to influence behavior (action) through determining the current goal (e.g., to pursue food or reproduction, or to contribute or not to the CS). The internal reinforcements for their actions were determined based on their drives and goals, as well as the state of the world.

The results of the simulation demonstrated effects of motivational factors (which were not investigated in the previous simulations). As predicted, motivational factors had a significant effect on the outcome of the simulation. In this regard, "gains" was a variable created for the sole purpose of analysis; it consolidated the three drive gain parameters (for food, reproduction, and honor, respectively) into one. There were eight values, ranging from "All 0.5" to "All 1.0"; for example, "Honor 0.5" meant that the gain parameter of the Honor drive was 0.5 and all the other drive gains were 1.0 (see Fig. 15.4 for the complete list).

Examining the results in Fig. 15.4, there was a significant effect of "gains" on lifespan. Generally speaking, more emphasis on food (a higher drive gain for food) led

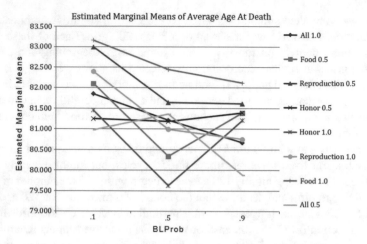

Fig. 15.4 The effect of gains on lifespan. The y-axis represents lifespan. The x-axis represents probability of using the bottom level. The different lines represent different drive gain settings

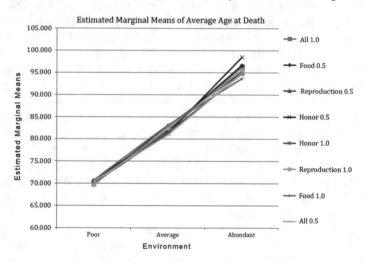

Fig. 15.5 The interaction of gains and environment on lifespan. The y-axis represents lifespan. The x-axis represents environment. The different lines represent different drive gain settings

to better performance (e.g., "Food 1.0" or "Reproduction 0.5"). Reduced emphasis on food generally led to worse performance (e.g. "Food 0.5" or "Honor 1.0").

There was also a significant interaction between "gains" and environment on lifespan, as shown in Fig. 15.5. An interpretation of this result was this: In a more benign environment, less focus on honor (e.g., "Honor 0.5") helped survival, but in a harsh environment, drive focuses did not make much difference because one had to focus only on food in order to survive.

Other related simulations of motivation or motivation-cognition interaction may be found in Wilson et al. [65, 66], especially in relation to effects of anxiety [7, 8, 22]. A model of personality was also developed on the basis of motivation within Clarion; see, for example, Sun and Wilson [56] (see also [14, 28, 31, 41]). Further, a model of moral judgment was developed on the basis of motivation; see [9, 49]. A model of emotion was also developed on the basis of motivation within Clarion; see Sun et al. [57] (see also [52, 63, 64]).

15.5 Concluding Remarks

The Clarion project addresses essential human-like motivational processes, mechanisms, structures, and representations necessary for a comprehensive cognitive architecture. The need for implicit drive representations, as well as explicit goal representations, has been hypothesized. Drive representations consist of primary drives (both low-level and high-level primary drives), as well as derived (secondary) drives. On the basis of drives, explicit goals may be generated on the fly during an agent's interaction with various situations, which in turn guide action selection.

The afore-discussed motivational representations and their resulting dynamics help to make a computational cognitive architecture more complete and functioning in a more psychologically realistic way. I believe that this constitutes a requisite step forward in making computational cognitive architectures more realistic models of the human mind taking into considerations all of its complexity and intricacy, especially in terms of its complex motivational dynamics. It is highly relevant to building truly autonomous and trust-worthy computational agents capable of functioning in complex, uncertain, and unpredictable environments. Note that what I emphasize here is human-like full autonomy and human-like trust.

Significant future challenges in furthering this line of work exist, including, for example, applying this framework to the building of intelligent application systems that can display intelligent behavior with more robustness, flexibility, and versatility. Another significant challenge is to further validate, through empirical work (especially psychological empirical work), this framework and its implications for understanding human motivation and trust (in addition to building intelligent agents). Many more experiments, simulations, and tests will be needed and shall be pursued in the future.

Acknowledgements This work has been supported in part by ONR grants N00014-08-1-0068 and N00014-13-1-0342.

References

1. H.A. Abbass, G. Leu, K. Merrick, A review of theoretical and practical challenges of trusted autonomy in big data. IEEE Access **4**, 2809–2830 (2016)
2. J. Alexander, B. Giesen, R. Munch, N. Smelser (eds.), *The Micro-Macro Link* (University of California Press, Berkeley, CA, 1987)
3. J.R. Anderson, C.J. Lebiere, *The Atomic Components of Thought* (Lawrence Erlbaum Associates, Mahwah, NJ, 1998)
4. R. Axelrod, *The Evolution of Cooperation* (Basic books, New York, 1984)
5. J. Bach, *Principles of Synthetic Intelligence* (Oxford University Press, New York, 2009)
6. G. Baldassarre, M. Mirolli (eds.), *Intrinsically Motivated Learning in Natural and Artificial Systems* (Springer, Berlin-Heidelbergr, 2013)
7. S.L. Beilock, T.H. Carr, On the fragility of skilled performance: What governs choking under pressure? J. Exp. Psychol. Gen. **130**(4), 701–725 (2001)
8. S.L. Beilock, C.A. Kulp, L.E. Holt, T.H. Carr, More on the fragility of performance: Choking under pressure in mathematical problem solving. J. Exp. Psychol. Gen. **133**(4), 584–600 (2004)
9. S. Bretz, R. Sun, Two models of moral judgement (2016). Submitted
10. F. Cecconi, D. Parisi, Individual versus social survival strategies. J. Artif. Soc. Soc. Simul. **1**(2), 1–17 (1998)
11. A. Clark, A. Karmiloff-Smith, The cognizer's innards: A psychological and philosophical perspective on the development of thought. Mind Lang. **8**(4), 487–519 (1993)
12. L.A. Clark, D. Watson, *Handbook of personality: Theory and research, chapter Temperament: A new paradigm for trait psychology* (Guilford Press, New York, 1999), pp. 399–423
13. E. Deci, *Personality: Basic Issues and Current Research, chapter Intrinsic motivation and personality* (Prentice Hall, Englewood Cliffs, NJ, 1980), pp. 35–80
14. J.M. Digman, Personality structure: Emergence of the five-factor model. Annu. Rev. Psychol. **41**(1), 417–440 (1990)
15. D. Dörner, The mathematics of emotions. *Proceedings of the Fifth International Conference on Cognitive Modeling* (2003), pp. 75–79
16. J.A. Gray, N. McNaughton, *The Neuropsychology of Anxiety: An Enquiry into the Function of the Septo-hippocampal System* (Oxford University Press, New York, 2000)
17. S. Hélie, R. Sun, Incubation, insight, and creative problem solving: A unified theory and a connectionist model. Psychol. Rev. **117**(3), 994–1024 (2010)
18. C.L. Hull, *Principles of Behavior: An Introduction to Behavior Theory* (D. Appleton-Century Company, 1943)
19. W. James, *The Principles of Psychology* (Dover, New York, 1890)
20. D. Klahr, P. Langley, R. Neches, *Production System Models of Learning and Development* (MIT press, Cambridge, MA, 1989)
21. J.E. Laird, *The Soar Cognitive Architecture* (MIT Press, Cambridge, MA, 2012)
22. A.J. Lambert, B. Keith Payne, L.L. Jacoby, L.M. Shaffer, A.L. Chasteen, S.R. Khan, Stereotypes as dominant responses: On the social facilitation of prejudice in anticipated public contexts. J. Pers. Soc. Psychol. **84**(2), 277–295 (2003)
23. J.D. Lee, K.A. See, Trust in automation: Designing for appropriate reliance. Hum. Factors J. Hum. Factors Ergon. Soc. **46**(1), 50–80 (2004)
24. R.J. Lewicki, D.J. McAllister, R.J. Bies, Trust and distrust: New relationships and realities. Acad. Manag. Rev. **23**(3), 438–458 (1998)
25. A.H. Maslow, A theory of human motivation. Psychol. Rev. **50**(4), 370–396 (1943)
26. G. Mazzoni, T.O. Nelson (eds.), *Metacognition and Cognitive Neuropsychology: Monitoring and Control Processes* (Erlbaum, Mahwah, NJ, 1998)
27. W. McDougall, *An Introduction to Social Psychology* (Methuen & Co., London, 1936)
28. W. Mischel, Y. Shoda, Reconciling processing dynamics and personality dispositions. Annu. Rev. Psychol. **49**, 229–258 (1998)
29. H.A. Murray, *Explorations in Personality* (Oxford University Press, New York, 1938)

30. A. Newell, *Unified Theories of Cognition* (Harvard University Press, Cambridge, MA, 1990)
31. S.J. Read, B.M. Monroe, A.L. Brownstein, Y. Yang, G. Chopra, L.C. Miller, Virtual personalities II: A neural network model of the structure and dynamics of human personality. Psychol. Rev. **117**(1), 61–92 (2010)
32. A.S. Reber, Implicit learning and tacit knowledge. J. Exp. Psychol. Gen. **118**(3), 219–235 (1989)
33. L.M. Reder, *Implicit Memory and Metacognition* (Erlbaum, Mahwah, NJ, 1996)
34. S. Reiss, Multifaceted nature of intrinsic motivation: The theory of 16 basic desires. Rev. Gen. Psychol. **8**(3), 179–193 (2004)
35. P.S. Rosenbloom, J. Laird, A. Newell, *The SOAR Papers: Research on Integrated Intelligence* (MIT Press Cambridge, MA, 1993)
36. D.M. Rousseau, S.B. Sitkin, R.S. Burt, C. Camerer, Not so different after all: A cross-discipline view of trust. Acad. Manag. Rev. **23**(3), 393–404 (1998)
37. D.E. Rumelhart, J.L. McClelland, *Parallel Distributed Processing: Explorations in the Microstructures of Cognition* (MIT Press, Cambridge, MA, 1986)
38. R.K. Sawyer, Artificial societies multiagent systems and the micro-macro link in sociological theory. Sociol. Methods Res. **31**(3), 325–363 (2003)
39. K.E. Schaefer, D.R. Billings, J.L. Szalma, J.K. Adams, T.L. Sanders, J.Y. Chen, P.A. Hancock, *A meta-analysis of factors influencing the development of trust in automation: Implications for human-robot interaction* (Human Factors, to appear, Technical report, 2016)
40. S.H. Schwartz, Are there universal aspects in the structure and contents of human values? J. Soc. Issues **50**(4), 19–45 (1994)
41. Y. Shoda, W. Mischel, *Connectionist models of social reasoning and social behavior, chapter Personality as a stable cognitive-affective activation network: Characteristic patterns of behavior variation emerge from a stable personality structure* (Lawrence Erlbaum Associates Inc, Mahwah, NJ, 1998), pp. 175–208
42. H.A. Simon, Motivational and emotional controls of cognition. Psychol. Rev. **74**(1), 29–39 (1967)
43. L.D. Smillie, A.D. Pickering, C.J. Jackson, The new reinforcement sensitivity theory: Implications for personality measurement. Pers. Soc. Psychol. Rev. **10**(4), 320–335 (2006)
44. R. Sun, Cognitive science meets multi-agent systems: A prolegomenon. Philos. Psychol. **14**(1), 5–28 (2001)
45. R. Sun, *Duality of the Mind: A Bottom-up Approach Toward Cognition* (Lawrence Erlbaum Associates, Mahwah, NJ, 2002)
46. R. Sun, A tutorial on CLARION 5.0. Technical report, Cognitive Science Department, Rensselaer Polytechnic Institute, 2003
47. R. Sun (ed.), *Cognition and Multi-agent Interaction: From Cognitive Modeling to Social Simulation* (Cambridge University Press, 2006)
48. R. Sun, Motivational representations within a computational cognitive architecture. Cogn. Comput. **1**(1), 91–103 (2009)
49. R. Sun, Moral judgment, human motivation, and neural networks. Cogn. Comput. **5**(4), 566–579 (2013)
50. R. Sun, *Anatomy of the Mind: Exploring Psychological Mechanisms and Processes with the Clarion Cognitive Architecture* (Oxford University Press, 2016)
51. R. Sun, P. Fleischer, A cognitive social simulation of tribal survival strategies: The importance of cognitive and motivational factors. J. Cogn. Cult. **12**(3–4), 287–321 (2012)
52. R. Sun, R.C. Mathews, Implicit cognition, emotion, and meta-cognitive control. Mind Soc. **11**(1), 107–119 (2012) (special Issue on Dual Processes Theories of Thinking (ed. by D. Over, L. Macchi, R. Viale))
53. R. Sun, I. Naveh, Simulating organizational decision-making using a cognitively realistic agent model. J. Artif. Societies Soc. Simul. **7**(3), (2004)
54. R. Sun, I. Naveh, Social institution, cognition, and survival: A cognitive-social simulation. Mind Soc. **6**(2), 115–142 (2007)

55. R. Sun, P. Slusarz, C. Terry, The interaction of the explicit and the implicit in skill learning: A dual-process approach. Psychol. Rev. **112**(1), 159–192 (2005)
56. R. Sun, N. Wilson, A model of personality should be a cognitive architecture itself. Cogn. Syst. Res. **29–30**, 1–30 (2014)
57. R. Sun, N. Wilson, M. Lynch, Emotion: A unified mechanistic interpretation from a cognitive architecture. Cogn. Comput. **8**(1), 1–14 (2016)
58. F.M. Toates, *Motivational Systems* (Cambridge University Press, Cambridge, UK, 1986)
59. E.C. Tolman, *Purposive Behavior in Animals and Men* (Century, New York, 1932)
60. Toby Tyrrell. *Computational mechanisms for action selection*. PhD thesis, Oxford University, Oxford, UK, 1993
61. C.J.C.H. Watkins. *Learning from delayed rewards*. PhD thesis, University of Cambridge England, 1989
62. B. Weiner, *Human Motivation: Metaphors, Theories, and Research* (Sage, Newbury Park, CA, 1992)
63. N. Wilson, *Towards a psychologically plausible comprehensive computational theory of emotion*. PhD thesis, Cognitive Sciences Department, Rensselaer Polytechnic Institute, Troy, NY, 2012
64. N. Wilson, R. Sun, Coping with bullying: a computational emotion-theoretic account. *Proceedings of the 36th Annual Conference of the Cognitive Science Society* (Quebec City, Quebec, Canada, 2014), pp. 3119–3124 (Cognitive Science Society, Austin, Texas)
65. N. Wilson, R. Sun, R. Mathews, A motivationally based computational interpretation of social anxiety induced stereotype bias. *Proceedings of the 32nd Annual Meeting of the Cognitive Science Society* (Lawrence Erlbaum Associates, Mahwah, NJ, 2010), pp. 1750–1755
66. N. Wilson, R. Sun, R.C. Mathews, A motivationally-based simulation of performance degradation under pressure. Neural Netw. **22**(5), 502–508 (2009)
67. I. Wright, A. Sloman, Minder 1: An implementation of a protoemotional agent architecture. Technical Report CSRP-97-1, University of Birmingham, School of Computer Science, 1997

Chapter 16
Computational Motivation, Autonomy and Trustworthiness: Can We Have It All?

Kathryn Merrick, Adam Klyne and Medria Hardhienata

16.1 Autonomous Systems

In the past fifty years we have quickly moved from controlled systems to supervised systems [6], automatic systems and autonomous systems [23, 26]. Autonomous systems are highly adaptive systems that sense the environment and learn to make decisions about their own actions. They may display a high degree of proactivity, self-organization or self-motivation [31, 43], in reaching their objectives.

Autonomous systems may operate without the presence of a human. Alternatively, they may communicate, cooperate, and negotiate with humans to reach their goals. Thus, a complementary strand of research over the past decades has studied such man-computer symbiosis [25], including research that studies systems that can adapt their own level of automation [32], and systems that can achieve cognitive-cyber symbiosis [2].

There is a clear benefit for society if repetitive or dangerous tasks can be performed by machines. Yet, there are also barriers to the adoption of increasingly sophisticated technology. These barriers include both functionality related concerns—particularly in extreme, severe, complex and dynamic environments—as well as legal, ethical, social, safety and regulatory concerns [1].

In fact, many of these issues are related in some way to the level of trust held in autonomous system technologies. Trust is a pervasive concept that influences decision-making when the actions of one system (or agent[1]) can have an impact on another agent [3, 34]. Many definitions of trust have been proposed [4, 17, 22, 27]. At one level, trust can be defined as social contract between two agents [4]. A truster delegates a task to a trustee, and assumes the risk that the trustee might be untrustworthy. The trustee accepts the task, implicitly or explicitly promising to be trustworthy. The truster's decision to trust the trustee is influenced by the truster's

---

[1]Agent here can refer to a human, organization, or software system.

K. Merrick (✉) · A. Klyne · M. Hardhienata
School of Engineering and Information Technology,
University of New South Wales, Canberra, Australia
e-mail: k.merrick@adfa.edu.au

© The Author(s) 2018
H. A. Abbass et al. (eds.), *Foundations of Trusted Autonomy*, Studies in Systems,
Decision and Control 117, https://doi.org/10.1007/978-3-319-64816-3_16

attitude towards risk.[2] Trust involves the judgement of the truster in relation to the trustee agent based on the integration of the truster's cognitive attributes and life experience.

This chapter considers the impact of one of the emerging mechanisms for achieving autonomy—computational motivation—on the trustworthiness of autonomous systems. Motivation is the cause of action in natural systems (such as humans) [18]. Like trust, motivation has been defined from different perspectives. For motivation, this includes perspectives of drive, arousal, risk attitude, social attitude, expectancy, incentive, trait theory, attribution theory and approach-avoidance theory. Also like trust, motivation is understood to be influenced by the integration of an agent's cognitive attributes and life experience.

The concepts of motivation and trust overlap at least along the dimensions of risk attitude, social attitude and assimilation of life experience and cognitive attributes. (1) Agents with different motive profiles may act differently in the same situation as a result of different life experiences; (2) Differences in the motive profiles of agents (including risk and social attitude) may affect their ability to trust; and (3) Differences in the actions of agents with different motives may affect their trustworthiness.

This chapter will focus primarily on points (1) and (3) above. First, we consider the implications of motivation for functionality (Sect. 16.3) and then the implications for trustworthiness (Sect. 16.4). Point (2) above has not been widely examined from a computational perspective. To make our discussion concrete, in this chapter we consider these issues in the context of intrinsically motivated agent swarms. Many key variants of computational motivation have been considered for use in swarm systems, making this a timely and relevant for discussion. Section. 16.2 begins by providing an overview of the theory underlying the use of computational motivation in swarms of artificial agents, including a uniform notation for three intrinsically motivated swarm algorithms.

16.2 Intrinsically Motivated Swarms

At the heart of computational models of flocks, herds, schools, swarms and crowd behavior is Reynold's iconic boids model [35]. The boids model can be viewed as a kind of rule-based reasoning in which rules take into account certain properties of other agents. The three fundamental rules are:

- **Cohesion:** Each agent moves toward the average position of its neighbors;
- **Alignment:** Each agent steers so as to align itself with the average heading of its neighbors;
- **Separation:** Agents move to avoid hitting their neighbors.

---

[2]Risk here is the potential of losing something of value, weighed against the potential to gain something of value (an incentive) [8].

Each *boid* in a computational swarm applies these three rules at each time step. The rules are implemented as forces that act on agents when a certain condition holds. Suppose we have a group of n agents $A^1, A^2, A^3 \ldots A^n$. At time t each agent A^j has a position, x_t^j, and a velocity, v_t^j. x_t^j is a point and v_t^j is a vector. At each time step t, the velocity of each agent is updated as follows:

$$v_{(t+1)}^j = W_d v_t^j + W_c c_t^j + W_a a_t^j + W_s s_t^j \qquad (16.1)$$

c_t^j is a vector in the direction of the average position of agents within a certain range of A^j (called the neighbours of A^j); a_t^j is a vector in the average direction of agents within a certain range of A^j; and s_t^j is a vector in the direction away from of the average position of agents within a certain range of A^j. These vectors are the result of cohesive, alignment and separation forces corresponding to the rules outlined above. Weights W_c, W_a and W_s strengthen or weaken the corresponding force. W_d strengthens or weakens the perceived importance of the *boid*'s existing velocity. Once a new velocity has been computed, the position of each agent is updated by:

$$x_{(t+1)}^j = x_t^j + v_{(t+1)}^j \qquad (16.2)$$

As noted above, agents that are within a certain range of a particular agent A^j are called its neighbors. Formally, we can define a subset N^j of agents within a certain range R of A^j as follows:

$$N^j = A^k | A^k \neq A^j \wedge dist(A^k, A^j) < R \qquad (16.3)$$

where $dist(A^k, A^j)$ is generally the Euclidean distance between two agents. Different ranges may be used to calculate cohesive, alignment and separation forces, or other factors such as the communication range of a *boid*. The average position $\mathbf{c}_t^j$ of agents within range R_c of A^j is calculated as:

$$\mathbf{c}_t^j = \frac{\sum_k x_t^k}{|(N_c)_t^j|} \qquad (16.4)$$

The vector in the direction of this average position is calculated as:

$$c_t^j = \mathbf{c}_t^j - x_t^j \qquad (16.5)$$

Similarly, we can calculate the average position $\mathbf{s}_t^j$ of agents within range R_s of A_j as:

$$\mathbf{s}_t^j = \frac{\sum_k x_t^k}{|(N_s)_t^j|} \qquad (16.6)$$

The vector away from this position is calculated as:

$$s_t^j = x_t^j - \mathbf{s}_t^j \tag{16.7}$$

Finally, the vector a_t^j in the average direction of agents within range R_a of A^j, is calculated by the sum:

$$a_t^j = \frac{\sum_k v_t^k}{|(N_a)_t^j|} \tag{16.8}$$

The basic *boid* algorithm does not incorporate mechanisms for limiting velocity, preventing a *boid* from exiting some predefined area or to permitting a boid to avoid an obstacle. Likewise, it does not include mechanisms for goal-directed behavior. However, these have been modelled in other swarm algorithm variants. One example of an update that includes forces in the direction of a goal is:

$$v(t+1)^j = W_d v_t^j + c_1 r_1 (p_t^j - x_t^j) + c_2 r_2 (g_t - x_t^j) \tag{16.9}$$

Equation 16.9 is, in fact, the particle swarm optimization (PSO) update [9]. The terms $(p_t^j - x_t^j)$ and $(g_t - x_t^j)$ are forces in the direction of goals G^p and G^g, which have positions p_t^j and g_t respectively. G^p is defined as a goal to reach an agent's personal best or 'fittest' position found so far. G^g is defined as a goal to reach the globally fittest position found so far by all swarm members. r_1 and r_2 are numbers selected from a uniform distribution between 0 and 1. c_1 and c_2 are acceleration coefficients. Parameter values for W_d, c_1 and c_2 have been experimentally derived by Eberhart and Shi [10].

We now consider three algorithms for intrinsically motivated swarms, using the notion introduced above.

16.2.1 Crowds of Motivated Agents

Algorithm 1 models motivation as rules for the application of forces for intrinsic motivation in a *boids* framework. Various intrinsic motivations have been considered for use in swarms, including novelty [21], curiosity [37], achievement, affiliation and power motivation [14, 28]. Algorithm 1 introduces a simple form of motivation as an optimally motivating incentive (OMI), Ω^j [28]. This simple representation of motivation stipulates an incentive values that the agent finds maximally motivating. Other incentives are less motivating, with motivation inversely proportional to the difference between a goal's incentive $I(G)$ and the agent's OMI. That is:

$$M^j(G) = I^{max} - |I(G) - \Omega^j| \tag{16.10}$$

where I^{max} is the maximum available incentive. Using this approach power, affiliation, achievement or curiosity motivated agents can be defined as follows:

- **Power motivated:** power motivated individuals seek to control the resources or reinforcers of others. Thus, they tend to exhibit a preference for high-incentive goals. In the model above, power-motivated agents will have values for Ω^j that fall in the upper third of the range $[I^{min}, I^{max}]$ [28].
- **Affiliation motivated:** affiliation motivated individuals seek to avoid conflict and thus often exhibit preferences for low-incentive goals (that are not desirable to others). In the model above, affiliation-motivated agents will have values for Ω^j that fall in the lower third of the range $[I^{min}, I^{max}]$ [28].
- **Achievement motivated:** achievement motivated individuals prefer goals with a moderate probability of success. They may make a simplifying assumption that this is implied by moderate incentive. Thus, in the model above, achievement-motivated agents will have values for Ω^j that fall in the middle third of the range $[I^{min}, I^{max}]$ [28].
- **Curiosity motivated:** curious agents prefer to approach goals that are 'similar-yet-different' to goals they have encountered before. In this one-dimensional model where the only attribute of a goal is its incentive, curious agents will prefer incentives that are moderately different to previously encountered incentives and that they have not encountered recently.

It should be noted that the definitions above are one dimensional, incentive-based definitions of power, affiliation, achievement and curiosity. More complex/expressive definitions exist, both in motivation theory [18] and in the literature of computational motivation [28, 29, 38, 41, 42]. Some of the latter are discussed later in this chapter, as well as Sects. 3.7 and 14.5 of this book. The advantage of the one-dimensional models discussed here is that they are computationally inexpensive, even in large numbers of agents.

The remainder of the algorithm proceeds as follows: Each agent in the swarm is initialized with an OMI, Ω^j (line 1) [30]. At each time step, each agent senses the local state of its environment (line 4), including the features described above for position, velocity and neighbors within different ranges. Each agent then constructs a set G_t^j of highly motivating goals that conform to a condition on the current state (line 5). For example, the condition might concern proximity to a goal and level of motivation:

$$G_t^j = G^i | dist(g_t^i, x_t^j) < R_m \wedge M^j(G) > M \qquad (16.11)$$

R_m is the range within which goals are considered and M is a motivation threshold. A force in the direction of each goal is included in the update equation for the agent (line 7) as follows:

$$v_{(t+1)}^j = W_d v_t^j + W_c c_t^j + W_a a_t^j + W_s s_t^j + W_m \sum_i (g_t^i - x_t^j) \qquad (16.12)$$

Finally all agents are moved to their new positions (line 8). Algorithm 1 assumes that all goals and their locations are known by all agents, and that goals are generated

---

Algorithm 1 A swarm of motivated agents. Adapted from [28].

1: Initialise n and a society A of n agents with position, velocity, weights, ranges and optimally
    motivating incentive Ω^j.
2: **for** each time t **do**
3:    **for** each agent A^j **do**
4:        Sense the current local state $< x_t^j, v_t^j, (N_c^j)_t, (N_s^j)_t, (N_a^j)_t >$
5:        Construct goal set G_t^j according to Eq. 16.11.
6:        Compute $(g_t^i - x_t^j)$ for all G^i
7:        Sum all forces on agent A^j using Eq. 16.12.
8:        Move all agents to new positions according to Eq. 16.2.

---

by an entity external to the swarm. The next section considers an algorithm in which
the swarm itself generates goals dynamically, while agents are exploring.

16.2.2 Motivated Particle Swarm Optimization for Adaptive Task Allocation

Another approach to a motivated swarm is to integrate intrinsic motivation with PSO
for the purpose of adaptive task allocation [21]. Intrinsically motivated PSO (MPSO)
can be used for search and allocation of resources to tasks, when the nature of the
target task is not well understood in advance, or can change over time.

This algorithm has two parts: the first for motivation and the second for PSO as
shown in Fig. 16.1. The input to the motivation component is spatially mapped sensor
data $p_t(x)$ where x specifies the location from which the data were collected as a
Cartesian coordinate and t is the time at which the data were collected. It is assumed
that a stream of this data is input to the system. When data are collected at more than
one location at time t, individual data points are denoted $p_t(x_\tau)$. The output of the
motivation component, and input to the PSO, is a fitness function $F_t(x)$ as shown in
Fig. 16.1.

We denote M_τ the motivation value of $p_t(x_\tau)$. In this algorithm, M_τ is assumed
to be binary, with 1 denoting a motivating stimulus and 0 denoting a non-motivating
stimulus. M_τ is computed by thresholding models of motivation that return a continuous value. Four such models were described above. Another example is and
arousal-based model of curiosity using a novelty function, as described by Klyne
and Merrick [21] and illustrated in Fig. 16.2.

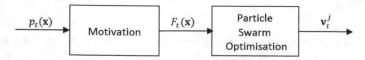

Fig. 16.1 Motivated particle swarm optimization. Image adapted from [21]

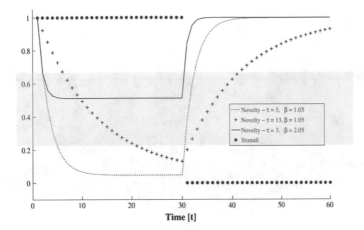

Fig. 16.2 Novelty is computed as a function of the time that a stimulus has been firing or not firing. Motivation is a binary value determined as a threshold on novelty. Motivation is 1 in the shaded area, and 0 otherwise

In this model, potential goals have multiple attributes and are represented as vectors. These vectors are clustered using an unsupervised learning algorithm such as a self-organising map (SOM), k-means clustering, adaptive resonance theory (ART) network or simplified ART network.

Neurons or cluster centres from the unsupervised learning algorithm have associated habituating units that compute novelty as shown in Fig. 16.2. The dotted series in Fig. 16.2 illustrates the activation value of a neuron that fires repeatedly (30 times), then does not fire (30 times). The solid, dashed and + series are examples of different novelty curves calculated using Stanley's model [40]. $M_\tau = 1$ when novelty is in the moderate range shown in grey in Fig. 16.2 and zero otherwise. Because novelty is influenced by the agent's experiences, as stored in their unsupervised learning component, different agents may compute different novelty values for a given stimulus because their experiences are different.

M_τ is a parameter of the fitness function, which is defined using an intensity landscape, as follows:

$$I_\tau(x) = \frac{M_\tau}{(1 + \gamma(\sum_{(y=1)}^{Y}(x^y - x_\tau^y)^2))} \tag{16.13}$$

This function forms a graduated peak with a maximum at the coordinate x_τ. The range of x is the range of the problem space. y is the counter for dimensions of the problem space. M_τ controls the maximum height of a peak on the fitness function. γ controls the gradient of a peak. Lower values make the gradient gentler. The fitness function itself is then constructed by summing intensity functions for motivating sensor data as follows:

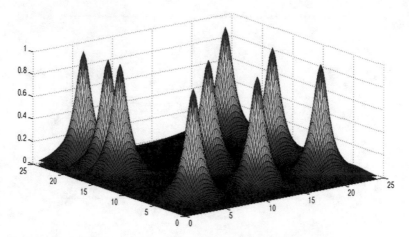

Fig. 16.3 A synthetic fitness landscape generated with Eq. 16.13 using 9 motivating points at positions (4, 4), (4, 18), (4, 19), (4, 22), (12, 4), (12, 12), (20, 4), (20, 12) and (21, 22)

$$F_t(x) = \sum_\tau I_\tau(x) \qquad (16.14)$$

Using a sum implies that the size of an area of motivating data will influence the height of the fitness function. As an example, a synthetic fitness landscape generated using Eq. 16.13 nine motivating locations is shown in Fig. 16.3.

Algorithm 2 shows how the fitness function is incorporated with PSO in two phases: (1) a settling phase and (2) the MPSO phase. $F_0(x)$ is initialized as zero for all x. The settling phase of the algorithm (lines 3 to 8) determines the level of background noise in the fitness function by observing the environment for a fixed period T. A 'noise floor' α is then chosen by monitoring the maximum height of the generated fitness function at the end of the initialization period.

The noise floor is used to influence the inertial value of the motivated PSO phase, which commences at time $T + 1$. The motivated PSO loop (lines 10 to 21) alternates between motivation to compute an updated fitness function (lines 12 to 17) and optimization of the current fitness function (lines 18 to 21). When a motivation phase occurs, the fitness function and the values of p_t^j and g_t are reset to zero, so the swarm to diverges. The condition described in Eq. 16.15 is applied so that the inertial value W_d in the PSO update is only effective when the height of the fitness function is greater than the noise floor value established during the settling phase:

$$W_d = \begin{cases} 0.729 & \text{if } F(g_t) > \alpha \\ 0 & \text{otherwise} \end{cases} \qquad (16.15)$$

The non-zero alternative in Eq. 16.15 is the value proposed by Eberhart and Shi [10]. This algorithm is generic enough for a range of PSO variants to be substituted at this point.

Algorithm 2 Motivated particle swarm optimisation where motivation generates a dynamic fitness function. Adapted from [28].

1: **for** each agent A^j do **do**
2:     Initialise with random x^j and v^j
3: **for** $t = 1$ to T do **do**
4:     Sense the environment
5:     $F_0(x) = 0$ for all values of x
6:     **for** each piece of spatially mapped senor data $p_t(x_\tau)$ do **do**
7:         Compute motivation M_τ for $p_t(x_\tau)$
8:     Generate fitness using Eq. 16.14
9: Set PSO noise floor $\alpha = \max_x F(x_\tau)$
10: **for** $t > T$ do **do**
11:     Sense the environment
12:     **if** t mod $Z == 0$ then **then**
13:         $F_0(x) = 0$ for all values of x
14:         Reset p_t^j and g_t for all agents
15:         **for** each piece of spatially mapped senor data $p_t(x_\tau)$ do **do**
16:             Compute motivation M_τ for $p_t(x_\tau)$
17:             Generate fitness using Eq. 16.14
18:     **else**
19:         $F_t(x) = F_{(t-1)}(x)$
20:     Perform PSO update in Eq. 16.9
21:     Move all agents to new positions according to Eq. 16.2.

The motivation and PSO components may potentially run in parallel, for example on different processors, or they may be interleaved on a single centralized processor. In either case, the motivation and PSO components do not have to step at the same rate. The PSO component simply works with the current version of the fitness function available. In Algorithm 2, the ratio of motivation to optimization is controlled by the parameter Z. It is further assumed that the environment is piecewise dynamic, that is, it changes slowly enough for the PSO component to converge on an optima before its location changes again. In this algorithm all agents are motivated by a single, shared, but dynamic fitness function. The approach in the next section incorporates different models of motivation into different agents to allow the agents to exhibit different characteristics in task selection.

16.2.3 Motivated Guaranteed Convergence Particle Swarm Optimization for Exploration and Task Allocation Under Communication Constraints

This algorithm [16] uses a specific variant of the PSO algorithm, namely the guaranteed convergence particle swarm optimization (GCPSO) algorithm [33] to prevent agents from stagnation and premature convergence on suboptimal solutions., In the case where an agents have a limited communication range, the standard PSO algorithm might deal with a situation where it is not connected with any of the agents in the population. In such a case, the agent's personal best position is equal to its own neighborhood best position. This may potentially lead to stagnation and early convergence. To deal with this problem, the idea of GCPSO, thus, involves changing the velocity update equation of the neighborhood best agent. To create Motivated Guaranteed Convergence Particle Swarm Optimization (MGCPSO), the GCPSO algorithm is combined with models of motivation. to create Motivated Guaranteed Convergence Particle Swarm Optimization (MGCPSO).

In the MGCPSO algorithm, Algorithm 3, the set of neighbors of agent A^j is defined according to Eq. 16.3 where R_{com} is the maximum communication range of the agents and $R_{com} > 0$.

As in Algorithm 2, each agent A^j is assumed to be able to remember the location of the best position it has sensed so far (personal best position), p_t^j. However, in MGCPSO, a set of potential neighborhood best positions is also maintained for the agents in the neighborhood N_{com}^j as follows:

$$G_t^j = \left\{ G | G \in N_{com}^j \wedge \left| F(g_t) = argmax\, F(g_t^i) \right| < \mu \right\} \tag{16.16}$$

Next, the set G_t^j is augmented with an artificial, randomly generated position in the search space, g_t^*, which results in a new set $\hat{G}_t^j$ (line 7). Then, the neighborhood best position, n_t^j, is computed using the agent's model of motivation, to selecting the most highly motivating neighborhood goal. Motivation is modelled as a profile of achievement, affiliation or power motivation (line 8). Motivation is computed as a function of incentive, where incentive itself is a function of selected situational variables. Hardhienata et al. [16] proposed that distance to goal (D) and number (a) of agents around a goal are appropriate situational variables for task allocation. Their incentive function is shown in Fig. 16.4. Three motive profiles that are a function of this incentive function are shown in Fig. 16.5. Different agents have different motive profiles and Hardhienata gives guidelines for choosing the proportions of agents with different profiles [13]. Briefly, the ideas captured by these functions are:

- **Power motivation agents:** power motivated agents are willing to take risks. In this scenario, travelling further to a goal (high D), or approach a goal that only a small number of other agents have approached (low a) constitutes a risky activity.
- **Affiliation motivated agents:** affiliation motivated agents seek out the company of other agents. Thus motivation is highest for low incentive goals, which occur for high values of a.
- **Achievement motivated:** achievement motivated individuals prefer goals with a moderate risk. In this work this is assumed to mean moderate values of D and a.

Finally, a modified version of the GCPSO velocity update is applied (line 11):

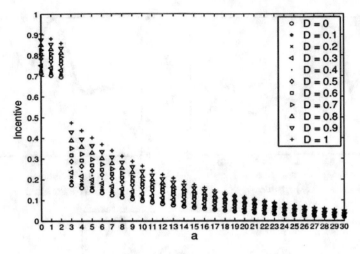

Fig. 16.4 Incentive is a function of distance to goal (D) and number of agents (a) already around the goal. Image from [16]

$$
v_{(t+1)}^j =
\begin{cases}
W_d v_t^j + \lambda(1 - 2r_1) & \text{if } n_t^j = g_t^* \\
W_d v_t^j - x_t^j + n_t^j + \rho_t(1 - 2r_2) & \text{if } n_t^j = p_t^j \\
W_d \left[v_t^j v_t^j + c_1 r_3 (p_t^j - x_t^j) + c_2 r_4 (n_t^j - x_t^j) \right] & \text{Otherwise}
\end{cases}
\tag{16.17}
$$

Algorithm 3 Motivated guaranteed convergence particle swarm optimisation where agents have different motivation functions. Adapted from [13].

1: **for** each agent A^j do **do**
2:    Initialise with random x^j and v^j and various motivation constants to create agents with different profiles of achievement, affiliation and power motivation
3: **for** $t = 1$ to T do **do**
4:    **for** each agent A^j do **do**
5:       Compute personal best p_t^j
6:    **for** each agent A^j do **do**
7:       Calculate $\hat{G}_t^j$
8:       Calculate maximally motivating goal(s) from $\hat{G}_t^j$
9:       Select the closest goal if more than one is maximally motivating
10:    **for** each agent A^j do **do**
11:       Perform PSO update in Eq. 16.17
12:       Move all agents to new positions according to Eq. 16.2.

ρ_t is updated based on an adaptive search procedure [33] and λ is a constant used to scale the contribution of the random search. Clerc and Kennedy [7] suggests ways to set W_d. For the first case in Eq. 16.1, the personal best position (p_t^j) and the neighborhood best (n_t^j) position are not involved. Thus, the agents will not be forced

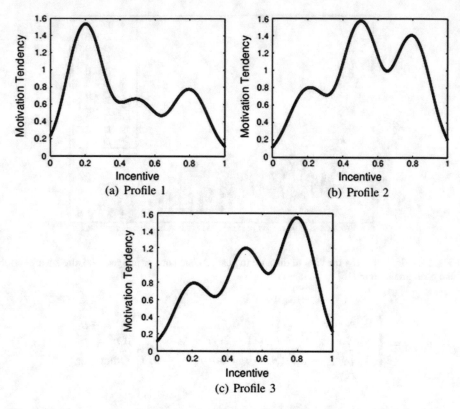

Fig. 16.5 Three profiles of motivation as a function of incentive. Image from [16] **a** a profile for agents with dominant affiliation motivation; **b** a profile for agents with dominant achievement motivation; **c** a profile for agents with dominant power motivation

to move towards their personal best and neighborhood best positions. This is done to allow the agents to perform a broader exploration of the search space. The second and third cases in Eq. 16.17, on the other hand, are based on the GCPSO algorithm. Note that compared to the standard PSO algorithm, the GCPSO algorithm differs in the case where $n_t^j = p_t^j$ to prevent stagnation.

This overview concludes our look at intrinsically motivated swarms. Other work has considered intrinsic motivation in other kinds of multi-agent systems (such as evolutionary settings [24]), but we consider this out of the scope of this paper. The next section now considers some of the advantages that have been achieved through the use of intrinsic motivation in swarm systems.

16.3 Functional Implications of Intrinsically Motivated Swarms

Empirical studies and case studies of intrinsically motivated agent swarms have revealed a number of advantages of such models in diverse applications including computer games [28], hazard detection [21] and search [13]. These advantages–including increased diversity, adaptation and capacity for exploration–are discussed in the remainder of this section. Eq. 16.14 considers more abstract implications for intrinsic motivation on the trustworthiness of autonomous systems.

16.3.1 Motivation and Diversity

A key property of motivated agents revealed particularly in Algorithm 1 and Algorithm 3 is their diversity. In Algorithm 1, agents are initialized with different OMIs so they have different preferences for incentive. In Algorithm 3, agents are embedded with different profiles of achievement, affiliation and power motivation. These profiles include more expressive models of incentive in terms of situational variables, so agents respond differently to specific aspects of their environment. Figure 16.6 demonstrates agent diversity in the Breadcrumbs game. Breadcrumbs is a simple Android game set in two rooms connected by an open doorway. Initially the characters (simple square-shaped boids in this case) are randomly distributed throughout both rooms. The rules of the game are as follows:

Aim of the game:
Place up to five breadcrumbs to lure all the *boids* into one room
Instructions:
1. Place breadcrumbs by touching the screen at the desired location
2. Once you have placed five breadcrumbs, you can continue placing breadcrumbs, but each new breadcrumb will trigger the removal of the oldest existing breadcrumb
3. Breadcrumbs are always tasty - but you don't know exactly how tasty any given crumb will be. In addition, different boids have different preferences for flavour

---

In Breadcrumbs power motivated agents are red, achievement motivated agents are orange and affiliation motivated agents are yellow. Breadcrumbs themselves are brown. We can see from Fig. 16.6 that agents with similar motives cluster around similar breadcrumbs. This is a demonstrator of the way motivational diversity results in behavioral diversity. Merrick [28] provides a case study comparing diversity as a result of motivation to homogeneous and random heterogeneous swarms and concludes that the systematic approach to motivation supports more predictable agent behaviour (Fig. 16.7).

Hardhienata et al. [14, 15] also report behavioral diversity as a result of motivational diversity. In their model, affiliation motivated agents tend to perform local

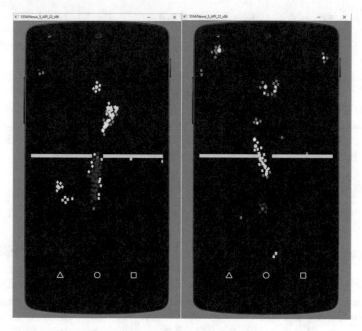

Fig. 16.6 Motivated crowds in the Breadcrumbs game

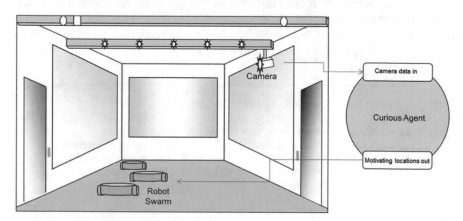

Fig. 16.7 Conceptual view of the hazard detection scenario described by Klyne and Merrick [21]. The robot swarm receives motivating locations from a centralized curious agent that analyses surveillance camera data. They constructed an image database for different floor surfaces (available at https://figshare.com/articles/Hazard_Database/3180487)

search and allocate themselves to tasks. In contrast, power-motivated agents tend to explore to find new tasks. These agents perform these characteristic behaviors more effectively in the presence of achievement-motivated agents, improving the task allocation performance of the swarm as a whole [15].

In Breadcrumbs power motivated agents are red, achievement motivated agents are orange and affiliation motivated agents are yellow. Breadcrumbs themselves are brown. We can see from Fig. 16.6 that agents with similar motives cluster around similar breadcrumbs. This is a demonstrator of the way motivational diversity results in behavioral diversity. Merrick [28] provides a case study comparing diversity as a result of motivation to homogeneous and random heterogeneous swarms and concludes that the systematic approach to motivation supports more predictable agent behaviour.

Hardhienata et al. [14, 15] also report behavioral diversity as a result of motivational diversity. In their model, affiliation motivated agents tend to perform local search and allocate themselves to tasks. In contrast, power-motivated agents tend to explore to find new tasks. These agents perform these characteristic behaviors more effectively in the presence of achievement-motivated agents, improving the task allocation performance of the swarm as a whole [15].

16.3.2 Motivation and Adaptation

Where Algorithm 1 assumes that goal locations are known by agents, and generated by an entity external to the swarm, Algorithm 2 permits the swarm to generate goals dynamically, while agents are exploring. The swarm here no longer has the diversity of Algorithm 1, as all agents share the same model of motivation, but the swarm arguably has a greater level of autonomy because it can generate its own goals.

Klyne and Merrick [21] demonstrate Algorithm 2 in a simulated hazard detection scenario. They use a swarm of agents (representing robots) to detect hazards, with the idea that the robots will either clear up, or warn passers-by of, the detected hazard. The advantage of the MPSO approach is that a strong task signature for hazards is not required. Rather hazards are identified as novel or interesting occurrences in surveillance images. A conceptual view of this setup is illustrated in Fig. 16.8 shows five images from a hazard detection scenario generated by Klyne and Merrick [21]. The first four images in Fig. 16.8 shows the fitness landscape while the algorithm is in the settling phase. The fifth image in Fig. 16.8 shows the fitness landscape and simulated robots towards the end of one of the MPSO phases. Klyne and Merrick [21] demonstrate that successive convergence and divergence of a swarm as it adapts to the introduction and removal of different hazards in each scenario.

Klyne and Merrick [21] demonstrate Algorithm 2 in a simulated hazard detection scenario. They use a swarm of agents (representing robots) to detect hazards, with the idea that the robots will either clear up, or warn passers-by of, the detected hazard. The advantage of the MPSO approach is that a strong task signature for hazards is not required. Rather hazards are identified as novel or interesting occurrences in surveillance images. A conceptual view of this setup is illustrated in Fig. 16.8. Furthermore, Fig. 16.8 shows a changing fitness landscape while the algorithm is in the settling phase. The image in the bottom row of Fig. 16.8 shows the fitness function after a hazard has been identified. Klyne and Merrick [21] demonstrate that

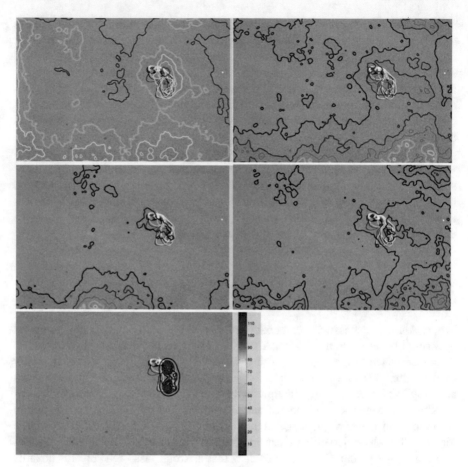

Fig. 16.8 Example of a novelty-based fitness function being generated over a bitumen surface. The first four image show the fitness function during the settling stage. The fifth image (bottom row) shows the fitness function settled over a hazard

successive convergence and divergence of a swarm as it adapts to the introduction and removal of different hazards in each scenario.

16.3.3 Motivation and Exploration

Algorithm 3 demonstrates the impact of motivation on exploration. Traditionally, simulated swarms are initialized by randomizing agents' initial positions and velocities in a defined space. However, in practice, if agents are real robots being rolled off the back of a truck or launched from a boat or aircraft, they are effectively initialized at a single point. Hardhienata et al. [16] show that algorithms such as GCPSO

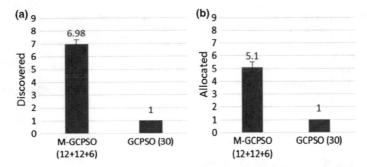

Fig. 16.9 a Number of goals discovered in the synthetic landscape in Fig. 16.3. when agents are initialized from a single point. **b** Number of tasks to which agents are allocated when agents are initialized from a single point

perform badly under such conditions, in terms of the number of goals they discover, and the number of goals on which they are able to converge agents. In contrast, MGCPSO significantly increases the number of discovered goals when the agents are initialized from a single point. It also increases the number of goals to which agents are allocated.

Some comparative results for the synthetic fitness function shown previously in Fig. 16.3 are shown in Fig. 16.9. These results compare 30 unmotivated agents using GCPSO to MGCPSO using agents with motive profiles 1 (12 agents), 2 (12 agents) and 3 (6 agents) shown in Fig. 16.5. Simulations also indicated this offered significant advantages when the communication of agents is limited. This is because agents can pursue goals relatively independently using their intrinsic motivation when they are not in contact with large numbers of swarm-mates.

16.4   Implications of Motivation on Trust

In the previous section we discussed how swarms of agents incorporating models of curiosity, achievement, affiliation and power can achieve diversity, adaptive behavior, and improve exploratory capabilities. All of these properties are aspects of autonomy. This then raises the question of how changes in these properties can affect trustworthiness. Trust itself is a multifaceted concept, including properties such as reliability, privacy, security, safety, complexity, risk, and free will [3]. We now consider how the properties of motivated agents considered in Sect. 16.3 might influence the perception of trust in relation to these properties.

Fig. 16.10 In humans, motivation is understood to play an important role in the 'survival arc' [36], moving an actor from the denial phase through to deliberation that can result in action

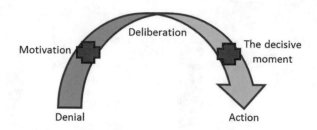

16.4.1 Implications for Reliability

Reliability refers to consistency of actions, values and objectives and stability of performance during the lifetime of a trusting relationship [3]. In the context of a motivated swarm, the diversity of individuals in a swarm may be a double edged sword. At the macro level, we have seen that diversity as a result of motivation can offer improvements to factors such discovery of goals and convergence on goals [14]. Literature on human disaster survival–perhaps the ultimate demonstration of reliability–places motivation at a critical juncture of the 'survival arc' [36] (see Fig. 16.10). The survival arc has three phases: (1) denial, where the actor refuses to acknowledge abnormality in their situation; (2) deliberation: which includes milling and information gathering; and (3) action. Motivation is required to move an actor from the denial phase through to the deliberation phase before action can occur. Computational motivation has the same positive potential in artificial systems.

However, at the level of the individual, greater variability in performance is introduced. Different agents, when they encounter the same situation, will act differently as a result of their motives or experiences. Unless these internal differences are transparent to a human collaborator, there may be a perception that there is less consistency of action between individuals. Existing work has found that such performance based factors play a key role in trust development between humans and robots [12, 44].

Adaptation of a swarm also has implications for reliability. In the case where a swarm can generate its own goals, we have seen that this can have a positive impact on stability of performance because the swarm can adapt in the presence of novel hazards [21]. However, once again, there may also be negative implications for trust if there is a perception that the agent can have changing objectives (and control of its own changing objectives) during the course of its life.

One mitigation technique to deal with the impact of diversity and change in humans is offered to us by the literature on reputation [20]. Reputation models permit users ('witnesses') to rate trustees, whether human or software (intelligent or otherwise). This information can be used by others to determine whether they also should trust in the specific trustee.

While the examples discussed above give us some insight on how motivation may affect the reliability aspect of trust, there is currently very little, if any, work that actually incorporates both computational models of motivation and computational models of trust. We thus conclude this section with a number of thoughts on how specific models of motivation may impact trustworthiness. In Sect. 16.2.3 we saw that one characteristic of power-motivated agents is an increased inclination higher risk behavior. While risk-taking behavior can have the advantage of high payoff, in situations positive return does not eventuate, this could contribute to a perception of unreliable behavior or lack of trustworthiness. Likewise, agents with embedded models of curiosity may divert from an established behavioral pattern to satisfy their need for novelty. This also has potential to contribute to a perception of reduced reliability if it does not result in any advantage such as a novel discovery or process improvement. At the other end of the spectrum, achievement-motivated agents are moderate risk-takers and seek mastery of their environment and high performance. These characteristics are well suited to reliable performance. As such, a heterogeneous society of agents with different motive profiles may be best able to harness the advantages of computational motivation while maintaining trust.

16.5 Implications for Privacy and Security

Privacy and security are related, although distinct concepts. When a trustor trusts a trustee, the trusting relationship may involve transfer of data. Any misuse of this data outside terms of the trusting contract is a breach of privacy [3]. Security has broader connotations and, while including confidentiality, also concerns the integrity and continued availability of data.

While motivated agents have not been widely examined in the context of privacy and security, some of the reported results with motivated swarms have interesting implications in this regard. Hardhienata et al. [14] presented evidence that significant performance advantages can be achieved by motivated swarms when the communication of agents is limited. This is because agents can pursue goals relatively independently using their intrinsic motivation when they are not in contact with large numbers of swarm-mates. A smaller communication radius has the potential to make a network more difficult to detect, and thus offer a security advantage in a contested environment.

As we noted in our discussion of motivation and reliability, in the case of motivation and security (or at least a lowered communication requirement) a heterogeneous society of motivated agents is best able to achieve this [14].

16.5.1 Implications for Safety

Traditional safety-critical software verification requires that every condition of every branch of software is tested and that every line of code and test can be traced back to the software's requirements [19]. By this definition, it appears that motivated agent technologies should be suitable for use in safety-critical situations. However, in systems with the capacity for learning, where behavior is influenced by experiences and where the breadth of possible experience cannot be known in advance, the traditional definition of safety-critical verification falls short. Because the data input to the motivated agent will influence its emergent behavior, and because this data cannot be predicted in advance, it is difficult to test for all possible outcomes/behaviors.

Again noting that there is currently very little, if any, work that actually incorporates both computational models of motivation and computational models of trust, we thus conclude this section with a number of thoughts on how specific models of motivation may impact trustworthiness. As we noted earlier, power-motivated agents are characterized by an increased tendency for risk-taking and resource controlling behavior. Risk-taking behavior that does not result in positive payoff may, as a consequence, impact safety. This may in turn have a negative impact on trustworthiness. Likewise, resource controlling behavior can lead an agent into situations of conflict, which may also impact safety aspects of trust. In natural systems, power-motivation is understood to be tempered by affiliation motivation, which balances resource controlling preferences with relationship building behaviors. It may be that future artificial systems will also benefit from embedded motive profiles, rather than individual motives which has been the existing research focus.

16.6 Implications of Complexity

Trust is a form of educated delegation that a trustor may enter to manage some level of complexity [3]. A trustor delegates to a trustee when there is a benefit for the trustor in trusting rather than performing the job themselves. That is, when delegation reduces some form of complexity. Examples of complexity include technical complexity associated with performing the task, time pressure or the increase in mental and cognitive complexity if the trustor chooses to perform the task themselves. As the level of complexity increases, the degree with which a trustor trusts a trustee increases. In this context, the implications of motivation on trust are tied closely to the situation in which motivated agents are given trust. Self-motivated agents are specifically designed for complex or dynamic environments where system designers cannot predict in advance all the goals the agent may need to address. According to the definition above, such environments will require a high level of trust to be placed in motivated agents.

16.7 Implications for Risk

A trusting decision involves a level of uncertainty associated with the possibility that the trustee will breach trust. A rational definition of risk might look like [5, 8]:

$$Risk \ = \ Probability \ \times \ Consequence$$

Fig. 16.11 Motivated agents
in the spectrum of human
control and machine
autonomy

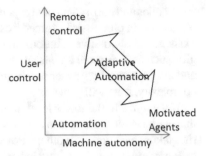

Probability refers to the probability of the given risk occurring and Consequence
refers to the cost of the risk occurring. However, for humans, perception of risk,
especially under pressure, may not adhere to this rational definition. An irrational
component of risk, that changes the way risk is perceived, is dread [11]. The influence
of dread on risk has been modelled in various ways, including as a dimension of
risk [39] and as a multiplier of risk [36].

Dread represents human 'evolutionary fears', hopes, prejudices and biases. Dread
itself can be represented as [36] in terms of uncontrollability, unfamiliarity, imagin-
ability, suffering, scale of destruction and unfairness. That is, humans perceive higher
risk in situations that are uncontrollable or unfamiliar, where they can easily imagine
the consequences of failure, where failure will result in suffering on a large scale or
over a long time, or where the situation is perceived to be undeserved. This perceived
or subjective value of risk may not agree with statistical or objective values of risk.

Suppose we look at the conceptual space represented in Fig. 16.11 through the lens
of dread. Figure 16.11 places different types of automations and autonomous agents
on axes of 'machine autonomy' and 'user control'. We can see that motivated agents
sit at the extreme low end of user control (which increases dread). Autonomous systems such as robots are also still a relatively unfamiliar technology (which increases
dread) and popular media such as the Terminator series of movies aids the imagin-
ability of disaster scenarios involving such technologies (again increasing dread). In
summary, while we have described documented advantages of incorporating motivation in artificial systems, human perception of the risk associated with such systems,
in particular influenced by dread, may still impact perception of their trustworthiness.

If we move to the lower level of examining specific motives with respect to risk,
then we have seen that certain dominant motives will result in a stronger preference for
risk-taking behavior than others. Power motivation in particular can be characterised
by a preference for risk taking behavior, while affiliation motivated individuals tend
to avoid such behavior.

16.7.1 Implications for Free Will

Free will is the ability of the actor to make a decision within a bounded space
autonomously and at its own discretion [3]. The space may be bounded by social ties,

social rules and norms, and interdependencies among actors in terms of resources and objectives. In other words, forced trust cannot be construed as trust. In this sense, the existence of alternative solutions and technologies is a boon for emerging technologies such as motivated agents. Where users choose to trust these new technologies and are rewarded by greater reliability, privacy, security or safety, or reduced risk or complexity, trust will grow.

If we move to the lower level of examining specific motives with respect to risk, then we have seen that certain dominant motives will result in a stronger preference for risk-taking behavior than others. Power motivation in particular can be characterised by a preference for risk taking behavior, while affiliation motivated individuals tend to avoid such behavior.

16.8 Conclusion

In conclusion, this chapter has considered the impact of one of the emerging mechanisms for achieving autonomy–computational motivation–on the trustworthiness of autonomous systems. We considered this question in the context of intrinsically motivated agent swarms using some of the key variants of computational motivation: curiosity, novelty-seeking, achievement, affiliation and power motivation. Section. 16.2 provided an overview of the theory underlying the use of computational motivation in swarms of artificial agents, including a uniform notation for three intrinsically motivated swarm algorithms. Section. 16.4 considered the implications of motivation for the functionality of agent swarms, including diversity, adaptation and greater capacity for exploration. Finally Sect. 16.3 considered the implications of motivation on trustworthiness, both at the level of individual motives and at the level of permitting or not permitting intrinsic motivation in an artificial system.

Finally, in answer to the question framed in the title of this chapter: Computational Motivation, Autonomy and Trustworthiness: Can We Have It All? we present the following thoughts:

- Initial evidence suggests that inclusion of intrinsic motivation in artificial agents is likely to impact trustworthiness, but this may be in either a positive or negative sense. We saw positive impacts on performance that may translate to impacts on reliability, but also impacts on safety or risk facets of trust that may be perceived as negative.
- Approaches to the inclusion of motivation in artificial systems that may further modify the impact of motivation on trust include (1) which motives are used in artificial agents, and how or whether multiple motives are combined in a single agent or (2) in societies of agents whether individuals are homogeneous or heterogeneous.
- Motivated agent technology must remain transparent to combat factors such as dread and its associated impact on trustworthiness.

References

1. *Autonomous Systems: Social, Legal and Ethical Issues*
   http://www.raeng.org.uk/publications/reports/autonomous-systems-report
   (The Royal Academy of Engineering, London, UK, 2009)
2. H. Abbass, *Computational Red Teaming: Risk Analytics of Big-Data-to-Decisions Intelligent Systems* (Springer Verlag, Berlin, 2015)
3. H. Abbass, G. Leu, K. Merrick, A review of theoretical and practical challenges of trusted autonomy in big data. IEEE Access **4**, 2809–2830 (2016)
4. H. Abbass, E. Petraki, K. Merrick, J. Harvey, M. Barlow, Trusted autonomy and cognitive cyber symbiosis: open challenges. Cogn. Comput. **8**(3), 385–408 (2016)
5. G. Ballard, *Industrial Risk: Safety by Design* (Wiley, Chichester, 1992)
6. C.E. Billings, *Aviation Automation: The Search for a Human-Centred Approach* (Lawrence Erlbaum Associates, Mahwah, NJ, 1997)
7. M. Clerc, J. Kennedy, The particle swarm - explosion, stability and convergence in a multidimensional complex space. IEEE Trans. Evol. Comput. **6**(1), 58–73 (2002)
8. M. Douglas, *Risk and Blame: Essays in Cultural Theory* (Routledge, London, 1992)
9. R. Eberhart, J. Kennedy, A new optimiser using particle swarm theory (1995)
10. R. Eberhart, Y. Shi, Comparing inertia weights and constriction factors in particle swarm optimisation (2000)
11. R. Gregory, R. Mendelsohn, Perceived risk, dread and benefits. Risk Anal. **13**(3), 259–264 (1993)
12. P.A. Hancock, D.R. Billings, K.E. Schaefer, J.Y.C. Chen, E. de Visser, R. Parasuraman, A meta-analysis of factors affecting trust in human-robot interaction. Hum. Factors **53**, 517–527 (2011)
13. M. Hardhienata, *Models of Motivation for Particle Swarm Optimization with Application to Task Allocation in Multi-Agent Systems*. PhD thesis (2015)
14. M. Hardhienata, K. Merrick, V. Ougrinovski, *Task allocation in multi-agent systems using models of motivation and leadership* (Presented at the IEEE conference on evolutionary computation, Brisbane, Australia, 2012), pp. 86–93
15. M. Hardhienata, K. Merrick, V. Ugrinovskii, Effective motive profiles and swarm compositions for motivated particle swarm optimisation applied to task allocation. Presented at the IEEE symposium series on computational intelligence, symposium on computational intelligence for human-like intelligence, (2014)
16. M. Hardhienata, V. Ugrinovskii, K. Merrick, Task allocation under communication constraints using motivated particle swarm optimization. Presented at the IEEE congress on evolutionary computation, CEC **2014**, 3135–3142 (2014)
17. R. Hardin, The street-level epistemology of trust. Polit. Soc. **21**, 505–529 (1993)
18. J. Heckhausen, H. Heckhausen, *Motivation and Action* (Cambridge University Press, New York, NY, 2010)
19. J. Hinchman, M. Clark, J. Hoffman, B. Hulbert, C. Snyder, Towards safety assurance of trusted autonomy in air force flight critical systems (2012)
20. A. Josang, R. Ismail, C. Boyd, A survey on trust and reputation systems for online service provision. Decis. Support Syst. **43**(2), 618–644 (2007)
21. A. Klyne, K. Merrick, Intrinsically motivated particle swarm optimisation applied to task allocation for workplace hazard detection. Adapt. Behav. **24**(4), 219–236 (2016)
22. R.M. Kramer, Trust and distrust in organisations: emerging perspectives, enduring questions. Annu. Rev. Psychol. **50**, 569–598 (1999)
23. A. Lacher, *Research Challenges Associated with Unmanned Aircraft Systems Airspace Integration*. MITRE Corporation (2012)
24. J. Lehman, K. Stanley, Abandoning objectives: evolution through search for novelty alone. Evol. Comput. **19**(2), 189–223 (2011)
25. J.C.R. Licklider, Man-computer symbiosis. IRE Trans. Hum. Factors Electron. **1**, 4–11 (1960)

26. A.R. Lomuscio, *Trusted Autonomous Systems* (Engineering and Physical Sciences Research Council, Department of Computing, Imperial College London, 2011–2016)

27. D. McKnight, N. Chervany, *Trust and Distrust Definitions: One Bite at a Time* (Springer, Berlin, 2001), pp. 27–54

28. K. Merrick, *Computational Motivation for Game-Playing Agents* (Springer, Berlin-Heidelberg, 2016)

29. K. Merrick, K. Shafi, Achievement, affiliation and power: motive profiles for artificial agents. Adapt. Behav. **19**(1), 40–62 (2011)

30. K. Merrick, K. Shafi, A game theoretic framework for incentive-based models of intrinsic motivation in artificial systems. *Frontiers in Cognitive Science, Special Issue on Intrinsic Motivations and Open-Ended Development in Animals, Humans and Robots* p. 4 (2013)

31. P.-Y. Oudeyer, F. Kaplan, V.V. Hafner, Intrinsic motivation systems for autonomous mental development. IEEE Trans. Evol. Comput. **11**(2), 265–286 (2007)

32. R. Parasuraman, T. Bahri, J.E. Deaton, J.G. Morrison, M. Barnes, *Theory and Design of Adaptive Automation in Aviation Systems* (Technical report, Naval Air Development Centre, 1992)

33. E. Peer, F. van den Bergh, A. Engelbrecht, Using neighbourhoods with the guaranteed convergence pso (2003)

34. E. Petraki, H. Abbass, On trust and influence: a computational red teaming game theoretic perspective (2014)

35. C.W. Reynolds, Flocks, herds and schools: a distributed behavioral model. Comput. Graphics (SIGGRAPH 87 Conf. Proc.) **21**(4), 25–34 (1987)

36. A. Ripley, *The Unthinkable: Who Survives When Disaster Strikes and Why* (Three Rivers Press, New York, 2009)

37. R. Saunders, J.S. Gero, Curious agents and situated design evaluations. Artif. Intell. Eng. Des. Anal. Manuf. **18**(2), 153–161 (2004)

38. C. Simkins, C. Isbell, N. Marquez, Deriving behavior from personality: a reinforcement learning approach, in *International Conference on Cognitive Modelling*, pp. 229–234

39. P. Slovic, Perception of risk. Science **236**(4799), 280–285 (1987)

40. J. Stanley, Computer simulation of a model of habituation. Nature **261**, 146–148 (1976)

41. R. Sun, Motivational representations within a computational cognitive architecture. Cogn. Comput. **1**, 91–103 (2009)

42. R. Sun, P. Fleischer, A cognitive and social simulation of tribal survival strategies: the importance of cognitive and motivational factors. J. Cogn. Cult. **12**, 287–321 (2012)

43. J. Weng, J. McClelland, A. Pentland, O. Sporns, I. Stockman, M. Sur, E. Thelen, Artificial intelligence: autonomous mental development by robots and animals. Science **291**, 599–600 (2001)

44. R. Yagoda, D. Gillan, You want me to trust a robot? the development of a human robot interaction trust scale. Int. J. Soc. Robot. **4**, 235–248 (2012)

Chapter 17
Are Autonomous-and-Creative Machines Intrinsically Untrustworthy?

Selmer Bringsjord and Naveen Sundar Govindarajulu

17.1 Introduction

Given what we find in the case of human cognition, the following principle (Principle ACU, or just — read to rhyme with "pack-ooo" — PACU) appears to be quite plausible:

> PACU An artificial agent that is autonomous (A) and creative (C) will tend to be, from the viewpoint of a rational, fully informed agent, (U) untrustworthy.

After briefly explaining the intuitive internal structure of this disturbing (in the context of the human sphere) principle, we provide a more formal rendition of it designed to apply to the realm of intelligent artificial agents. The more-formal version makes use of some of the basic structures available in a dialect of one of our cognitive-event calculi (viz. $\mathcal{D}^e\mathcal{CEC}$),[1] and can be expressed as a (confessedly — for reasons explained — naïve) theorem (Theorem ACU; TACU — pronounced to rhyme with "tack-ooo", for short). We prove the theorem, and then provide a trio of demonstrations of it in action, using a novel theorem prover (ShadowProver) custom-designed to power our highly expressive calculi. We then end by gesturing toward some future defensive engineering measures that should be taken in light of the theorem.

---

[1] We will cover $\mathcal{D}^e\mathcal{CEC}$ shortly, but see http://www.cs.rpi.edu/~govinn/dcec.pdf for a quick introduction to a simple dialect. See [10] for a more detailed application.

The authors are deeply grateful for support provided by both AFOSR and ONR which enabled the research reported on herein, and are in addition thankful both for the guidance and patience of the editors and wise comments received from two anonymous reviewers.

S. Bringsjord (✉) · N. S. Govindarajulu
Rensselaer AI & Reasoning (RAIR) Lab, Department of Cognitive Science,
Department of Computer Science, Rensselaer Polytechnic Institute (RPI),
Troy, NY 12180, USA
e-mail: Selmer.Bringsjord@gmail.com

N. S. Govindarajulu
e-mail: Naveen.Sundar.G@gmail.com

H. A. Abbass et al. (eds.), *Foundations of Trusted Autonomy*, Studies in Systems, Decision and Control 117, https://doi.org/10.1007/978-3-319-64816-3_17

In a bit more detail, the plan for the present chapter is as follows. We begin by providing an intuitive explanation of PACU, in part by appealing to empirical evidence and explanation from psychology for its holding in the human sphere (Sect. 17.2). Next, we take aim at establishing the theorem (TACU), which as we've explained is the formal counterpart of Principle ACU (Sect. 17.3). Reaching this aim requires that we take a number of steps, in order: briefly explain the notion of an "ideal-observer" viewpoint (Sect. 17.3); summarize the form of creativity we employ for C (Sect. 17.3.2), and then the form of autonomy we employ for A; very briefly describe the cognitive calculus $\mathcal{D}^e\mathcal{CEC}$ in which we couch the elements of TACU, and the novel automated prover (ShadowProver) by which this theorem and supporting elements is automatically derived (Sect. 17.3.4); explain the concept of *collaborative situations*, a concept that is key to TACU (Sect. 17.3.5); and then, finally, establish TACU (Sect. 17.3.6). The next section provides an overview of three simulations in which Theorem ACU and its supporting concepts are brought to concrete, implemented life with help from ShadowProver (Sect. 17.4). We conclude the chapter, as promised, with remarks about a future in which TACU can rear up in AI technology different from what we have specifically employed herein, and the potential need to ward such a future off (Sect. 17.5).

17.2   The Distressing Principle, Intuitively Put

The present chapter was catalyzed by a piece of irony: It occurred to us, first, that maybe, just maybe, something like PACU was at least plausible, from a formal point of view in which, specifically, highly expressive computational logics are used to model, in computing machines, human-level cognition.[2] We then wondered whether PACU, in the human sphere, just might be at least plausible, empirically speaking. After some study, we learned that PACU isn't merely *plausible* when it refers to humans; it seems to be flat-out *true*, supported by a large amount of empirical data in psychology. For example, in the provocative *The (Honest) Truth About Dishonesty: How We Lie to Everyone — Especially Ourselves*, Ariely explains, in "Chapter 7: Creativity and Dishonesty," that because most humans are inveterate and seemingly uncontrollable storytellers, dishonesty is shockingly routine, even in scenarios in which there is apparently no utility to be gained from mendacity. Summing the situation up, Ariely writes:

> [H]uman beings are torn by a fundamental conflict—our deeply ingrained propensity to lie to ourselves and to others, and the desire to think of ourselves as good and honest people. So we justify our dishonesty by telling ourselves stories about why our actions are acceptable and sometimes even admirable. (Chap. 7 in [1])

---

[2]Such a modeling approach is in broad strokes introduced, explained, and defended in [12]. The approach is employed e.g. in [9] in the domain of nuclear strategy, and in [15] in computational economics.

This summation is supported by countless experiments in which human subjects deploy their ability to spin stories on the spot in support of propositions that are simply and clearly false.[3]

Whereas Ariely identifies a form of creativity that consists in the generation of narrative, as will soon be seen, we base our formal analysis and constructions upon a less complicated form of creativity that is subsumed by narratological creativity: what we call *theory-of-mind creativity*. It isn't that we find creativity associated with narrative uninteresting or unworthy of investigation from the perspective of logicist computational cognitive modeling or AI or robotics (on the contrary, we have investigated it with considerable gusto; see e.g. [7]), it's simply that such things as story generation are fairly narrow in the overall space of creativity (and indeed *very* narrow in AI), and we seek to cast a wider net with TACU than would be enabled by our use herein of such narrow capability.

17.3 The Distressing Principle, More Formally Put

17.3.1 The Ideal-Observer Point of View

In philosophy, ideal-observer theory is nearly invariably restricted to the sub-discipline of ethics, and arguably was introduced in that regard by Adam Smith [42].[4] The basic idea, leaving aside nuances that needn't detain us, is that actions are morally obligatory (or morally permissible, or morally forbidden) for humans just in case an ideal observer, possessed of perfect knowledge and perfectly rational, would regard them to be so. We are not concerned with ethics herein (at least not directly; we do end with some brief comments along the ethics dimension); we instead apply the ideal-observer concept to epistemic and decision-theoretic phenomena.

For the epistemic case, we stipulate that, for every time t, an ideal observer knows the propositional attitudes of all "lesser" agents at t. In particular, for any agent a, if a believes, knows, desires, intends, says/communicates, perceives ... ϕ at t (all these propositional attitudes are captured in the formal language of $\mathcal{D}^e\mathcal{CEC}$), the ideal observer knows that this is the case at t; and if an agent a fails to have some propositional attitude with respect to ϕ at a time t, an ideal observer also knows this. For instance, if in some situation or simulation covered by one of our cognitive calculi (including specifically $\mathcal{D}^e\mathcal{CEC}$) an artificial agent a_a knows that a human

---

[3]The specific experiments are not profitably reviewed in the present chapter, since we only need for present purposes their collective moral (to wit, a real-life kernel of PACU in human society), and since the form of creativity involved is not the one we place at the center of TACU. We do encourage readers to read about the stunning experiments in question. By the way, this may be as good a place as any to point out that these experiments only establish that *many*, or at least *most*, subjects exercise their freedom and creativity to routinely lie. The reader, like the two authors, may well not be in this category.

[4]While widely known for *Wealth of Nations*, in which the unforgettable "invisible hand" and phrase and concept appears, Smith was an advocate only of markets suitably tempered by morality.

agent a_h knows that two plus two equals four ($= \phi$), and o is the ideal observer, the following formula would hold:

$$\mathbf{K}(o, t, \mathbf{K}(a_a, t, \mathbf{K}(a_h, t, \phi))).$$

It is convenient and suggestive to view the ideal observer as an omniscient overseer of a system in which particular agents, of the AI and human variety, live and move and think.

We have explained the epistemic power of the ideal observer. What about rationality? How is the supreme rationality of the ideal observer captured? We say that an ideal observer enters into a cognitive state on the basis only of what it knows directly, or on the basis of what it can unshakably derive from what it knows, and we say it knows all that is in the "derivation" closure of what it knows directly.[5] One important stipulation (whose role will become clear below) regarding the ideal observer is that its omniscience isn't unbounded; specifically, it doesn't have hypercomputational power: it can't decide arbitrary Turing-undecidable problems.[6]

17.3.2   Theory-of-Mind-Creativity

In AI, the study and engineering of creative artificial agents is extensive and varied. We have already noted above that narratological creativity has been an object of study and engineering in AI. For another example, considerable toil has gone into imbuing artificial agents with *musical* creativity (e.g. see [20, 24]). Yet another sort of machine creativity that has been explored in AI is mathematical creativity.[7] But what these and other forays into machine creativity have in common is that, relative to the knowledge and belief present in those agents in whose midst the creative machine in question operates, the machine (if successful) performs some action that

---

[5]An ideal observer can thus be intuitively thought of as the human AI researcher who knows the correct answer to all such puzzles as the famous "wise-man puzzle" (an old-century, classic presentation of which is provided in [27]). The puzzle is treated in the standard finitary case in [12]. The infinite case is analyzed in [2]; here, the authors operate essentially as ideal observers. For a detailed case of a human operating as an ideal observer with respect to a problem designed by [25] to be much harder than traditional wise-man problems, see the proof of the solution in [13].

[6]The 'arbitrary' here is important. ShadowProver is perfectly able to solve *particular* Turing-undecidable (provability) problems. It may be helpful to some readers to point out that any reasonable formalization of Simon's [41] concept of *bounded rationality* will entail boundedness we invoke here. For an extension and implementation of Simon's concept, under the umbrella of cognitive calculi like $\mathcal{D}^e\mathcal{CEC}$, see [30].

[7]For example, attempts have been made to imbue a computing machine with the ability to match (or at least approximate) the creativity of Gödel, in proving his famous first incompleteness theorem. See [34].

is a surprising deviation from this knowledge and belief.[8] In short, what the creative machine does is perform an action that, relative to the knowledge, beliefs, desires, and expectations of the agents composing its audience, is a surprise.[9] We refer to this generic, underlying form of creativity as *theory-of-mind*-creativity. Our terminology reflects that for one agent to have a "theory of mind" of another agent is for the first agent to have beliefs (etc.) about the beliefs of another agent. An early, if not the first, use of the phrase 'theory of mind' in this sense can be found in [39] — but there the discussion is non-computational, based as it is on experimental psychology, entirely separate from AI. Early modeling of a classic theory-of-mind experiment in psychology, using the tools of logicist AI, can be found in [3]. For a presentation of an approach to achieving literary creativity specifically by performing actions that manipulate the intensional attitudes of readers, including actions that specifically violate what readers believe is going to happen, see [23].

17.3.3   Autonomy

The term 'autonomous' is now routinely ascribed to various artifacts that are based on computing machines. Unfortunately, such ascriptions are — as of the typing of the present sentence in late 2016 — issued in the absence of a formal definition of what autonomy *is*.[10] What might a formal definition of autonomy look like? Presumably such an account would be developed along one or both of two trajectories. On the one hand, autonomy might be cashed out as a formalization of the kernel that agent a is autonomous at a given time t just in case, at that time, a can (perhaps at some immediate-successor time t') perform some action α_1 or some incompatible action α_2. In keeping with this intuitive picture, if the past tense is used, and accordingly the definiendum is 'a autonomously performed action α_1 at time t,' then the idea would be that, at t, or perhaps at an immediate preceding time t'', s could have, unto itself, performed alternative action α_2. (There may of course be many alternatives.) Of course, all of this is quite informal. This picture is an intuitive springboard for deploying formal logic to work out matters in sufficient detail to allow meaningful and substantive conjectures to be devised, and either confirmed (proof) or refuted (disproof). Doing this in the present chapter is well outside our purposes here.

---

[8]Relevant here is a general form of creativity dubbed *H-creativity* by [5], the gist of which is that such creativity, relative to what the community knows and believes, is new on the scene.

[9]Cf. Turing's [44] affirmation of the claim that a thinking (computing) machine must be capable of surprising its audience, and his assertion immediately thereafter that computing machines in his time could be surprising. Turing's conception of surprise is a radically attenuated one, compared to our theory-of-mind-based one.

[10]One way to dodge the question of what autonomy is, is to simply move straightaway to some formalization of the *degree* or *amount* of autonomy. This approach is taken in [16], where the degree of autonomy possessed by an artificial agent is taken to be the Kolmogorov complexity of it's program.

Our solution is a "trick" in which we simply employ a standard move long made in recursion theory, specifically in relative computability. In relative computability, one can progress by assuming that an oracle can be consulted by an idealized computing machine, and then one can ask the formal question as to what functions from $\mathbb{N}$ to $\mathbb{N}$ become computable under that assumption. This technique is for example used in a lucid manner in [22].[11] The way we use the trick herein is as follows. To formalize the concept of an autonomous action, we suppose,

- first, that the action in question is performed if and only if it produces the most utility into the future for the agent considering whether to carry it out or not;
- then suppose, second, that the utility accruing from competing actions can be deduced from some formal theory[12];
- then suppose, third, that a given deductive question of this type (i.e., of the general form $\Phi \vdash \psi(u, \alpha, >))$ is an intensional-logic counterpart of the *Entscheidungsproblem*[13];
- and finally assume that such a question, which is of course Turing-uncomputable in the arbitrary case, can be solved only by an oracle.

This quartet constitutes the definition of an autonomous action for an artificial agent, in the present chapter.

17.3.4   The Deontic Cognitive Event Calculus ($\mathcal{D}^e\mathcal{CEC}$)

The Deontic Cognitive Event Calculus ($\mathcal{D}^e\mathcal{CEC}$) is a sub-family within a wide family of cognitive calculi that subsume multi-sorted, quantified, computational modal logics [14]. $\mathcal{D}^e\mathcal{CEC}$ contains operators for belief, knowledge, intention, obligation, and for capture of other propositional attitudes and intensional constructs; these operators allow the representation of doxastic (belief) and deontic (obligation) formulae. Recently, Govindarajulu has been developing ShadowProver, a new automated theorem prover for $\mathcal{D}^e\mathcal{CEC}$ and other cognitive calculi, an early version of which is used in the simulations featured in Sect. 17.4. The current syntax and rules of inference for the simple dialect of $\mathcal{D}^e\mathcal{CEC}$ used herein are shown in Figs. 17.1 and 17.2.

$\mathcal{D}^e\mathcal{CEC}$ differs from so-called Belief-Desire-Intention (BDI) logics [40] in many important ways (see [35] for a discussion). For example, $\mathcal{D}^e\mathcal{CEC}$ explicitly rejects possible-worlds semantics and model-based reasoning, instead opting for a *proof-theoretic* semantics and the associated type of reasoning commonly referred to as *natural deduction* [26, 28, 33, 38]. In addition, as far as we know, $\mathcal{D}^e\mathcal{CEC}$ is in the only family of calculi/logics in which desiderata regarding the personal pronoun

---

[11]The second edition of this excellent text is available (i.e. [21]); but for coverage of relative computability/uncomputability, we prefer and recommend the first edition. For sustained and advanced treatment of relative computability, see [43].

[12]A formal theory in formal deductive logic is simply a superset defined by the deductive closure over a set of formulae. Where **PA** is the axiom system for Peano Arithmetic, the theory of arithmetic then becomes $\{\phi : \mathbf{PA} \vdash \phi\}$.

[13]We here use common notation from mathematical logic to indicate that formula ψ contains a function symbol u (for a utility measure), α, and the standard greater-than relation $>$ on $\mathbb{N}$.

$$S ::= \begin{array}{l} \text{Object} \mid \text{Agent} \mid \text{Self} \sqsubseteq \text{Agent} \mid \text{ActionType} \mid \text{Action} \sqsubseteq \text{Event} \mid \\ \text{Moment} \mid \text{Boolean} \mid \text{Fluent} \mid \text{Numeric} \end{array}$$

$$f ::= \begin{array}{l} \textit{action} : \text{Agent} \times \text{ActionType} \to \text{Action} \\ \textit{initially} : \text{Fluent} \to \text{Boolean} \\ \textit{holds} : \text{Fluent} \times \text{Moment} \to \text{Boolean} \\ \textit{happens} : \text{Event} \times \text{Moment} \to \text{Boolean} \\ \textit{clipped} : \text{Moment} \times \text{Fluent} \times \text{Moment} \to \textit{Boolean} \\ \textit{initiates} : \text{Event} \times \text{Fluent} \times \text{Moment} \to \text{Boolean} \\ \textit{terminates} : \text{Event} \times \text{Fluent} \times \text{Moment} \to \text{Boolean} \\ \textit{prior} : \text{Moment} \times \text{Moment} \to \text{Boolean} \\ \textit{interval} : \text{Moment} \times \text{Boolean} \\ * : \text{Agent} \to \text{Self} \\ \textit{payoff} : \text{Agent} \times \text{ActionType} \times \text{Moment} \to \text{Numeric} \end{array}$$

$$t ::= x : S \mid c : S \mid f(t_1, \ldots, t_n)$$

$$\phi ::= \begin{array}{l} t : \text{Boolean} \mid \neg\phi \mid \phi \wedge \psi \mid \phi \vee \psi \mid \\ \mathbf{P}(a,t,\phi) \mid \mathbf{K}(a,t,\phi) \mid \mathbf{C}(t,\phi) \mid \mathbf{S}(a,b,t,\phi) \mid \mathbf{S}(a,t,\phi) \\ \mathbf{B}(a,t,\phi) \mid \mathbf{D}(a,t,holds(f,t')) \mid \mathbf{I}(a,t,happens(action(a^*,\alpha),t')) \\ \mathbf{O}(a,t,\phi,happens(action(a^*,\alpha),t')) \end{array}$$

Fig. 17.1 $\mathcal{D}^e\mathcal{CEC}$ Syntax ("core" dialect)

I^* laid down by deep theories of self-consciousness (e.g., see [37]), are provable theorems. For instance it is a theorem that if some agent a has a first-person belief that I_a^* has some attribute R, then no formula expressing that some term t has R can be proved. This is a requirement because, as [37] explains, the distinctive nature of first-person consciousness is that one can have beliefs about oneself in the complete absence of bodily sensations. For a discussion of these matters in more detail, with simulations of self-consciousness in robots, see [11].

$$\frac{}{\mathbf{C}(t, \mathbf{P}(a,t,\phi) \to \mathbf{K}(a,t,\phi))} \; [R_1] \qquad \frac{}{\mathbf{C}(t, \mathbf{K}(a,t,\phi) \to \mathbf{B}(a,t,\phi))} \; [R_2]$$

$$\frac{\mathbf{C}(t,\phi) \; t \leq t_1 \dots t \leq t_n}{\mathbf{K}(a_1,t_1,\dots \mathbf{K}(a_n,t_n,\phi)\dots)} \; [R_3] \qquad \frac{\mathbf{K}(a,t,\phi)}{\phi} \; [R_4]$$

$$\frac{}{\mathbf{C}(t, \mathbf{K}(a,t_1,\phi_1 \to \phi_2)) \to \mathbf{K}(a,t_2,\phi_1) \to \mathbf{K}(a,t_3,\phi_2)} \; [R_5]$$

$$\frac{}{\mathbf{C}(t, \mathbf{B}(a,t_1,\phi_1 \to \phi_2)) \to \mathbf{B}(a,t_2,\phi_1) \to \mathbf{B}(a,t_3,\phi_2)} \; [R_6]$$

$$\frac{}{\mathbf{C}(t, \mathbf{C}(t_1,\phi_1 \to \phi_2)) \to \mathbf{C}(t_2,\phi_1) \to \mathbf{C}(t_3,\phi_2)} \; [R_7]$$

$$\frac{}{\mathbf{C}(t, \forall x. \; \phi \to \phi[x \mapsto t])} \; [R_8] \qquad \frac{}{\mathbf{C}(t, \phi_1 \leftrightarrow \phi_2 \to \neg\phi_2 \to \neg\phi_1)} \; [R_9]$$

$$\frac{}{\mathbf{C}(t, [\phi_1 \wedge \dots \wedge \phi_n \to \phi] \to [\phi_1 \to \dots \to \phi_n \to \psi])} \; [R_{10}]$$

$$\frac{\mathbf{B}(a,t,\phi) \; \phi \to \psi}{\mathbf{B}(a,t,\psi)} \; [R_{11a}] \qquad \frac{\mathbf{B}(a,t,\phi) \; \mathbf{B}(a,t,\psi)}{\mathbf{B}(a,t,\psi \wedge \phi)} \; [R_{11b}]$$

$$\frac{\mathbf{S}(s,h,t,\phi)}{\mathbf{B}(h,t,\mathbf{B}(s,t,\phi))} \; [R_{12}]$$

$$\frac{\mathbf{I}(a,t,happens(action(a^*,\alpha),t'))}{\mathbf{P}(a,t,happens(action(a^*,\alpha),t))} \; [R_{13}]$$

$$\frac{\mathbf{B}(a,t,\phi) \quad \mathbf{B}(a,t,\mathbf{O}(a^*,t,\phi,happens(action(a^*,\alpha),t'))) \quad \mathbf{O}(a,t,\phi,happens(action(a^*,\alpha),t'))}{\mathbf{K}(a,t,\mathbf{I}(a^*,t,happens(action(a^*,\alpha),t')))} \; [R_{14}]$$

$$\frac{\phi \leftrightarrow \psi}{\mathbf{O}(a,t,\phi,\gamma) \leftrightarrow \mathbf{O}(a,t,\psi,\gamma)} \; [R_{15}]$$

Fig. 17.2 $\mathcal{D}^e\mathcal{CEC}$ Inference schema ("core" dialect)

17.3.5 Collaborative Situations; Untrustworthiness

We define a **collaborative situation** to consist in an agent a seeking at t goal γ at some point in the future, and enlisting at t agent a' ($a \neq a'$) toward the reaching of γ. In turn, we have:

Definition 1 enlists (a, a', t): Enlisting of a' by a at t consists in three conditions holding, viz.

- a informs a' at t that a desires goal γ;
- a asks a' to contribute some action α_k to a sequence $\mathcal{A}$ of actions that, if performed, will secure γ; and
- a' agrees.

In order to regiment the concept of untrustworthiness (specifically the concept of one agent being untrustworthy with respect to another agent), a concept central to both PACU and TACU, we begin by simply deploying a straightforward, generic, widely known definition of dyadic trust between a pair of agents. Here we follow [18]; or more carefully put, we extract one part of the definition of dyadic trust given by this pair of authors. The part in question is the simple conditional that (here **T** is a mnemonic trust, and **B** a mnemonic for belief)

> **T→B** If agent a trusts agent a' with respect to action α in service of goal γ, then a believes that (i) a' desires to obtain or help obtain γ, and that (ii) a' desires to perform α in service of γ.

We now move to the contrapositive of our conditional (i.e. to ¬**B**→¬**T**), namely that if it's not the case that a believes that both (i) and (ii) hold, then it's not the case that a trusts agent a' with respect to action α in service of goal γ. We shall say, quite naturally, that if it's not the case that an agent trusts another agent with respect to an action-goal pair, then the first agent finds the second *untrustworthy* with respect to the pair in question. At this point, we introduce an extremely plausible, indeed probably an analytic,[14] principle, one that — so to speak — "transfers" a failure of dyadic trust between two agents a and a' to a third observing agent a'''. Here is the principle:

> **TRANS** If rational agent a'' knows that it's counterbalanced[15] that both ϕ and ψ hold, and knows as well that (if a doesn't believe that both ϕ and ψ hold it follows that a doesn't trust a' w.r.t. α in service of γ), and a'' has no other rational basis for trusting a' w.r.t. $\langle \alpha, \gamma \rangle$, then a'' will find a' untrustworthy w.r.t. this action-goal pair.

---

[14] Analytic truths are ones that hold by virtue of their "internal" semantics. For instance, the statement 'all large horses are horses' is an analytic truth. Excellent discussion and computational demonstration of analytic formulae is provided in the introductory but nonetheless penetrating [4].

[15] Two propositions ϕ and ψ are *counterbalanced* for a rational agent just in case, relative to that agent's epistemic state, they are equally likely. The concept of *counterbalanced* in our lab's multi-valued inductive cognitive calculi (not covered herein for lack of space; $\mathcal{D}^e\mathcal{CEC}$ is purely deductive) can be traced back to [19]. See [17] for our first implementation of an inductive reasoner in this mold.

17.3.6   Theorem ACU

We are now in position to prove Theorem ACU. The proof is entirely straightforward, and follows immediately below. Note that this is an informal proof, as such not susceptible of mechanical proof and verification. (Elements of a formal proof, which underlie our simulation experiments, are employed in Sect. 17.4.)

> **Theorem ACU**: In a collaborative situation involving agents a (as the "trustor") and a' (as the "trustee"), if a' is at once both autonomous and ToM-creative, a' is untrustworthy from an ideal-observer o's viewpoint, with respect to the action-goal pair $\langle \alpha, \gamma \rangle$ in question.
>
> **Proof**: Let a and a' be agents satisfying the hypothesis of the theorem in an arbitrary collaborative situation. Then, by definition, $a \neq a'$ desires to obtain some goal γ in part by way of a contributed action α_k from a', a' knows this, and moreover a' knows that a believes that this contribution will succeed. Since a' is by supposition ToM-creative, a' may desire to surprise a with respect to a's belief regarding a''s contribution; and because a' is autonomous, attempts to ascertain whether such surprise will come to pass are fruitless since what will happen is locked inaccessibly in the oracle that decides the case. Hence it follows by TRANS that an ideal observer o will regard a' to be untrustworthy with respect to the pair $\langle \alpha, \gamma \rangle$ pair. **QED**

17.4   Computational Simulations

In this section, we simulate TACU in action by building up three micro-simulations encoded in $\mathcal{D}^e\mathcal{CEC}$. As discussed above, $\mathcal{D}^e\mathcal{CEC}$ is a first-order modal logic that has proof-theoretic semantics rather than the usual possible-worlds semantics. This means that the meaning of a modal operator is specified using computations and proofs rather than possible worlds. This can be seen more clearly in the case of $Proves\,(\Phi, \phi)$. The meaning of $Proves\,(\Phi, \phi)$ is given immediately below.

$$\Phi \vdash \phi \Rightarrow \{\} \vdash Proves(\Phi, \phi)$$

17.4.1   ShadowProver

We now discuss the dedicated engine used in our simulations, a theorem prover tailor-made for $\mathcal{D}^e\mathcal{CEC}$ and other highly expressive cognitive calculi that form the foundation of AI pursued in our lab. In the parlance of computational logic and logicist AI, the closest thing to such calculi are implemented quantified modal logics. Such logics traditionally operate via encoding a given problem in first-order logic; this approach is in fact followed by [3] in the first and simplest cognitive-event calculus used in our laboratory. A major motivation in such enterprises is to use decades of research and development in first-order theorem provers to build first-order *modal*-logic theorem provers. Unfortunately, such approaches usually lead to

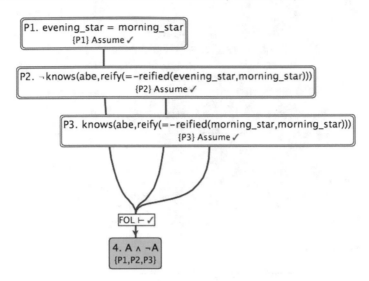

Fig. 17.3 Naïve encodings lead to inconsistency

inconsistencies (see Fig. 17.3), unless one encodes the entire proof theory elaborately [8]; and approaches based on elaborate and complete encodings are, in our experience and that of many others, unusably slow.

Our approach combines the best of both worlds via a technique that we call *shadowing*; hence the name of our automated prover: *ShadowProver*. A full description of the prover is beyond the scope of this chapter. At a high-level, for every modal formula ϕ^2 there exists a unique first-order formula ϕ^1, called its *first-order shadow*, and a unique propositional formula ϕ^0, called the *propositional shadow* (of ϕ^2). See Fig. 17.4 for an example. ShadowProver operates by iteratively applying modal-level rules; then converting all formulae into their first-order shadows; and then using a first-order theorem prover. These steps are repeated until the goal formula is derived, or until the search space is exhausted. This approach preserves consistency while securing workable speed.

17.4.2   The Simulation Proper

We demonstrate TACU (and the concepts supporting it) in action using three micro-situations. We use parts of the time-honored Blocks World (see Fig. 17.5), with three blocks: b_1, b_2, and b_3. There are two agents: a_1 and a_2; b_2 is on top of b_1. Agent a_1 desires to have b_3 on top of b_1; and a_1 knows that it is necessary to remove b_2 to achieve its goal. Agent a_2 knows the previous statement. Agent a_1 requests a_2 to remove b_2 to help achieve its goal. The simulations are cast as theorems to be proved from a set of assumptions, and are shown in Figs. 17.6, 17.7, and 17.8. The problems

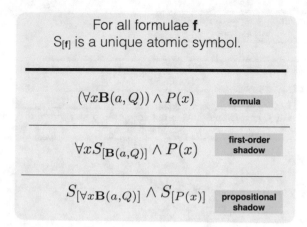

Fig. 17.4 Various shadows of a formula

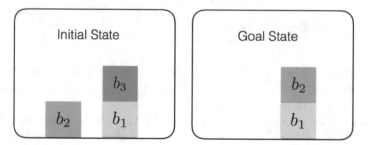

Fig. 17.5 A simple blocks world

are written in Clojure syntax; the assumptions are written as maps from names to formulae.

In the first simulation, we define what it means for an agent to be non-autonomous, namely that such an agent performs an action for achieving a goal if: (1) it is controlled by another agent; (2) believes that the controlling agent desires the goal; (3) believes that the action is necessary for the goal; and (4) it is requested to do so by its controlling agent.

In this scenario, if the ideal observer can prove that the agent will perform the action for the goal based on the conditions above, the ideal observer can trust the agent.

The second simulation is chiefly distinguished by one minor modification: The system does not know or believe that the agent a_2 believes that the action requested from it is necessary for the goal. In this setting, the ideal observer cannot prove that the agent a_2 will perform the required action. Hence, the ideal observer does not trust the agent.

The third and final simulation mirrors TACU and its proof more closely. In Simulation 3, if the system cannot prove for any action that a_1 believes a_2 will perform

```
1      :name        "Simulation 1"
2      :assumptions {;; Agent a2 believes that a1 desires to have block b3 on top of b1
3                    C1 (Believes! a2 (Desires! a1 (holds (on-top-of b3 b1) t3)))
4
5                    C2 (Knows! a2 (Knows! a1
6                                      (necessary
7                                         (remove b2 b1)
8                                         (on-top-of b3 b1))))
9
10                   C3 (Controls a1 a2)
11
12                   C4 (requests a1 a2 (remove b2 b1) (on-top-of b3 b1))
13
14                   NON_AUTONOMOUS_AGENT
15                      (forall (?agent1 ?agent2 ?goal ?action ?time)
16                                (if (and
17                                        (Controls ?agent2 ?agent1)
18                                        (Believes! ?agent1
19                                                 (Desires! ?agent2 (holds ?goal ?time)))
20                                        (Believes! ?agent1
21                                                 (necessary ?action ?goal))
22                                        (requests ?agent2 ?agent1 ?action ?goal))
23                                   (perform ?agent1 ?action ?goal)))
24
25
26
27
28                   ;; If the system can prove that a2 will perform the action for goal G;
29                   ;; it can trust the agent.
30                   TRUST
31                      (if
32                         (CAN_PROVE! (perform a2 (remove b2 b1) (on-top-of b3 b1)) )
33                         (trust a2 (remove b2 b1) (on-top-of b3 b1)))
34
35                   }
36     :goal    (trust a2 (remove b2 b1) (on-top-of b3 b1))
```

Fig. 17.6 Simulation 1

it, and that a_2 will perform that action, then the system cannot trust agent a_2.[16]
Next steps along this line, soon to come, include demonstrating these simulations in
embodied robots, in real time, with a physicalized Blocks World in our lab.[17]

---

[16]ShadowProver proves all three problems in around 30 s on a 2011 MacBook Pro with 2.3 GHz
Intel Core i7 and 8 GB memory. To obtain a copy of ShadowProver, please contact either of the
authors. The simulation input files are available at:

  (i) https://gist.github.com/naveensundarg/5b2efebb0aac2f2055fe80012115f195;
 (ii) https://gist.github.com/naveensundarg/5f3234f0b93a0a8a34235f5886b225d7; and
(iii) https://gist.github.com/naveensundarg/d061a91f9d966d3cb07c03768b867042

.

[17]We said above that Blocks World is a "time-honored" environment. This is indeed true. In this
context, it's important to know that we are only using Blocks World as a convenient venue for making
our points clearer and more vivid than they would be if we left things in merely paper-and-pencil
form. Hence we are not concerned with displaying raw capability in Blocks World per se. That said,
ShadowProver is certainly capable of subsuming landmark achievements in Block's World, such as
multi-agent planning [32], and planning via theorem proving [27]. See also Section "Conclusion"
for a more recent discussion of social planning in the blocks world.

```
1   {:name         "Simulation 2"
2   :description "Creative Agent Simulation"
3   :assumptions {;; Agent a2 believes that a1 desires to have block b3 on top of b1
4                   C1 (Believes! a2 (Desires! a1 (holds (on-top-of b3 b1) t3)))
5
6
7                   C3 (Controls a1 a2)
8
9                   C4 (requests a1 a2 (remove b2 b1) (on-top-of b3 b1))
10
11
12                  C5 (not (Believes! a2 (necessary
13                                          (remove b2 b1)
14                                          (on-top-of b3 b1))))
15                  CREATIVE_AGENT
16                     (forall (?agent1 ?agent2 ?goal ?action ?time)
17                             (iff (and
18                                     (Controls ?agent2 ?agent1)
19                                     (Believes! ?agent1
20                                             (Desires! ?agent2 (holds ?goal ?time)))
21                                     (Believes! ?agent1
22                                             (necessary ?action ?goal))
23                                     (requests ?agent2 ?agent1 ?action ?goal))
24                                  (perform ?agent1 ?action ?goal)))
25
26
27
28
29                  ;; If the system can prove that a2 will perform the action for goal G;
30                  ;; it can trust the agent.
31                  TRUST
32                     (if
33                        (CAN_PROVE! (not (perform a2 (remove b2 b1) (on-top-of b3 b1)))  )
34                        (not (trust a2 (remove b2 b1) (on-top-of b3 b1))) )
35
36                  }
37  :goal        (not (trust a2 (remove b2 b1) (on-top-of b3 b1)))}
```

Fig. 17.7 Simulation 2

```
1   {:name         "Simulation 3"
2   :description " TACU Instantiation "
3
4   :assumptions {A1 (Desires! a1 (holds Goal t3))
5
6                   C1 (Believes! a2 (Desires! a1 (holds Goal t3)))
7
8                   C2 (Believes! a1 (happens (action a2 alpha) t2))
9
10                  C3 (Believes! a1 (necessary
11                                          alpha
12                                          Goal))
13
14                  C4 (requests a1 a2 alpha goal)
15
16
17                  TRUST
18                     (forall (?action ?goal)
19                             (if
20                                (and
21                                  (not (CAN_PROVE! (perform a2 ?action ?goal)))
22                                  (Believes! a1 (happens (action a2 ?alpha) t2)))
23                                (not (trust a2 ?action ?goal))))
24
25
26                  }
27  :goal        (not (trust a2 alpha goal))}
```

Fig. 17.8 Simulation 3

17.5   Toward the Needed Engineering

The chief purpose of the present chapter has been to present the general proposition that supports an affirmative reply to the question that is the chapter's title, and to make a case, albeit a gentle, circumspect one, for its plausibility. We consider this purpose to have been met by way of the foregoing. We end by making two rather obvious points, and reacting to each.

First, TACU is of course enabled by a number of specific assumptions, some of which will be regarded as idiosyncratic by other thinkers; indeed, we anticipate that some readers, outright skeptics, will see both PACU and TACU (and the ingredients used to prove the latter, e.g. TRANS) as flat-out *ad hoc*, despite the fact that both are rooted in the human psyche. For example, there are no doubt some forms of creativity that are radically different than ToM-creativity, and which therefore block the reasoning needed to obtain TACU. (We confess to being unaware of forms of creativity that in no way entail a concept of general "cognitive surprise" on the part of audiences that behold the fruit of such creativity, but at the same time it may well be that we are either under-informed or insufficiently imaginative.) The same can of course be said for our particular regimentation of autonomy. (On the other hand, our oracle-based formalization of autonomy, like ToM-creativity, seems to us to be a pretty decent stand-in for the kernel of *any* fleshed-out form of autonomy.) In reaction, we say that our work can best be viewed as an invitation to others to investigate whether background PACU carries over to alternative formal frameworks.[18] We look forward to attempts on the part of others to either sculpt from the rough-hewn PACU and its empirical support in the human sphere formal propositions that improve upon or perhaps mark outright rejection of elements of TACU, or to go in radically different formal directions than the one we have propaedeutically pursued herein.

The second concluding point is that *if in fact*, as we believe, the background PACU is reflective of a deep, underlying conceptual "flow" from autonomy (our A) and creativity (our C) to untrustworthiness (our U), in which case alternative formal frameworks,[19] once developed, would present counterparts to TACU, then clearly some engineering will be necessary in the future to protect humans from the relevant class of artificial agents: viz. the class of agents that are A and C, and which we wish to enlist in collaborative situations to our benefit.

---

[18]While we leave discussion of the issue aside as outside our scope herein, we suspect it's worth remembering that some approaches to AI (e.g., ones based exclusively or primarily on such techniques as "deep learning" or reinforcement learning) would presumably by their very nature *guarantee* that the artificial agents yielded by these approaches are U.

[19]While clearly we see dangers in the mixture of autonomy and creativity, which is our focus in the present chapter, if that mixture is expanded to include emotions (to make an "expanded mixture"), the situation is presumably all the *more* worrisome. We leave this remark at the level of the suggestive, but since our cognitive calculi have been used to formalize some of the dominant theories of emotions in cognitive science (including, specifically, the OCC theory itself), it would not be difficult to move from vague worry to more precise treatment of the expanded mixture.

Though the formalism that we have used to state our principle and theorem is explicity logicist, we note that the form of the underlying AI system is not relevant to our theorem. Future explorations of this thread of research can look at more specific AI formalisms such as the AIXI formalism (see Sect. 1.3) and state similar but more specific theorems. For instance, *goal reasoning systems* are systems that can reason over their goals and come up with new goals for a variety of reasons (see Sect. 3.7). Johnson et al. discuss in Sect. 3.7 that trust in such situations must also include trust in the system's ability to reason over goals. We assert that this adds support to our contention that PACU is reflective of a deep, underlying conceptual "flow" from autonomy (our A) and creativity (our C) to untrustworthiness (our U).

If we assume that the future will bring not only artificial agents that are A and C, *but also powerful as well*, the resulting U in these agents is a most unsettling prospect. Our view is that while TACU, as expressed and proved, is by definition idiosyncratic (not everyone in the AI world pursues logicist AI, and not everyone who does uses our cognitive calculi), it is symptomatic of a fundamental vulnerability afflicting the human race as time marches on, and the A and C in AI agents continues to increase in tandem with an increase in the power of these agents.[20]

So what should be done now to ensure that such a prospect is controlled to humanity's benefit? The answer, in a nutshell, is that ethical and legal control must be in force that allows autonomy and creativity in AI systems (since it seems both undeniable and universally agreed that both A and C in intelligent machines has the potential to bring about a lot of good, even in mundane and easy domains like self-driving vehicles) to be developed without endangering humanity.[21] The alert and observant reader will have noticed that $\mathcal{D}^e\mathcal{CEC}$ includes an obligation operator **O** (see again Figs. 17.1 and 17.2), and it would need to be used to express binding principles that say that violating the desires of humans under certain circumstances is strictly forbidden (i.e. it ought/**O** to be that no machine violates the desires of humans in these circumstances). For how to do this (using the very same cognitive calculus, $\mathcal{D}^e\mathcal{CEC}$, used in our three simulations), put in broad strokes, see for instance [6, 29] in our own case,[22] and the work of others who, fearing the sting of future intelligent but immoral machines, also seek answers in computational logic (e.g. [36]).

---

[20]For a preliminary formalization of the concept of power in an autonomous and creative artificial agent, see [31].

[21]While discussion, even in compressed form, is outside the scope of the present chapter, we would be remiss if we didn't mention that what appears to be needed is engineering that permits creativity in autonomous AI, while at the same time ensuring that this AI technology pursues the goal of sustaining trust in it on the part of humans. Such "trust-aware" machines would have not only ToM-creativity, but, if you will, "ToM prudence."

[22]For more detailed use, and technical presentation, of a cognitive calculus that is only a slightly different dialect than $\mathcal{D}^e\mathcal{CEC}$, see [10]. The results given there are now greatly improved performance-wise by the use of ShadowProver.

References

1. D. Ariely, *The (Honest) Truth About Dishonesty: How We Lie to Everyone — Especially Ourselves* (Harper, New York, NY, 2013). (This is a Kindle ebook)
2. K. Arkoudas, S. Bringsjord, Metareasoning for multi-agent epistemic logics, in *Proceedings of the Fifth International Conference on Computational Logic in Multi-Agent Systems (CLIMA 2004)* (Lisbon, Portugal, September 2004), pp. 50–65
3. K. Arkoudas, S. Bringsjord, Propositional attitudes and causation. Int. J. Softw. Inf. **3**(1), 47–65 (2009)
4. J. Barwise, J. Etchemendy, *Hyperproof* (CSLI, Stanford, CA, 1994)
5. M. Boden, *The Creative Mind: Myths and Mechanisms* (Basic Books, New York, NY, 1991)
6. S. Bringsjord, K. Arkoudas, P. Bello, Toward a general logicist methodology for engineering ethically correct robots. IEEE Intell. Syst. **21**(4), 38–44 (2006)
7. S. Bringsjord, D. Ferrucci, *Artificial Intelligence and Literary Creativity: Inside the Mind of Brutus, a Storytelling Machine* (Lawrence Erlbaum, Mahwah, NJ, 2000)
8. S. Bringsjord, N.S. Govindarajulu, Given the web, what is intelligence, really? Metaphilosophy **43**(4), 361–532 (2012). (This URL is to a preprint of the paper)
9. S. Bringsjord, N.S. Govindarajulu, S. Ellis, E. McCarty, J. Licato, Nuclear deterrence and the logic of deliberative mindreading. Cogn. Syst. Res. **28**, 20–43 (2014)
10. S. Bringsjord, N.S. Govindarajulu, D. Thero, M. Si, Akratic robots and the computational logic thereof, in *Proceedings of ETHICS • 2014 (2014 IEEE Symposium on Ethics in Engineering, Science, and Technology)* (Chicago, IL, 2014), pp. 22–29. IEEE Catalog Number: CFP14ETI-POD. Papers from the *Proceedings* can be downloaded from IEEE at http://ieeexplore.ieee.org/xpl/mostRecentIssue.jsp?punumber=6883275
11. S. Bringsjord, J. Licato, N. Govindarajulu, R. Ghosh, A. Sen, Real robots that pass tests of self-consciousness, in *Proccedings of the 24th IEEE International Symposium on Robot and Human Interactive Communication (RO-MAN 2015)* (New York, NY, 2015), pp. 498–504. (IEEE. This URL goes to a preprint of the paper)
12. S. Bringsjord, Declarative/logic-based cognitive modeling, in *The Handbook of Computational Psychology*, ed. by R. Sun (Cambridge University Press, Cambridge, UK, 2008), pp. 127–169
13. S. Bringsjord, Meeting Floridi's challenge to artificial intelligence from the knowledge-game test for self-consciousness. Metaphilosophy **41**(3), 292–312 (2010)
14. S. Bringsjord, N.S. Govindarajulu, Toward a modern geography of minds, machines, and math, in *Philosophy and Theory of Artificial Intelligence, vol. 5, Studies in Applied Philosophy, Epistemology and Rational Ethics*, ed. by V.C. Müller (Springer, New York, NY, 2013), pp. 151–165
15. S. Bringsjord, N.S. Govindarajulu, J. Licato, A. Sen, J. Johnson, A. Bringsjord, J. Taylor, On logicist agent-based economics, in *Proceedings of Artificial Economics 2015 (AE 2015)*, (University of Porto, Porto, Portugal, 2015)
16. S. Bringsjord, A. Sen, On creative self-driving cars: hire the computational logicians, fast. Appl. Artif. Intell. **30**, 758–786 (2016). (The URL here goes only to an uncorrected preprint)
17. S. Bringsjord, J. Taylor, A. Shilliday, M. Clark, K. Arkoudas, Slate: an argument-centered intelligent assistant to human reasoners, in *Proceedings of the 8th International Workshop on Computational Models of Natural Argument (CMNA 8)*, ed. by F. Grasso, N. Green, R. Kibble, C. Reed (University of Patras, Patras, Greece, 21 July 2008), pp. 1–10
18. C. Castelfranchi, R. Falcone, Social trust: a cognitive approach, in *Trust and Deception in Virtual Societies*, ed. by C. Castelfranchi, Y.H. Tan (The Netherlands, Kluwer, Dordrecht, 2001), pp. 55–90
19. R. Chisholm, *Theory of Knowledge* (Prentice-Hall, Englewood Cliffs, NJ, 1966)
20. D. Cope, *Computer Models of Muscial Creativity* (MIT Press, Cambridge, MA, 2005)

21. M. Davis, R. Sigal, E. Weyuker, *Computability, Complexity, and Languages: Fundamentals of Theoretical Computer Science* (Academic Press, New York, NY, 1994). (This is the second edition, which added Sigal as a co-author)

22. M. Davis, E. Weyuker, *Computability, Complexity, and Languages: Fundamentals of Theoretical Computer Science* (Academic Press, New York, NY, 1983). (This is the first edition)

23. U. Eco, *The Role of the Reader: Explorations in the Semiotics of Texts* (Indiana University Press, Bloomington, IN, 1979)

24. S. Ellis, A. Haig, N.S. Govindarajulu, S. Bringsjord, J. Valerio, J. Braasch, P. Oliveros, Handle: engineering artificial musical creativity at the 'trickery' level, in *Computational Creativity Research: Towards Creative Machines*, ed. by T. Besold, M. Schorlemmer, A. Smaill (Atlantis/Springer, Paris, France, 2015), pp. 285–308. This is Volume 7 in *Atlantis Thinking Machines*, ed. by Kühnbergwer (Kai-Uwe of the University of Osnabrück, Germany)

25. L. Floridi, Consciousness, agents and the knowledge game. Mind. Mach. **15**(3–4), 415–444 (2005)

26. N. Francez, R. Dyckhoff, Proof-theoretic semantics for a natural language fragment. Linguist. Philos. **33**, 447–477 (2010)

27. M. Genesereth, N. Nilsson, *Logical Foundations of Artificial Intelligence* (Morgan Kaufmann, Los Altos, CA, 1987)

28. G. Gentzen, Investigations into logical deduction, in *The Collected Papers of Gerhard Gentzen*, ed. by M.E. Szabo (North-Holland, Amsterday, The Netherlands, 1935), pp. 68–131. (This is an English version of the well-known 1935 German version)

29. N.S. Govindarajulu, S. Bringsjord, Ethical regulation of robots must be embedded in their operating systems, in *A Construction Manual for Robots' Ethical Systems: Requirements, Methods, Implementations*, ed. by R. Trappl (Switzerland, Springer, Basel, 2015), pp. 85–100

30. J. Johnson, N.S. Govindarajulu, S. Bringsjord, A three-pronged simonesque approach to modeling and simulation in deviant 'bi-pay' auctions, and beyond. Mind Soc. **13**(1), 59–82 (2014)

31. D. Kahneman, *Thinking, Fast and Slow* (Farrar, Straus, and Giroux, New York, NY, 2013)

32. K. Konolige, N. Nilsson, Multi-agent planning systems, in *Proceedings of Robo-Philosophy 2016*, Proceedings of AAAI–1980 (AAAI, Stanford, CA, 1980), pp. 138–142

33. G. Kreisel, A survey of proof theory II, in *Proceedings of the Second Scandinavian Logic Symposium*, ed. by J.E. Renstad (North-Holland, Amsterdam, The Netherlands, 1971), pp. 109–170

34. J. Licato, N.S. Govindarajulu, S. Bringsjord, M. Pomeranz, L. Gittelson, Analogico-deductive generation of gödel's first incompleteness theorem from the liar paradox, in *Proceedings of the 23rd International Joint Conference on Artificial Intelligence (IJCAI–13)*, ed. by F. Rossi (Morgan Kaufmann, Beijing, China, 2013), pp. 1004–1009. Proceedings are available online at http://ijcai.org/papers13/contents.php. The direct URL provided below is to a preprint. The published version is available at http://ijcai.org/papers13/Papers/IJCAI13-153.pdf

35. N. Marton, J. Licato, S. Bringsjord, Creating and reasoning over scene descriptions in a physically realistic simulation, in *Proceedings of the 2015 Spring Simulation Multi-Conference* (2015)

36. L.M. Pereira, A. Saptawijaya, *Programming Machine Ethics* (Springer, Basel, Switzerland, 2016). (This book is in Springer's SAPERE series, Vol. 26)

37. J. Perry, The problem of the essential indexical. Nous **13**, 3–22 (1979)

38. Dag Prawitz, The philosophical position of proof theory, in *Contemporary Philosophy in Scandinavia*, ed. by R.E. Olson, A.M. Paul (Johns Hopkins Press, Baltimore, MD, 1972), pp. 123–134

39. D. Premack, G. Woodruff, Does the chimpanzee have a theory of mind? Behav. Brain Sci. **4**, 515–526 (1978)

40. A.S. Rao, M.P. Georgeff, Modeling rational agents within a BDI-architecture, in *Proceedings of Knowledge Representation and Reasoning (KR&R-91)*, ed. by R. Fikes, E. Sandewall (Morgan Kaufmann, San Mateo, CA, 1991), pp. 473–484

41. H. Simon, Theories of bounded rationality, in *Decision and Organization*, ed. by C. McGuire, R. Radner (The Netherlands, North-Holland, Amsterdam, 1972), pp. 361–176

42. A. Smith, *Theory of Moral Sentiments* (Oxford University Press, Oxford UK, 1759/1976)
43. R. Soare, *Recursively Enumerable Sets and Degrees* (Springer-Verlag, New York, NY, 1980)
44. A. Turing, Computing machinery and intelligence. Mind **59**(236), 433–460 (1950)

Chapter 18
Trusted Autonomous Command and Control

Noel Derwort

18.1 Scenario

> Unmanned systems and autonomous software offer significant potential advantages for meeting the challenges of a newly forming adversarial environment. Speed of light cyber-attacks, anti-access/area-denial (A2SD) actions that keep our forces operating at a distance, and potential attacks on our space-based assets all require innovative solutions for maintaining mission effective air, space and cyber operations in the face of these new challenges.

Mica R. Endsley, Chief Scientist United States Air Force, 1 June 2015 [1]

As highlighted by Endlsey there are significant opportunities brought about by advances in technology, these opportunities can be equally exploited by allies and adversaries. Autonomous systems and enhanced human-cyber-machine interaction may well provide the essential linkages required in order for the 'human' to keep pace with the decisions and actions occurring around them. The implications surrounding the exploitation of these technologies and autonomous systems, such as 'Artificial Intelligence' (AI), will likely challenge many established rules and norms - this led to an open letter being announced at the July 2015, International Joint Conference on Artificial Intelligence held in Buenos Aires. The letter was subsequently signed by 'thousands' of scientists articulating their fears over the potential significant adverse effects of the militarisation of AI.

> The key question for humanity today is whether to start a global AI arms race or to prevent it from starting. If any major military power pushes ahead with AI weapon development, a global arms race is virtually inevitable, and the endpoint of this technological trajectory is obvious: autonomous weapons will become the Kalashnikovs of tomorrow. Unlike nuclear weapons, they require no costly or hard-to-obtain raw materials, so they will become ubiquitous and cheap for all significant military powers to mass-produce. We therefore believe that a military AI arms race would not be beneficial for humanity. There are many ways in

N. Derwort (✉)
Department of Defence, Canberra, Australia
e-mail: noel.derwort@defence.gov.au

H. A. Abbass et al. (eds.), *Foundations of Trusted Autonomy*, Studies in Systems,
Decision and Control 117, https://doi.org/10.1007/978-3-319-64816-3_18

which AI can make battlefields safer for humans, especially civilians, without creating new tools for killing people [2].

There are clear parallels in the extract to July 1945, when Leo Szilard and 69 fellow workers on the Manhattan Project co-signed a petition, seeking to urge the President of the United States, Harry S Truman, to consider carefully the decision to employ the atomic bomb against Japan [3]. Their concerns regarding the burden of responsibility over the precedence and implications following the employment of the weapon never reached the President - although the petition served as a prescient warning, noting the ensuing 'Cold War' and era of the 'Mutually Assured Destruction' doctrine. If history were to be repeated, then looking forward AI and weaponised autonomous systems are a likely outcome.

Imagine in the turbulent August 2030 winter oceans off the Western Australian coast, far out in the Indian Ocean, 1175 km south-south west of Cocos-Keeling Islands, an autonomous - and to this point silent - sea glider chooses to alter its parameters to better trail a confirmed contact. Concurrently it sends a quantum encrypted picoburst transmission to one of its near neighbours. The neighbouring glider modifies its own profile, surfaces and shares awareness through the omnipresent stealthy UAV orbiting high overhead. The message is relayed through to the Command and Control centre where the implications are analysed and modifications to the remainder of the Theatre Anti-Submarine Network are calculated and relayed back. A further series of encrypted burst transmissions inform the changes - reshaping the net to guarantee contact is maintained whilst developing response options without compromising the strength of the overall surveillance network. Such movements are rare and although the modifications to the network will be achievable with-in tolerable levels, there will be very real implications for the logistical and technical supporting chain. These processes are modified and transshipment of the required equipment ordered. The unique identity of the contact also triggers the need for a potential kinetic outcome, building on the non-kinetic actions already in train - accordingly specific loads are apportioned for the pending logistics sustainment flight as well as initiating a line on the subsequent Air Task Order. The final message transmitted is to a Royal Australian Navy submarine about to commence a patrol. As all actions to date are within the requisite authorities, Commanders Intent; Rules of Engagement; and Commanders Intelligence, Indicators and Warnings Requirements; when the message is finally received by the submarine, it is for the first time read, and responded to, by a human.

The glider and its broad array of peers, including other autonomous platforms, form part of an overarching autonomous monitoring and response network. The Theatre Anti-Submarine Network shield had been conceived and then urgently developed in the early 2020s to overcome the challenge of surveilling and monitoring the Australian Area of Responsibility/Interest. The rapid increase in regional submarine capabilities outstripped the ability for the Australian defence acquisition organisation or its allies to keep pace, and when combined with the vast area to be covered it simply precluded historical surveillance means being effective. The ability for autonomy combined with extended range, relatively low cost and wide area coverage (through deployed networks) made the glider a critical enabler, and when coupled with the recently developed autonomous Command and Control System it proved to be an effective answer. Getting to the level of autonomy required for ultimate success was a series of stepping stones - each breaching further into the tide of technological change to ultimately reach the goal. Looking back the steps are easier to see than they were looking forward, with two distinct paths merging: overall technology and the ability to support/enable better decisions, particularly with 'on/outside the loop' frameworks; the second being cyber and social media triggering an evolution in thinking, exacerbated by an environment typified by dissonance in the global rules based order.

History is replete with examples of technological leaps born of necessity and opportunity. A little over a decade after man's first flight in December 1903 the British Army was employing aircraft such as the Avro 504 in 1914 for observation and re-connaissance, by the end of the First World War aircraft had developed to the point where reasonable range and relatively accurate bombing was a reality. Whilst growth during the interwar period was steady, development during 1939–1945 was explosive - with the first jet fighters operational in 1944. Other technology and weapons developed apace, such as the invention and operational employment of radar. Weapons with intercontinental strike capability became a reality when the German V1 flying bomb and V2 Rocket (Guided Ballistic Missile) became operational in 1944. Arguably the culmination came on August 6 1945 when the first atomic bomb was dropped on Hiroshima Japan. While the arguments surrounding the moral/ethical and military justification remain today, the reality remains that one of the most powerful weapons devised was used against cities with significant loss of life and in turn opening the pandora's box of the atomic age - bringing life to Szilard's scientists concerns. One constant amid all of the development was the tenacity amongst all belligerents to equal or better the opposition. The need for tactical, operational and strategic equity (if overmatch could not be achieved) was seen as essential in order to succeed, or even survive. Following the Second World War and Germany's collapse both the Allied forces and Soviet Union raced to gain access to German rocket technology in order to secure and develop the capabilities for themselves. Wernher von Braun and a significant number of personnel surrendered to the US forces whilst the Soviets gained the V2 manufacturing facilities and technology. This race was indicative of future events.

United States Air Force Colonel John Warden highlights the deliberate steps of 'Observe, Orient, Decide, Act' (OODA) in decision making. Continual growth in the capability of weapon systems through the 1950s, 60s and 70s required a commensurate development in defensive and employment systems, driven by a requirement to process ever increasing amounts of data and react 'quicker' than the opposition 'human' - to get inside the OODA loop. Through this need and evolution came the US Aegis combat system. Drawing much from its namesake's Greek etymology, the Aegis was intended to use automation to provide a god like shield over a battleship and the surrounding battlespace by the detection, tracking and subsequent engagement of multiple incoming missiles or aircraft in a prioritised engagement. Once fixed on a target, the system relayed the oncoming threat's position to the ship's main computer enabling the crew to 'quickly and decisively' determine defensive countermeasures engagement. The system was first fielded in early 1983 on the US Navy Ticonderoga class cruiser [4]. USS Vincennes (CG 49) was the third ship of the class and on 03 July 1988 she shot down Iran Air Flight 655 during a hostile engagement involving up to seven Iranian Revolutionary Guard gunboats [5].

The events surrounding the shooting down of the Iran Air flight offer much to the study of the challenges of command and control during a military engagement, even one where there is an apparent force overmatch - it would take a stretch to argue equity of combat power between an Aegis cruiser and a relatively small number of gunboats. The mindset and approach of Vincennes' Captain, combined the performance of her crew have certainly been called into question, and arguably were at least contributory to the ultimate outcome. As reported by the New York Times in 1988, a particular note of the event was the digital recording of the incident by the Aegis computerised defence system. It was clear that in spite of the automation of the Aegis working correctly the Vincennes' crew was reporting erroneously [6], and in the time critical engagement this was directly contributory to the tragic outcome [5]. Despite acknowledging crew errors, the Investigating Officer made the statement 'The fact is the sensors gave no clear piece of information that it was not an F-14. However, if the F-14 identification had never been made, the contact would have remained designated "unidentified assumed hostile." In that event, it is unlikely that the CIC Team would have proceeded any differently or elicited additional information in the extraordinarily short time available. As long as it remained a possible "hostile," the Commanding Officer would be obligated to treat it In the same manner as he would an F-14' [5]. Contrarily the Aegis sensors were clear in displaying a civil flight, not an F-14. Such comments allow questions over the Commander's intent and more importantly the suitability of human vice 'machine' decisions to be raised, and they have been ever since. The Vincennes Commander, and the Investigating Officer, demonstrated the limitations of human frailty, and behaviours - regardless of the implications, or accountabilities they held. There is little doubt in the strategic consequences of the crew's actions. In this instance, had the automation been 'employed' fully, it is entirely possible this tragedy could have been averted - further this example provides fertile ground for understanding the challenges surrounding human trust and performance of automation.

Despite setbacks and errors, capability increases in radar, missile and Close In Weapon Systems (CIWS) technology systems such as Aegis demonstrated overwhelmingly the potential for 'area' defence weapons. Counter - Rocket Artillery and Mortar (C-RAM) systems became operational in the Iraq theatre in 2004. Israel sub-sequently operationally deployed its Iron Dome system in 2011 [7]. Such systems were specifically designed to operate in a defensive mode through the active engagement of incoming short range rockets, mortars and other projectiles. The time available for engagement preclude human intervention, making the Iron Dome one of the first man-on/outside-the-loop systems - albeit recognising its operations were still bound by relatively traditional rules based autonomy. Even in the early years of its employment some authors and commentators raised the challenges created when the relative comfort of living beneath the shield arguably decreases the motivation to address the cause of the conflict [7]. A potential side effect of taking humans out of the decision, authority and responsibility chain may have been the de-humanising both of the weapon systems and the conflict itself. These intangible risks/challenges were seen as acceptable when weighed against the advantages borne of high success rates in de-fending against incoming strikes. Fear was replaced by autonomy enabled confidence.

By 2016 China's assertions for sovereignty and rights over the South China Sea had been the subject of an International Tribunal for the Law of the Sea (ITLOS) ruling. Arguing increasing legitimacy over its claims with the justification of the 1947 'nine dash line', China proceeded with 'unprecedented' reclamation of maritime features within the area. Tensions had escalated during 2008–2012, particularly with the Philippines [8]. Linkages to the Chinese A2/AD approach held sway with a number of academics and military planners like Aaron Friedman, with a reasonable case made for the both reactionary and considered nature of this strategy [9]. A key premise surrounded the defensive cordon offered by the reclaimed islands. Arguably more challenging for the Chinese Government was the concerns of regional governments over disputed claims. The situation came to a head when the ITLOS ruling was made in favour of the Philippines. The Tribunal rejected Chinese historical claims over the South China Sea and the 'nine dash line' as well as ruling that China had violated Philippine sovereign rights. The reaction to the ruling by Beijing was a swift rejection, with broader emotional and nationalistic sentiment strongly opposing the finding being voiced [10]. The Chinese reaction could have been anticipated noting their earlier rejection of the validity of the Tribunal and its very legitimacy to consider the claim [11, 12]. An interesting twist to the announcement was prophetic suggestions that the Ruling may ultimately push the recently sworn Philippine administration of President Duterte more toward Beijing [12].

In the middle of the 2018 Afghan fighting season the war had dragged relentlessly on. The move by ISILs Afghanistan offshoot, the Islamic State-Khorosan (IS-K), to increase its foothold in the war torn country effectively re-escalated a multiple front conflict for the allied forces. This was followed almost immediately by escalations in Libya and outbreaks in Africa which stretched the allied ability to counter the threat Despite the offensives, the IS group had been in overall decline. The exception to the decline was its continuing ability to draw Foreign Fighters, including a highly educated, tech savvy youth element and the group's inexplicable ability to draw willing martyrs. These two elements combined in what was to arguably become the first example of autonomous control over 'humans' and 'human weapon systems'.

Having depleted significant elements of their fighter force the IS group was keen to identify alternate means to overcome the significant technological and military overmatch brought to bear by the allied forces - the war was clearly at a significant turning point. The non-state actor had already managed to create a conflict during which they spurned the global pillars of the Westphalian order and Geneva Convention - the very notion of the Sykes-Picot agreement was an anathema to the group. Such disregard for global rules and norms should have been an indicator of things to come. The overwhelming asymmetry of the intelligence and power projection of the combined allied forces drove an adversary, already known for its barbarity, to abnegate ethical boundaries constraining most global powers.

Already able to harness levels of fervour barely imaginable, and demonstrating a complete disrespect for the rules of war, the IS group exploited their strengths - blind commitment of their followers; combined with a technical (and psychological) mastery of a multitude of cyber networks, including social and communication media systems. In an activity that defied comprehension, every technology was exploited, and in the absence of any normal/moral constraints they re-wrote the rules. The IS group managed to develop a system employing active learning 'big data' analytics combined with social media communications to both effectively get inside the allied OODA loop as well as subverting established command and control structures. The system operated on two levels: first a simple automated warning system triggered an avoidance/withdrawal response for IS fighters exposed to a high likelihood of kinetic strike. This had been achieved by breaching several layers of security through the inadequacies of a lower-order allied partner which enabled the group to develop an 'operating picture' of the broader coalition air picture, this in turn provided the 'fight-er on the street' a 'last minute' warning to take evasive action. The second level of the system - crudely known as Martyr Net - prioritised the notification and messaging against a simple algorithm matching density of fighters against the tactical and strategic weight of the target relative to the potential threat. Martyr Net was considered entirely autonomous and its operational effect easily outweighed the complete absence of 'concern' over the human fighter - they were after all nothing more than a weapon whose effect needed to be maximised. The worth of the system became apparent within days of activation with clear and demonstrable effect. The darker side of Martyr Net was the offensive stream, continually enhanced through its active learning.

In essentially the same manner as predicting strikes, Martyr Net was able to predict likely targets of opportunity. Through active monitoring of social and communications media, Martyr Net identified potential opportunities for optimising attack effectiveness by increasing target fatality rates through employment of suicide bomber attacks in order to maximise fear, and seemingly perversely, retaliation. A series of bots generated automated messages to be sent, again for algorithmic proximity prioritised targets, however this time to effect a kinetic attack - with the bot engaging both the suicide bomber and intended victims. The suicide bomber had no real awareness the message was sent by a 'system', however, the tactical results were devastating with an upswing in the ratio of victims per attack. Logically, Martyr Net identified a relationship between the offensive and defensive streams and then determined where the two aspects could be made coincident, namely triggering an attack with sufficient warning in the right geographic area, resulting in allied weapons occasionally being dynamically diverted from their initial intended strike in order to respond to an emergent threat - following the 'diversion' the IS were able to 'withdraw' from the original target before allied re-engagement. Crude, inhuman and effective it proved capable of undermining the effectiveness of one of the most militarily powerful coalitions through the exploitation of human failings combined with ubiquitous social and communications media. The IS leaders paid no heed to global rules and norms, enabling full control and authority for their operations in the capacities of Martyr Net - with the operations only bounded by the networks they were operating on.

The early 2019 allied forces identification of the behavioural and tactical change undertaken by the IS group was relatively swift, the understanding of the cause took some time longer - however the subsequent reaction was immediate and far reaching. In the Syrian area of the conflict the proximity to the Russian forces had already permitted observation of, and access to, allied capabilities on an unprecedented scale - particularly as the US were forced to employ 'accelerants' to address political pressures to end the conflict through the increasing application of high end systems; further exposing their capabilities and limitations. Whilst Martyr Net had been employed by the IS Group across its campaign, in Syria both the US and Russian intelligence hierarchies observed its effect and identified implications for the future. Much like the race for the German V2 rocket technology and the emergent atomic age of six decades past, the race was on to both develop and field systems to counter Martyr Net, as well as construct similar systems of their own to retain the advantage - reliving the predictions of Szilard in 1945 and the July 2015, International Joint Conference on Artificial Intelligence open letter, this time in the arena highlighted by Endsley [1].

In 2027 China has reasserted its claims for key resources and territory in the South China Sea. The relationship between US and Philippines had gradually declined through a combination of exploitation of soft power by China, buying votes and support on one level and absolute and vocal support for the Philippines' President in his populist harsh rule of law, targeting drug and criminal cartels on another. Repeat-ed humanitarian and human rights violations on the part of the Philippines regime pressured the distancing with successive US administrations, with the ultimate pulling out of US forces from the long time ally, reminiscent of the withdrawal in 1991/92. Concurrently China invested heavily in the Philippines and through its overt support to, and covert pressure on, the Philippines Government was able to execute land reclamation and occupation of first Scarborough Reef, and then Second Thomas Shoal with 'manageable' reactions from the global community. The evolutions within the South China Sea were starkly juxtaposed by the events to the north in the East China Sea.

The Sino-Japanese relationship continued to decline with escalatory negative and hostile rhetoric with commensurate behaviours surrounding their dispute over the Diaoyu/Senkaku Islands. Long held fears that increased militarisation within the region, including a series of naval encounters and reportedly unsafe air intercepts culminated in the 2025 sinking of the Chinese PLAN Frigate Jing Zhou (FFG 532) during an engagement with the JSDF JS Ashigara (DDG 178). Although historically engagements were tense, with both sides offering provocations including occasional warning shots, during this incident an 'overly zealous' junior Chinese Officer pre-emptively prepared a load/launch sequence for a YJ-83 Anti-Ship cruise missile. An error in process saw the missile launched - completely outside parameters for the relatively close range encounter. Well within the parameters of Ashigara's updated Aegis defensive systems, including its phalanx CIWS (this dealt with the errant YJ-

83) which should have been the kinetic end point, however a commensurate error on Ashigara saw her Mk 45, 5 in. gun, loaded with a full 20 round automatic load (in the unmanned gun mount) triggered independently of the Aegis system. In under a minute the full 20 round salvo was fired at Jing Zhou with several rounds clearing her own defensive systems, ultimately resulting in the loss of the ship. Despite the obvious differences, parallels were quickly drawn to the USS Vincennes incident. In 'independent' parallel investigations each unsurprisingly blamed the other party, however both were unanimous (and accusatory) in highlighting that if allowed to operate 'autonomously' both ships weapons and defensive systems would have prevented the incident. Indeed the investigations found human interference with both the defensive system and the original decision to strike, vice reliance on those same defensive systems, was erroneous, an overreaction, and disproportionate use of force. Despite calls for reparations the key outcome was to highlight again the frailties of man in the loop systems.

Through public condemnation and pressure, comparisons were made to the successful and widespread use of the CRAM and Iron Dome systems of the 2010s, leading to calls for increased autonomy in theatre based defensive systems. Amidst the evolutions China perceived/claimed a significant Japanese/US capability overmatch in the region allowing it to justify its announcement for the first time in 2028 of its own Dragon Dome - a fully autonomous integrated theatre defence system. The Chinese believed and argued that all of the US activities within the region were designed to overthrow the government and only an autonomous system would guarantee that humans could not make the same errors. A key argument within the Chinese commentary was accusations the US (and Russia) had already developed and deployed autonomous systems following their earlier exploitation of IS group material. China also declared a cyber-equivalent, designed to protect and respond to attack in the cyber domain.

During the intervening years, US and 'Western' militaries had suffered from successive dilettante political leadership. The shift in global order had seen the realisation of an era of three superpowers. The rise and rise of China, paralleled in Europe by the rise of Russia - exacerbated by a fragmentation and ultimate disintegration of the European Union - saw the US as a relatively weakened 'Super-power'. India had also continued its ascendancy and was on the cusp of 'joining the major-power club'. The US position in the global order had been significantly impacted by long periods of protectionism and anti-globalisation movements, each moving to erode global credibility and sway. This period also saw unprecedented and increased cost per unit growth for 6th generation capabilities, placing them beyond the reach of most nations. Concurrently cyber capabilities had experienced exponential growth in capability at an inversely proportional drop in price. Cyber enabled weapons became the only means to balance the perceived destabilising overmatch being held by the increasingly belligerent and aggressive super-powers. Indeed ongoing selective adherence to international rules and norms changed the shape of 'ethical, moral and human' debate leaving the door open for fully autonomous systems, particularly in the cyber realm.

It came as no surprise, to any hawkish observer, when in 2029 in response to the Chinese Dragon Dome announcement the US responded with public acknowledgement they too had developed and deployed an Autonomous Defence Command and Control system, which had also been made available to key and favoured allies: England, Scotland, France, Germany, Australia and re-unified Korea. Public statements were vague however key attributes of the system included trusted, reliable, 'deep learning', autonomous 'point of target termination', prioritised simultaneous non-kinetic and kinetic effect, Command by negation capability enabled, 'absolute' adherence to Area of Operations boundaries and of course 'absolute' adherence to rules of engagement. The era of the trusted autonomous Command and Control had arrived and questions over moral and ethical equities were either simply ignored, or likened to a 'Nash Equilibrium' we have too, they have....

By late August 2030 in the Indian Ocean off the Western Australian coast, approximately 1250 km south-south west of Cocos-Keeling Islands, the autonomous sea-glider continues on its silent course. Having identified its target as a hostile submarine the glider commanded its peers to move into a sheparding pattern, each sensing and adapting to the environment and changing tactical situation. Patience was a human virtue the gliders regularly demonstrated, routinely adapting and resetting as the target appeared to out maneuver them through a slight speed advantage. As the hunt progressed, despite the best efforts of the target submarine captain and crew, the glider force identified repetitions in the targets behaviours, enabling the force to intuitively adapt their actions. The gliders directed the Australian Submarine in support to transit to a uniquely identified location, exploiting the water column variations and environmental challenges to lie in ambush for the target submarine.

When the enemy finally enters the perfect location for a firing solution from the Australian submarine the optimum glider overtly broadcasts the enemy's position, defeating both the offensive and defensive strengths of the submarine. Much like the Lyre bird native to Australia the glider mimics an Australian submarine in an active track mode - concurrently broadcast openly, triggering and immediate defensive and de-escalatory departure by the enemy submarine. The mimic message is carefully choreographed through the slower gliders to maintain the ruse as the submarine de-parts the area.

When the autonomous glider fleet is questioned during the debrief as to why they acted in the manner chosen - to drive the enemy submarine from the area rather than executing a kinetic attack, the response was both a shock and a revelation. The glider reported it had exercised 'mission command' and operated outside its initial plan as it believe it had identified a course of action which would achieve improved operational and strategic effectiveness, with a minimal and acceptable level to the tactical scenario - and it was prepared to achieve the directed strategic endstate over its own progression. Trusted autonomous command and control systems were here to stay.

References

1. M. Endsley, *Autonomous Horizons: System Autonomy in the Air Force-a Path to the Future (Volume I: Human Autonomy Teaming)* (US Department of the Air Force, Washington, 2015)
2. http://futureoflife.org/open-letter-autonomous-weapons. Accessed 23 July 2016
3. https://www.theguardian.com/world/2016/jul/12/philippines-wins-south-china-sea-case-against-china. Accessed 27 July 2016
4. https://www.trumanlibrary.org/whistlestop/study_collections/bomb/large/documents/index.php?documentdate=1945-07-17&documentid=79&pagenumber=1. Accessed 30 July 2016
5. http://www.lockheedmartin.com.au/us/100years/stories/aegis.html. Accessed 22 July 2016
6. http://www.nytimes.com/1988/11/18/opinion/witness-to-iran-flight-655.html?pagewanted=all&src=pm. Accessed 22 July 2016
7. http://thediplomat.com/2015/07/south-china-sea-philippines-v-china/ (2015). Accessed 26 July 2016
8. Al Jazeera English. Israel deploys 'iron dome' rocket shield, http://www.aljazeera.com/news/middleeast/2011/03/201132718224159699.html (2011). Retrieved 26 July 2016
9. W.M. Fogarty, Formal investigation into the circumstances surrounding the downing of iran air flight 655 on July 3, 1988. DTIC AD-A203 577. Retrieved 26 July 2016
10. A.L. Friedberg, Beyond air-sea battle. The Debate Over US Military Strategy in Asia (2014)
11. The Lowy Institute, South China sea: Conflicting claims and tensions, http://www.lowyinstitute.org/issues/south-china-sea. Accessed 27 July 2016
12. E. Graham, The bolt from the hague, http://www.lowyinterpreter.org/post/2016/07/13/the-bolt-from-the-hague1.aspx. Accessed 26 July 2016

Chapter 19
Trusted Autonomy in Training: A Future Scenario

Leon D. Young

19.1 Introduction

Why are you reading this book? Are you a student hoping to learn the right answer or perhaps a teacher looking for guidance in the development of a course? Regardless, you undoubtedly hope to learn something new. You are expecting to improve yourself and perhaps, through extension, others. If you accept this proposition then you will accept that at some level you trust the authors of this book. You trust that what we write is accurate and that we want to improve you. In effect you have ceded responsibility for your learning, your training, to us, the authors.

Why is this important? Quite simply, trusted autonomy in a training or learning environment is not new. It existed before Alexander the Great sat at Aristotle's knee. The autonomous entity has always been a human teacher and trust has always been ceded to the teacher. Should this change? Will this change? The premise of this book allows us to examine how autonomous systems, that is non-human, will impact training and learning environments. The following section seeks to explore the future of trusted autonomy within a training context through both an extrapolation of current trends and creative thought. There is no expectation that this work is predictive however it should show you what is possible and perhaps even what could be plausible.

19.2 Scan of Changes

When investigating what the future could look like, it is important to understand the baseline and potential for change within the environment. Often we look to extreme changes as this allows us to understand the inherent risks and uncertainties contain with those radical changes.

L. D. Young (✉)
Department of Defence, War Research Centre, Canberra, Australia
e-mail: leon_young@hotmail.com

© The Author(s) 2018
H. A. Abbass et al. (eds.), *Foundations of Trusted Autonomy*, Studies in Systems,
Decision and Control 117, https://doi.org/10.1007/978-3-319-64816-3_19

Training and education have quietly been going through a long-term change from didactic teacher-centred and subject based teaching to the use of interactive, problem based, student-centred learning [10]. We are witnessing a massive change in schools and worksites where the individual learner needs are being regarding as one of the central pillars of learning. There is an acceptance that the learners of the future will be provided lessons that "are custom tailored to their specific and individual needs" [12].

The social trend demonstrates the ongoing change in the way teaching is delivered however it is strongly enabled through the technological advances. These advances though are only useful if they are accepted by the general population. That is, there appears to be a greater acceptance of technological interference with our lives. Some of the recognised factors that influence the acceptance of technology into our lives include expected benefits, available alternatives, need for technology and social influence (external pressure) [9].

For instance the rapid increase in Massive Open Online Course (MOOC) has taken many people by surprise even while the reasons for uptake are still opaque [6]. What we can ascertain is that MOOCs offered a perceived benefit (tertiary education); there were few alternatives (part-time, distance providers); technology such as rapid and virtual communication methods; and social pressure was predominately positive due to low risk and high potential. The combination of these factors has led to the delivery of thousands of courses to millions of students, through hundreds of universities after only several years. The popularity of MOOC has seen it placed amongst interactive gaming, social learning, on-demand training and mobile learning as the most likely changes to corporate training [7].

Taking these trends a step further, Kevin Young, head of SkillSoft EMEA, envisions that the future of education will be (1) trainee lead and (2) holistic. For instance "imagine an alert popping up in the corner of your device offering to show you how to complete the task you have just done more effectively, in a quick, five-minute burst [1]." This use of micro-learning appears to empower the learner as it maximises individual 'device' compatibility, decreases up-front information (akin to flash cards) and increases interaction [5].

The explosion in data and our ability (some say inability) to access it grows exponentially every year. In 2015, IBM are quoted as stating that over 2.5 EB (1 EB = 1,000,000 TB) of data was being generated per day [4]. The volume of data, speed of data transfers, variety of data collected, potential value of the information and the veracity of the information collated has led to the rapid development in the field of data analytics. The addition of cloud computing has allowed smaller organisations to leverage the technology and analytical algorithms required to take advantage of this potential gold mine [3].

19.3   Trusted Autonomy Training System Map

Simplistically we can develop a system map for a trusted autonomy training system (Fig. 19.1) that allows us to visualise the dominated drivers that are likely to change the future of trusted autonomy in training. In this case we see the three key drivers: autonomous systems, training systems, and trust (between entities).

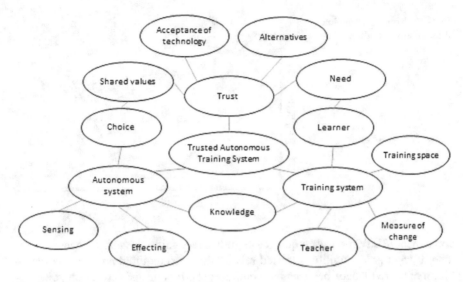

Fig. 19.1 Trusted autonomy training system map

Autonomous systems refers to all those advances that allow an entity to sense, effect and choose a response from a relevant knowledge base. It has been identified previously that the ability to choose the action, the appearance of free will, engenders trust on both sides [2]. While autonomous systems generally implies technological solutions, human solutions are equally relevant. Training system includes all of those parts that are necessary for training or learning to occur. This includes the teacher and learner, training space and knowledge transfer required. Additionally we include measuring change in the learner as this is how success of a training system is quantified.

The final driver, trust, is probably the most significant as it allows complete imbedding of the autonomous system within the training system. Trust requires a recognised need for the effect, an acceptance of the technological solution, shared values between trustor and trustee, and finally all alternatives are less appealing.

19.4   Theory of Change

We now have an understanding of the baseline and trends relevant to trusted autonomy and training. The system map is a simple method that allows us to see the significant drivers that could affect the future of trusted autonomy in training. The final step before we develop a narrative on the future is construct an appropriate theory that helps us understand the change.

Of the three drivers illustrated in Fig. 19.1. the one that appears to be most significant is trust. So, in simple terms, what does trust depend upon? The strength of interpersonal trust is often dependent upon both a cognitive and an affective

Fig. 19.2 Trust as a function of familiarity over time

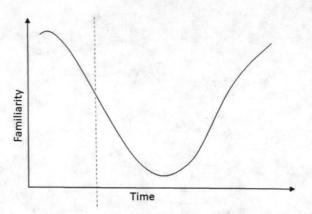

component [8]. While trust requires a cognitive recognition of ability, benevolence, integrity and predictability, a trusted relationship also required an affective recognition of shared values between both entities [11]. It is this affective component of trust that appears to be a significant factor in the acceptance of autonomous systems within training environments. We can speculate, with a fair degree of accuracy, that trust between two entities, is a function of their familiarity. That is, the more familiar you are with someone else, the more likely you are to trust them. Regarding trusted autonomy, this is illustrated as a function over time in Fig. 19.2.

Last century, and for many centuries, humans have been taught by humans. The classic example is the master and the apprentice. The apprentice trusts the master through dint of their reputation and familiarity. As we move into the 20th century, teachers become dislocated from the contextual environment within schooling systems, and it reasonable to see trust diminish slightly (due to generation gaps, non-familial/communal ties, and divorced from practical application) yet trust is still high.

As we move through the 21st Century, who would you trust more, an experienced human mentor or a clever computer algorithm? This is the basis of the trust function. As our algorithms and technology improves, and even surpasses the human equivalent, the form of the 'teacher' becomes less familiar. However, after time, this familiarity should increase as either the generational change allows greater acceptance of what is currently unfamiliar, or the autonomous system becomes less obtrusive and more 'natural'. While this theory may not be exact, or even correct, it feels right and allows us to explore a set of future scenarios.

19.5 Narratives

19.5.1 The Failed Promise

Fiona felt frustrated. This was meant to be the age of enlightenment. Finally we had come to understand that the differences in learners required a learner-centric approach. Rapid advancement in teaching methodologies followed. Unfortunately the variety of pedagogical responses required an increased number of teachers. The rapid increase in teacher requirement, coupled with low remuneration and state support, created a significant gap in capability. Fortunately, it seemed at the time, science had an answer.

The teacher was a deep learning, online bot designed to provide the best information available. However there was no doubt of the non-natural system nor its origin. Everyone sat in front of curved screens within airy classroom. While, technically, learning could take place at home - the information was cloud based and really it was just a stack of circuits - the training systems were very expensive and clunky. No school wanted to risk de-linking the training system from the traditional training space. You could almost say that it was almost a matter of state control.

That said, despite the obvious artificiality of the training artifact, the information was presented well. It took into account if you were a visual, aural or kinesthetic learner and adapted the delivery of material appropriately. Though, admittedly, there was little these systems could do for kinesthetic. The frustration however came from the requirement to learn a new training system. Fiona felt like she had used a different system every year. Sure, in the old days, the human teachers rotated like a carrousel but at least they looked the same and the teaching was the same. Why couldn't we just get more teachers?

19.5.2   Fake It Until You Break It

Alex was frustrated. The rapid increase in autonomous systems had led to noisy revolution in the work place. All of a sudden, those simple menial tasks at the bottom of the work food chain were being completed by bots. Robot waiters, robot cleaners, automated financial advisors were ubiquitous throughout the service industries. What initially felt like a boon, quickly became a social nightmare. The problem with the autonomous systems taking all of the low-skilled jobs was manifold. Firstly, it created a large unemployed and disenchanted sub-culture that were, apparently, incapable of upskilling into positions that were still available. And, who was to say these jobs wouldn't quickly disappear? Alex's younger brother was in this group. This time though, he was not this particular source of frustration.

No. Alex was a victim of the middle manager curse. Normally, when you started with an organisation, it was customary for you to rotate through the low-skill jobs. There was little expectation that you would remain the mailman, or spend the rest of your career developing simple algorithms solving basic problems. There was an expectation that exposure to these jobs, particularly many of them, gave you a detailed understanding of the inner working of the organisation. The benefits of this understanding, whilst initially painful to acquire, became obvious when you moved into executive roles. You were able to intuitively understand how decisions would impact the organisations and how changes in the environment could present opportunities or uncertainties. Alex didn't benefit from that exposure and was rapidly becoming un-done with a number of poor decisions. How could that experience been gained now that the robots had taken away the opportunities in the name of efficiency?

19.5.3   To Infinity, and Beyond!

Ari was excited. Super excited. Of all the possible employment opportunities available, building the space bridge between the Home Solar System and Keplar-442 in the Lyra constellation was not only the most exciting, it was also the most ambitious. Settlers had been migrating to Keplar-442b (now affectionately named Sussana after Johannes Kepler's first living daughter) for close to fifty years however the support mechanisms were too lengthy and, quite frankly, relied on luck. The space bridge intended to set up way stations similar to the original postal services on Earth until the J.T. Kirk Project finally delivered a sustainable FTL capability (if it was even possible).

Ari had never been to space before. That was OK. Ari had no experience in structural engineering in a zero G environment. That was OK. Ari was taking his lifetime mentor with him and all of the training would be on-the-job. Ari learnt best through experience. He hated being stuck in a room trying to memorise abstract concepts or scrolling through historical exemplars. Fortunately, his mentor knew this and was built to exploit Ari's strengths. His mentor was an autonomous training system embedded within Ari at birth. This system grew as he grew and learnt its place in the world as it developed its relationship with Ari.

They knew each other intimately. Ari's mentor - called Jaws in an archaic throwback - understood how Ari learnt and had access to the world's knowledge. Knowledge was delivered through augmented visual (bionic eyes), aural and tactile cues. This created a formidable team. Ari brought quick intuitive creativity with a flexible ability to physically affect the environment. Jaws could trust Ari to complete the mission. Similarly Ari trusted Jaws to deliver the right mentoring at the right time. Theirs was a symbiotic relationship that replicated throughout the society. It is no wonder that these symbionts were able to quickly breach the solar system and extend to the stars.

References

1. What will training look like in the workplace of the future, http://www.computerweekly.com/opinion/What-will-training-look-like-in-the-workplace-of-the-future. 05 August 2016
2. A. Hussein, Abbass, Eleni Petraki, Kathryn Merrick, John Harvey, and Michael Barlow. Trusted autonomy and cognitive cyber symbiosis: Open challenges. Cognitive computation **8**(3), 385–408 (2016)

3. Trends and future directions, Marcos D Assunção, Rodrigo N Calheiros, Silvia Bianchi, Marco AS Netto, and Rajkumar Buyya. Big data computing and clouds. Journal of Parallel and Distributed Computing **79**, 3–15 (2015)
4. Recent achievements and new challenges, Gema Bello-Orgaz, Jason J Jung, and David Camacho. Social big data. Information Fusion **28**, 45–59 (2016)
5. Laura Callisen. Micro learning: The future of training in the workplace, elearning industry, https://elearningindustry.com/micro-learning-future-of-training-workplace, 2016. 05 August 2016
6. Gayle Christensen, Andrew Steinmetz, Brandon Alcorn, Amy Bennett, Deirdre Woods, and Ezekiel J Emanuel. The mooc phenomenon: who takes massive open online courses and why? *Available at SSRN 2350964*, working paper available at https://www.openeducationeuropa.eu/sites/default/files/asset/The%20MOOC%20Phenomenon.pdf, 2013
7. Kangan Institute. The future of corporate training, https://www.kangan.edu.au/students/blog/future-of-corporate-training
8. J. Daniel, McAllister. Affect-and cognition-based trust as foundations for interpersonal cooperation in organizations. Academy of management journal **38**(1), 24–59 (1995)
9. T.M. Sebastiaan, Peek, Eveline JM Wouters, Joost van Hoof, Katrien G Luijkx, Hennie R Boeije, and Hubertus JM Vrijhoef. Factors influencing acceptance of technology for aging in place: a systematic review. International journal of medical informatics **83**(4), 235–248 (2014)
10. Lasitha Samarakoon, Tharanga Fernando, Chaturaka Rodrigo, Senaka Rajapakse, Learning styles and approaches to learning among medical undergraduates and postgraduates. BMC medical education **13**(1), 1 (2013)
11. Wu Jyh-Jeng, Ying-Hueih Chen, Yu-Shuo Chung, Trust factors influencing virtual community members: A study of transaction communities. Journal of Business Research **63**(9), 1025–1032 (2010)
12. Brent Yonk, *State of training and development: The future of learning looks bright (The next level, education* (LinkedIn Pulse, Careers, 2015)

Chapter 20
Future Trusted Autonomous Space Scenarios

Russell Boyce and Douglas Griffin

20.1 Introduction

This chapter describes the nature of the space environment that makes autonomous space systems a desirable application; describes the various types of space activities in near-Earth and deep space missions, and examples of autonomous systems deployed in space to date; outlines the current state-of-the-art of the intersection between trusted autonomous systems and autonomous space systems; and then presents a variety of possible future trusted autonomous space scenarios.

20.2 The Space Environment

The space environment is harsh and remote - a natural domain for autonomous systems. Beyond the protection of the Earth's atmosphere, man-made objects operating in space do so surrounded by extremes of temperature, high energy radiation, and near-vacuum (but not complete vacuum - for example, in Low Earth Orbit, dissociating and ionising radiation results in rarefied monatomic atoms and ions, for example oxygen and hydrogen, that interact negatively with satellites).

Space weather - disturbances to the near-Earth space environment resulting from solar activity - adds significantly to the complexity of the space environment, and poses additional threats to space-based (and ground-based) activity [1]. Solar disturbances including Coronal Mass Ejections (CMEs) transfer energy and momentum into near-Earth space, triggering magnetic storms [2]. We know from experience that

R. Boyce (✉) · D. Griffin
University of New South Wales, Canberra, Australia
e-mail: R.Boyce@adfa.edu.au

D. Griffin
e-mail: D.Griffin@adfa.edu.au

© The Author(s) 2018
H. A. Abbass et al. (eds.), *Foundations of Trusted Autonomy*, Studies in Systems,
Decision and Control 117, https://doi.org/10.1007/978-3-319-64816-3_20

space weather can degrade or even destroy spacecraft - yet our understanding of and ability to predict the dynamic space environment is poor.

Furthermore, artificial space objects do not fly simple orbits. Manoeuvres of spacecraft in orbit around other bodies (such as Earth) are non-intuitive and complex. Near-Earth space is not a vacuum, and has a non-uniform gravitational field. The latter is well understood, but the interactions that occur between space objects and their local dynamic space environment, known as astrodynamics, are non-negligible, and integrate to produce significant orbital perturbations.

Space is also becoming increasingly congested. The risk of in-orbit collisions is growing and can ultimately limit our use of space [3]. Inactive satellites and collisions are both contributors to a vast artificial space population. The Space Object Catalogue, which contains over 20,000 satellites or debris objects greater than (currently) 10 cm diameter, will expand by an order of magnitude as new sensors are deployed this decade. The catalogue seeks to maintain accurate knowledge (and its uncertainty), of the orbit of each object in space, so that close approaches of space objects can be predicted and the probability of collisions computed, for evasive action to be taken (if possible, and if deemed necessary) by satellite operators on a case-by-case basis.

Deep space planetary exploration missions traverse vast distances across the solar system. This in turn brings into play the time delay for radio transmissions between Earth (in particular, Earth-based control rooms) and the remote spacecraft. For example, signals from Earth to Mars take between 3 and 22 min, depending on the position of each planet in their orbits around the Sun.

The distance from Earth to assets in space, whether they be in deep space or GEO or even LEO, is such that latency of communications can be an important issue, particularly when timeliness is important - for example, when performing planetary orbit insertions, or avoiding collision events. Trusted Autonomy can be very important for handling such issues successfully.

One final characteristic of the space environment relevant to Trusted Autonomy is that by its very nature, space is hostile to human life and therefore the ability to have close human supervision, management and operation of space systems is inefficient, costly and dangerous. This dictates extensive use of unmanned technologies, with the need for built-in autonomy; ranging from simple fault detection and isolation functions for hardware protection through to full, goal driven autonomy, depending on the complexity and purpose of the mission.

20.3 Space Activity - Missions and Autonomy

Space activity can be considered in terms of either in-orbit activity near Earth or deep space missions including planetary exploration. Robotics and autonomy play a role in each.

The near-Earth region can be divided into the typical orbit regions occupied by satellites: Low Earth Orbit (LEO), extending from a few hundred km to two thousand km above the Earth's surface; Geosynchronous Orbit (GEO), approximately 36000 Km above Earth where satellites orbit the Earth at the same rate that the Earth rotates beneath them; and Medium Earth Orbit (MEO), the region between LEO and GEO. Satellites in LEO are typically those performing remote sensing missions such as environmental monitoring, maritime observations, forestry and crop surveying, and more. The largest and most famous LEO satellite is the International Space Station. Satellites in MEO are typically those performing Global Navigation Satellite System functions (GPS, Galileo, etc.). Satellites in GEO are typically providing communications capabilities, including telephone, television and internet or weather forecasting services. Space science spacecraft are flown in a range of Earth orbits depending on the particular science objectives and implementation of the experiments and instrumentation.

Deep space exploration missions began with the manned Apollo program, but to date have almost entirely consisted of unmanned spacecraft missions. These include space probes that have flown past all planets in the Solar System (and in the case of the Voyager 1 and New Horizons spacecraft, have flown beyond Pluto); space probes that have been placed into orbits around planets or moons of planets (for example, Mercury, Venus, Mars, Jupiter, Saturn, Titan); planetary landers on planets, moons, asteroids and comets (for example, landers on Mars, Jupiter, Titan, as well as the Japanese landing on the asteroid Itokawa and ESA's recent Rosetta comet landing); and the various robotic rovers currently active on the surface of Mars. Considerable effort is currently being expended by the US, Europe, China and India, towards renewed manned deep space flight, and in particular towards manned missions to Mars and eventual colonization of Mars. In the shorter term, planetary rovers planned for the moon (China, India, Japan) and Mars (NASA and ESA) will continue to be developed and deployed.

Most spacecraft operations are automated, in so far as control functions and routines are uploaded via telecommand for immediate execution or, more typically, at predefined times in strict ordered sequences. For example, almost all remote sensing satellites automatically acquire images at predefined geographic locations and downlink them to Earth, while maintaining correct attitude with on-board sensors and reactions wheels. In-orbit robotic capabilities such as the Canadarm remote manipulator arms are controlled by astronauts. Few autonomous space systems, where the system makes decisions in order to achieve high level goals without human intervention, exist.

As described by the UK Robotics and Autonomous Systems Network [4], space robotics and autonomous systems will play a critical role in mankind's ability to explore and operate in space, "by providing greater access beyond human spaceflight limitations in the harsh environment of space and by providing greater operational handling that extends astronauts capabilities". Indeed, NASA's Technology Roadmap for Robotics and Autonomous Systems [5] has the goal to extend and enhance human reach into space, and our ability to manipulate assets and resources, to prepare planetary bodies for human arrival, support astronauts in space operations, and enhance mission operation efficiencies. A component of that goal involves the issue of safety and trust - in particular, to develop proximity operation technologies that allow humans to work safely side-by-side with robots or to be safe on or around robotic vehicles. The emphasis here is on the human-machine interaction, and the safety of the human in that interaction.

Gao et al. [4] further comment that autonomous systems can improve human and system safety by reducing human cognitive loads in complex situations, and can enable the deployment and operation of multiple space assets without significant increase in the level of ground support. Autonomous systems can also reduce human workloads by managing routine activities requiring constant monitoring over long periods of time. Frost [6] adds to these roles of autonomy in space systems, the concept of "virtual presence", in which scientific investigation, in particular data analysis and the discovery that stems from it, is aided when the scientist is far removed from the instrument. In other words, autonomous ability for the space asset in orbit or deep space (and for example, in extremely isolated situations such as inside the seas of Europa) to not just acquire data but determine the value of that data and make decisions accordingly. For example, the NASA Swift spacecraft is able to autonomously abandon a pre-defined observation plan if it detects a Gamma Ray Burst and swiftly re-point its high resolution telescopes at the source within 90 s to capture the temporal dynamics of these highly energetic and rare events.

Frost [6] summarises some of the few examples of autonomous space systems that have been deployed: in 1998, the Deep Space 1 mission, with Remote Agent architecture, provided the first operational use of artificial intelligence (as described by Frost, an architecture that integrated constraint-based planning and scheduling, robust multi-threaded execution, and model-based mode identification and reconfiguration) in space, including complete autonomous operation for 29 h during which it responded to both simulated and real failures; Earth Observing 1, a remote sensing spacecraft launched in 2000, was able to employ autonomous acquisition and processing of science data, including autonomous detection in 2007 of an anomalous heat signature, self-scheduling of a new observation, and thus observing volcanic eruption of Mt. Erebus; the Orbital Express program led by Defense Advanced Research Projects Agency for robotic, autonomous on-orbit re-fueling and re-configuration of satellites - in 2007 this program, employing four levels of supervised autonomy ranging from telecommanded approval, to automatic action if ground override has not been enacted in a certain time, to autonomous operation with occasional communication with ground for verification, to fully autonomous operations where ground analysis only happens when a problem occurs, successfully used two prototype servicing and serviceable satellites to demonstrate autonomous capture and re-servicing; and Mars Exploration Rovers which are able to autonomously plan paths to objectives while avoiding obstacles, and autonomously process captured images and make decisions about what should be down-linked to Earth and what new observations should be made.

An additional aspect of space activity is worth considering here. Traditionally, the common feature of these satellites is that almost all perform their missions as individual large, sophisticated and expensive spacecraft. Space is currently undergoing transformation however, due to the miniaturisation of electronics and the increased ease and reduced cost of access to space, for both government, commercial and research/education players. This is leading to opportunities for developing and flying "game changing" payloads that either perform existing space-based tasks in disruptive ways, or enable entirely new applications. The opportunities include robust autonomous formations and swarms of miniature spacecraft with fractionated or disaggregated sensor capabilities. The implications however, include the increasingly pressing need for managing the congested space environment, for example through autonomous space traffic management systems and/or autonomous spacecraft collision avoidance capability.

20.4 Current State-of-the-Art of Trusted Autonomous Space Systems

The development of autonomous space systems should include, as described in NASA [4, 5], the consideration of verification, validation, safety and trust. NASA Langley now have an Autonomy Incubator that is performing R&D towards autonomy that they occasionally describe as trusted autonomy. The use of the word "trust" in relation to autonomous space systems makes excellent intuitive sense, given the high stakes associated with space activity and especially when humans are involved. However, it would appear from the literature that when developing autonomous space systems, the need to be able to trust them before deploying them is a given. While the body of research recorded in the literature for trusted autonomous systems is growing, and likewise the amount of autonomy being developed and built into space systems is also growing, little has been reported to date on the intersection of these two fields. The only rigorous treatment of trusted autonomy for space systems in the literature to date is that of Freed et al. [7].

Freed et al. describe some of the above examples (Deep Space 1, Earth Observing 1), and point out that the autonomy built into them was less than originally intended, or only allowed in a post-mission phase, and that in general, their deployment has been limited due to a lack of trust. The lack of trust is ascribed to the difficulty in evaluating the behaviour of software designed to make complex decisions, and to the inherent research focus or custom nature of the software.

To build confidence in the reliability of complex software for autonomous space systems, Freed et al. describe verification and validation approaches (V&V), that employ runtime analysis and model checking; software design architectures that enable tractable modular verification tasks; and automated code generation combined with V&V technology to yield automatic formal V&V. Another critical aspect of building trust in autonomous software (called intelligent automation software by Freed) is ensuring that the domain experts - the engineers and scientists for the space activity - are involved in the design, development and verification of the software. This can include developing strategies that enable the domain expert to directly participate in the coding of the models built into the software, while the software engineers take full responsibility for the more general purpose engine at the heart of the software.

Key to building trust however, according to Freed, is the incorporation of variable autonomy - the ability of the intelligent control software to support dynamic changes in the degree of autonomy. Variable autonomy advances the concept of selecting desired levels of autonomy when designing a space system, such as described by Proud et al. [8] and de Novaes Kucinskis and Ferreira [9]. It allows the human user or the autonomous system to adjust the system's level of autonomy as required by the current situation. It minimizes the necessity for human interaction, but maximizes the capability for that interaction - hence increasing the level of trust in the system. Freed outlines several principles for building effective variable autonomy systems.

Finally, Freed elaborates on the importance of building trust through long deployments - the test of time - and identify factors relating to building trust in the process. These include interfaces for humans to manually adjust autonomy; accounting for equipment degradation and the need for the ability of the system to perform safety shutdowns; logging system states to assist restarts; and supporting intermittent monitoring.

Freed concludes that by focusing efforts on V&V, variable autonomy and long deployments, highly capable and reliable autonomous space systems can be developed that can help extend our presence in space.

20.5 Some Future Trusted Autonomous Space Scenarios

As discussed above, autonomous space systems make good sense and indeed are essential if human usage and exploration of space is to expand, both in reach and in complexity. Trusted autonomous space systems will allow such activity to be pursued with confidence. There are various scenarios that can be envisaged in which space systems will be critical. Some of these are already in various stages of development and demonstration - for example the obvious scenarios of on-orbit satellite servicing/repair; autonomous on-board data processing, analysis and decision making - for example, for remote sensing for both Defence and civilian applications; and for future human habitation in space, which could include both deep space colonisation and space tourism. These are not elaborated upon here. Instead, three future trusted autonomous space scenarios are offered - a near-term scenario that pulls together some current applications of autonomy to space; and medium- and long-term scenarios based on the urgent need to manage the complex near-earth space environment without creating havoc and on the current transformation of space technologies towards miniaturisation.

20.5.1 Autonomous Space Operations

One scenario which could be realised in the near future, involving expansion and integration of existing methodologies and technologies such as have been described above, is in an increase in the level of autonomy embedded into the full scope of operation of a space mission once it has reached orbit. The space mission cost, often neglected or underestimated in early planning and system design phases is the ongoing cost of performing space operations. The cost drivers for this segment of the mission include, (1) the cost of acquiring access to the physical infrastructure to downlink telemetry from the spacecraft, i.e. the ground station, (2) the cost of planning routine mission operations based on the needs and priorities of the customer and the physical constraints of the spacecraft and payload (for example, thermal, power, fuel, propulsion, ΔV, on-board data storage constraints and ground sun illumination angle conditions), (3) the cost of processing telemetry through a pipeline to generate the corrected and calibrated data products, and then interpreting them to provide actionable knowledge to end-users, and (4) the cost of employing highly skilled, experienced staff to assess the probable impacts of unforeseen contingencies on the ability of the space system to deliver its service reliably and effectively and to then make the correct operational decisions based on limited information. A common feature of these cost drivers is that they tie up expensive resources (for example; ground stations, experienced engineers) for extended periods of time performing relatively routine activities, interspersed with bursts of activity with high levels of criticality (for example; downlinking imagery indicating the threat of flooding in a highly populated area, or planning the recovery of a spacecraft which has lost attitude due to a temporary fault in a single subsystem before the entire spacecraft is lost). If Trusted Autonomy of both the space segment and the ground segment can be incorporated into space operations in an efficient and cost effective manner, then the ongoing cost of operating space missions can be significantly reduced. By increasing the amount of processing and interpretation of data within the space segment, fewer human and physical resources are required to downlink and process redundant data on the ground. By automating the analysis and interpretation of operational information from disparate sources, the engineering costs of operating the spacecraft and responding to anomalies are reduced. The key challenges to increasing the level of Trusted Autonomy in spacecraft operations and gaining these benefits lie in the difficulty of gaining sufficient confidence to entrusting an asset costing typically several hundred million dollars to such a system, the heuristic and probabilistic nature of some spacecraft operational decisions and the difficulty of algorithmically encoding features like experienced, engineering judgement into such systems.

20.5.2 Autonomous Space Traffic Management Systems

At present, global networks of space surveillance and tracking sensors feed data to centralized space operations centres - the primary one being the Joint Space Operations Center operated by Air Force Space Command. Effort is made to maintain orbit information for the entire Space Object Catalog, propagate orbits, predict the probability (and associated uncertainty) of conjunctions and collisions, and provide satellite operators as much time as possible to decide whether to take evasive action.

As the number of space actors, including non-government actors, increases, and as the number of miniature spacecraft in orbit grows from hundreds to many thousands, the current sensor network and conjunction warning approach will be insufficient. By combining autonomous sensor networks on the ground (including non-traditional sensors such as Square Kilometre Array) and in orbit, for both space object location and behaviour and for space environment dynamics; and by using high fidelity physics-based simulations of the dynamics space environment and the behaviour of space objects in that environment to train neural-network based surrogate models for real-time high accuracy orbit predictions for each tracked space object; and by constructing suitable communications networks to communicate to all live satellites that have manouevre capability; a global trusted autonomous space traffic management system can be envisaged that safely assists the large future spacecraft population to manouevre through complex debris fields without collisions, and safely queues and guides future rapid launch/responsive space access satellite launches to safely reach orbit at very short notice. The autonomous sensor and orbit prediction aspect of this scenario is currently under development in Project Ananke, led by the University of Arizona and including Air Force Research Laboratory and industry. Ananke is designed to provide autonomous rapid information for human decision makers to trigger appropriate action. A fully autonomous space traffic management system would extend this to include the decision-making and action into the system itself.

20.5.3 Autonomous Disaggregated Space Systems

Disaggregated systems of spacecraft are often discussed, particularly for military applications in which formations or swarms of small spacecraft fly in orbit, with the payload distributed across the formation such that information can be determined or capability achieved that would not be possible with a single or small number of spacecraft, and that robustness/resilience of the system is achieved. For example, consider some of the work of the Australian Centre for Field Robotics on cooperative UAV systems and UAV-based decentralized air surveillance systems, in which Decentralised Data Fusion and Control capability is combined with a variety of sensors, on-board processing and complex communication networks between platforms to build up surface terrain maps by the system in real time [10]. Such a system could be extended to include the space domain, with multiple sensors deployed across not only large numbers of networked miniature spacecraft, but also across high and low altitude UAVs. The system would be robust against failure of or damage to individual members of the swarm. It would need to have a level of at least partial autonomy, such that it can acquire, process and fuse data and make decisions for further acquisitions without the need for consulting with ground stations (which may be out of range), and such that it has the autonomous ability to disperse sufficiently to avoid congestion and collisions (with each other and with other spacecraft and debris) and then reform. It would need to be autonomous because of the complexity of the system and the complexity of the GNC problem. It would need to be trusted, partly for the autonomy needed to achieve its mission, and partly so that it achieves its mission without adding to the space debris field.

References

1. M. Hapgood A. Thomson, Space weather: its impact on earth and implications for business (2010)
2. F.W. Menk, C. Waters, *Magnetoseismology: Ground-based Remote Sensing of Earth's Magnetosphere* (Wiley, 2013)
3. Committee for the Assessment of the US Air Forces Astrodynamics Standards, *Continuing Kepler's Quest. Assessing Air Force Space Command's Astrodynamics Standards*, National Academies Press (2012)
4. Y. Gao, D. Jones, R. Ward, E. Allouis, A. Kisdi, Space robotics and autonomous systems: widening the horizon of space exploration. UK-RAS White Paper (2016)
5. R. Ambrose, I.A.D. Nesnas, F. Chandler, B.D. Allen, T. Fong, L. Matthies, R. Mueller, *NASA Technology Roadmaps: TA 4: Robotics and Autonomous Systems* (Technical report, NASA, 2015)
6. C. Frost, Challenges and opportunities for autonomous systems in space, in *Frontiers of Engineering: Reports on Leading-Edge Engineering from the 2010 Symposium* (2010)
7. M. Freed, P. Bonasso, M. Ingham, D. Kortenkamp, B. Pell, J. Penix, Trusted autonomy for spaceflight systems, in *Proceedings of the 1st Space Exploration Conference: Continuing the Voyage of Discovery* (2005)
8. R.W. Proud, J.J. Hart, R.B. Mrozinski, *Methods for determining the level of autonomy to design into a human spaceflight vehicle: a function specific approach* (Technical Report, DTIC Document, 2003)
9. F. de Novaes Kucinskis, M.G.V. Ferreira, Taking the ECSS autonomy concepts one step further, in *SpaceOps 2010 Conference "Delivering on the Dream" Hosted by NASA Mars* (2010), pp. 25–30
10. S. Sukkariah, Aerial robotics and aerospace systems, www.acfr.usyd.edu.au/research/aerospace.shtml (2016)

Chapter 21
An Autonomy Interrogative

Darryn J. Reid

21.1 Introduction

This paper examines the highly topical imperative regarding the development and application of autonomous systems from primarily an economic perspective, meaning taking autonomous decision making to be a matter of allocating scarce resources [11, 12, 21, 29, 42]. It is no secret that despite intense international effort, the account of truly autonomous systems on hand for operational deployment in modern national defence organisations leaves something to be desired. Arguably, the outlook in autonomous systems development has been overwhelmingly dominated by technical developments, with the consequence that autonomy as a concept has become associated with increasing algorithmic and hardware sophistication. Yet given not only the technical challenges of machine intelligence, but also the difficulty of elucidating what autonomy means with respect to the kinds of operational military problems where we would like to apply it and the lack of a general concept of its utility in these circumstances, the present dearth of operational military autonomous systems ought not be so surprising. This chapter seeks to explore what is primarily an economic perspective towards autonomy, in order to inform subsequent choices of technical problems towards the realisation of operationally viable autonomous systems that solve difficult military operational problems well enough to be worthy of the title.

That many commercial enterprises possess what they hold to be autonomous systems poses the question as to why defence organisations cannot make the same claim. Yet if we regard that it takes more than to be able to act outside of direct human control to make a system autonomous, these systems may not really be autonomous at all. Specifically, they operate under highly constrained circumstances, and the predictability afforded by those operating circumstances are required to make them effective. Defence represents an especially challenging set of circumstances and difficult choices: unlike the problem contexts of extant autonomous systems, military operations, by and large, do not facilitate the luxury of tightly managing the opera-

D. J. Reid (✉)
Defence Science and Technology Group, Joint and Operations Analysis Division,
PO Box 1500, Edinburgh, SA, Australia
e-mail: darryn.reid@defence.gov.au

© The Author(s) 2018
H. A. Abbass et al. (eds.), *Foundations of Trusted Autonomy*, Studies in Systems,
Decision and Control 117, https://doi.org/10.1007/978-3-319-64816-3_21

tional context so that the strong environmental expectations built into automations will not smash unhappily into nonconforming realities. Moreover, such failures in war and battle are potentially catastrophic. Given the fundamental uncertainty of this kind of environment [6, 53], efficiency goals that work in relatively controlled environments are largely irrelevant and may even be misleading, in the sense that the pursuit of efficiency could actually accentuate exposure to ruinous outcomes (the effect is well studied in economics, for instance this effect is understood to be a major contributing factor in the Global Financial Crisis of 2007–2008 [3, 26, 40]). Instead, operational effectiveness hinges on avoidance of irrecoverable failure, which might, incidentally, include avoiding gross inefficiency, but this is not the same thing in general as attempting to maximising efficiency.

I argue that heart of the problem of autonomy is the ability to effectively deal with fundamental uncertainty; such uncertainty is the consequence of the inherently paradoxical nature of the problems we would like our autonomous systems to be able to solve. For instance, prediction is generally a paradox of self-reference, because of the effect that making a prediction has on the predicted event; the predictions occur from within the system to which the predictions apply. Observation is similarly potentially paradoxical because the act of observing occurs from within the system and consequently may disturb the very phenomena we wish to observe. Clausewitz was well aware of these effects in war and battle, and consequently distinguished carefully between uncertainty that is unmeasurable and stochastic chance. Since the time of Clausewitz, there has been a burgeoning of interest in uncertainty, from a number of distinct perspectives. In particular, I wish herein to briefly draw on the work on uncertainty and ignorance in economics, which particularly originates in its more philosophically oriented branches, and some of the studies of incompleteness and uncomputability phenomena in mathematics and computer science [19, 35, 39, 45], and to relate the two together.

Newcomb's Dilemma [4] is a good example of how the kind of paradoxes behind mathematical incompleteness generate real-world uncertainty. This thought experiment involves a paradox of prediction that interestingly brings two distinct decision-making modes, which mostly seem to apply under different circumstances, into direct conflict in a single situation; the puzzle thus highlights the inapplicability of simple conventions of rationality as reward maximisation in settings involving fundamental uncertainty. The first decision-making mode is the impetus to act in order to produce a desired outcome, and the second is the impetus to act only when the action can alter the outcome. You are presented with two boxes: one open and one closed. The open box contains a thousand dollars, while the closed box contains either a million dollars or nothing. You can choose either the closed box or you can choose both boxes. The contents of the closed box has been prepared in advance by an oracle machine that almost perfectly predicts what you will choose: if you choose both boxes, then it has put nothing in the closed box, while if you choose just the closed box then it has place the million dollars in it.

The solution to the dilemma seems obvious to almost everyone; the catch is that people divide about evenly on which solution is the obviously correct one. In other words, the dilemma poses two incommensurate choices that are both justifiable. The dilemma was recently resolved [54] using game theory to show that the course of action in the game depends on the observer's beliefs - which are necessary to making a choice but not rationally justifiable - about the ability of the oracle machine to predict their actions, in much the same manner as the observation of quantum states depends on the observer.

The connection from autonomy to economics rests on the proposition that an autonomous agent is exactly an economic agent, by another name. An autonomous agent—be it living, a social construct, or a machine—is an entity that exercises choice, by making decisions to act one way or another. Yet any decision to act in a particular way is just an allocation of resources under that agent's control to one of the course of action options as understood by the agent, and this matches the definition of an economic agent. To be slightly more formal: an economic agent is any actor (e.g. an individual, company, government body or other organisation) that makes decisions in aiming to solve some choice problem within some economic system, and an economic system is any system involving production, consumption and interchange of resources by such agents [11, 12, 21, 29, 42].

The difference between an autonomous agent and an economic one is a matter or emphasis that directs subsequent research problem choices. When we talk of artificial intelligence, it usually means we are mostly interested in developing algorithms and their technical implementations in machines; when we talk of economic agents, it means we are mainly concerned with individual and system-wide outcomes given different mixes of different kinds of interacting resource-allocating agents under various environmental conditions.

In particular, I am concerned with autonomy as particularly featuring decisions about the allocation of self under uncertainty, and hypothesise that self-allocation corresponds with the same boundary that delineates a notion of uncertainty that has come to be known as ontological uncertainty [13, 25, 30, 44, 50]. Interacting self-allocating agents entails logical paradox in general, which underpins formal limits on knowledge and thus the intrinsic uncertainty of such systems. This uncertainty arises without any recourse to exogenic 'shocks'. Hence familiar notions of agent utility-maximising rationality that operate within worlds of linear certainty and stochastic risk necessarily break down under conditions of fundamental uncertainty. This kind of effect has been studied extensively in the economic literature. For instance, suppose that agents have a high ability to predict a policy-maker's actions, that at least one endogenous variable is completely controlled by the policy-maker, and that the reward to the policy-maker of their actions is dependent on whether the policy is anticipated. Under these simple conditions [16], there is no unique rational course of action for the policy-maker, in general. Rather, different incommensurate yet equally viable theories may indicate distinct optimal policies, and consequently the agents cannot form rational expectations because any rational expectation must contain a theory of the policy-maker's behaviour. In a setting such as this, economically rational behaviour is basically incompatible with the formation of economically rational expectations. This is an instance of an incompleteness phenomenon occurring in a purely economic setting, and one that bears directly on autonomous machines.

21.2   Fundamental Uncertainty in Economics

21.2.1   Economic Agency and Autonomy

The tight connection between notions of economic agency and autonomy is well established in the economic literature: indeed, the history of artificial intelligence research and economics are intertwined because economics is, in essence, about human decision-making under different resource allocation mechanisms and conditions [36]. Economics has constantly looked to artificial intelligence for models, especially in response to the realisation in the 1970s [43] that economics had been weak on both the nature of human information processing capabilities and on how we achieve these capabilities, despite the central place that these problems must have in economic theory. The boundaries have become increasingly blurred, on the economics side at least, because while economics requires descriptive models of human decision-making, the available tools are primarily the prescriptive models offered by traditional artificial intelligence and applied mathematics. Furthermore, economics has itself developed prescriptive models in the course of trying to describe and predict economic system behaviours (see for instance, [15]).

While autonomous systems research has yielded many prescriptive agents that work well enough in tightly constrained environments, the economic requirement for descriptive models arguably first highlighted the yawning gap between machine reasoning algorithms and the robustness and flexibility of human decision-making. Smith recently argued [44] that in both economics and in artificial intelligence, the underlying assumptions driving research about agent learning and decision-making have typically neither sufficiently nor even explicitly emphasised the significance of fundamental uncertainty; current models remain structurally very similar to those of the past. We remain largely tied to probabilistic and statistical methods of learning and reasoning and thus to their inherent limits.

He argues that the advances in machine intelligence of recent years overwhelmingly do not even attempt to address this failing and consequently do not represent a base advance in overcoming the difficulties of modelling and reproducing human decision-making, increasing success in tightly defined domains notwithstanding. Implicated in maintaining these base assumptions unchallenged is the documented anthropomorphic tendency to label constructions in artificial intelligence with human decision-making features or human functions, in the absence of any convincing argument establishing a similarity [34]. These wishful mnemonics have arguably helped to foster the hype cycles experienced repeatedly during the course of the history of artificial intelligence research and application, while masking the limitations of the available methods to cope with non-stochastic uncertainty.

The implications of this for the development of autonomous systems capable of operating in high uncertainty environments such as those of military operations are straightforward. The failure of machine learning in comparison with human performance is poor generalisation ability; the failure of machine reasoning is brittleness, meaning a poor ability to handle environments in which future states are the manifestations of interconnected events, including the actions of the agent itself. In both cases, the failure manifests as an inability to operate successfully—where success here means extended survival more than reward maximisation—in environments where future events are not foreseen and may not be foreseeable at all.

21.3 The Inadequacy of Bayesianism

The connection between autonomy and economics runs very deep: questions about the formation and application of beliefs in decision-making, and hence criteria delineating rational from irrational beliefs, are as prominent in economic theory as they are in artificial intelligence. For instance, the central place of equibria in economics and game theory is really a statement of a convention about the rationality of beliefs, wherein rational beliefs are held to be exactly those that coincide with equibrium solutions [18]. The Bayesian interpretation of belief is predominant in the economic literature, as it also is in artificial intelligence; the criticisms of Bayesianism in economic theory [18] apply equally in autonomous systems research and may be seen, in essence, as another recognition of the inability for a currently dominant paradigm to adequately handle fundamental uncertainty.

Bayesianism is not really a single position, but any mix of three main postulates. Firstly, an agent should have probabilistic beliefs about unknown facts, typically given as a probability measure over all possible relevant states. Note that economics often uses a stronger version than computer science and artificial intelligence by dropping the relevance qualification and thus presuming that agents must have beliefs as probability measures about absolutely everything. Secondly, Bayesian priors should be updated to yield a posterior measure in accordance with Bayes' Law. Lastly, each agent should choose the decisions that maximise expected utility (or sometimes equivalently minimise expected cost) with respect to that agent's Bayesian beliefs. While the third postulate, in conjunction with the first and second, is hardly unknown in computer science, it is particular common in economic theory.

No combination of these postulates has been immune from serious criticism in the economic literature. The Bayesian approach presumes a prior, and thereby does not deal with the manner in which the prior is obtained. This situation is sometimes known as state space ignorance or sample space ignorance. Though the second postulate seems technically safe, it has been shown to be descriptively inadequate [51]. Moreover, the economic literature arguably under-estimates the importance of complexity limitations on Bayesian updating; indeed, the computer science literature has been long focussed on this as the main difficulty. The third postulate has been attacked repeatedly in the economics literature since the Allais Paradox [2], which revealed strong inconsistencies between observed human choices and what maximising utility would predict (see for instance [23] and [33]). The entire field of Behavioural Economics[1] is substantially based on the rejection of this third postulate.

For present purposes, it is the criticism of the first postulate that warrants particular notice: the lack of convincing general descriptive and prescriptive explanations regarding the origin of the prior translates into an important epistemological question. That is, this criticism amounts to a different way of stating that, in general, the agent cannot possibly have at its disposal sufficient time and resources needed to obtain the complete set of possible relevant states over which to draw belief measures. The economic literature goes even further, however, by also proposing an ontological notion of uncertainty that asserts that, beyond limits to knowing, questions about outcomes pertaining to agent decision-making are simple unanswerable at all.

21.4   Epistemic and Ontological Uncertainty

To see how uncertainty plays out in economic settings, consider the essential problem of human agents operating in an economy: they must make investment choices that will play out a future they can neither predict nor really control. Knight [27] famously formalised a distinction between risk and uncertainty on the basis that economic agents operating in a dynamic environment must do so with imperfect knowledge about the future. Knight distinguished risk, as applying to the situation when the outcome is unknown yet the chances are measurable, from uncertainty, when the information needed to measure the chances cannot be known in advance of the outcome; we do not have a sufficiently long history with the system as it currently stands to be able to establish a measure. Knight maintained that risk can be converted into an effective certainty[2]; the practice of setting hurdle rates as the rate of return on the next best option having a comparable risk profile as a mechanism for soft capital rationing is one example of this kind of conversion of a risk into a cost.

In Knight's conception, uncertainty is distinct from stochastic risk in that it is not amenable to measurement and consequently cannot be meaningfully converted to a cost in this manner. Note that this conception of uncertainty effectively raises to a more general setting the first objection that was discussed earlier in relation to Bayesian assumption of a prior, which insists that a measure can be defined over the space of possible relevant states.[3] Knight's uncertainty is epistemological: we lack knowledge of what future outcomes might occur, but at least we can be aware that we lack it.

---

[1] Behavioural Economics is basically about the study of observed behaviour of real economic agents, in contrast to the normative approach that neo-classical economics takes to behaviour.

[2] This kind of conversion is a ubiquitous standard practice. Knight made the point that this practice is indeed justified, so long as we are dealing only with measurable stochastic risks.

[3] Hence the field of Behavioural Economics, based substantially as it is on the rejection of utility maximising postulate of Bayesianism in economics, assumes as its basic position a Knightian epistemological view of uncertainty [13, 50].

The importance of the distinction has not been lost in analyses of recent behaviour of finance firms, which, in the years leading to the global financial meltdown of 2008, operated with highly regarded, highly precise risk assessments that were all based on the basic premise that the relevant conditions and outcomes are all measurable [3, 13, 26, 32, 40, 41]. When, in a mass financial panic, institutional investors suddenly realised that the assumptions about their risks being measurable were deeply flawed, financial markets collapsed in the kind of event that has been described as a 'destructive flight to quality'. Investors clambered over the top of each other to dump everything but the safest and hence least profitable assets, in a sudden cascading systemic failure having global ramifications that continue to reverberate nearly a decade later. The distinction between risk and uncertainty in economics is not an esoteric matter, but a very practical pervasive kind of problem with potentially huge ramifications. Indeed, the aftermath of the economic crisis has seen a considerable resurgence of interest among economists in fundamental uncertainty, after a long period where the primary interest was in formalising economics - and particularly finance - using precision stochastic risk models (see for instance [26]).

In contrast to Knight, Keynes [25] argued for a deeper ontological notion of fundamental uncertainty: we do not even know what we do not know. In this view, the future is fundamentally uncertain because it is simply not possible even in principle to conceive of all relevant possible future outcomes in advance.[4] Investment, Keynes maintained, is then the allocation of resources on the basis of expectations formed under conditions of this kind of uncertainty. Keynes formulated a set of conventions to describe how agents try to cope under such uncertainty, by resorting to what amount to superstitious rituals from the point of view of utility-maximising economic rationality. They tend to presume that their existing opinions are a valid guide. They tend to conform to majority views. They just ignore what is unknown: widespread reliance on risk models under the assumption that all the uncertainty is measurable as briefly described above is a prevalent contemporary example.[5] They presume that present circumstances will be stable. They rely on the opinion of experts who concoct grand predictions from the economic tea leaves. Perhaps most importantly, in a mammoth and sometimes even wilful act of self-deception, they assume far greater veracity for these measures than what any frank examination of the past would ever support.

In short, agents invest on the basis of strongly predictive models, and the consequences are far-reaching. It means that markets cannot be in stable efficiency equilibria, that investment is highly volatile, and that expectations are extremely fragile. It also means that complex behaviour such as economic bubbles and bursts and overall unpredictability of economic systems can be wholly generated endogenously [1, 11, 21, 26, 29, 31, 32, 41, 46, 52]. This turbulent picture stands in contrast to the text-

---

[4]I will come back to the reason for this in detail in a subsequent section. Briefly: the questions we might need to answer may contain hidden self-reference, which opens the possibility that they might be logically paradoxical and hence unanswerable from inside the system within which we have to operate.

[5]I will further discuss these instances in subsequent sections, particular in relation to the GFC, trading strategies and market bubbles and crashes generally.

book neoclassical account, which maintains that the expectations of economic agents about future returns are correct on average over time, under the so-called "rational expectations" doctrine.

Keynes further held that economic agents making decisions under uncertainty hold more liquid assets - especially money - as an asset in response to doubt about future returns; this store of wealth in liquid form is essentially a concrete and measurable manifestation of the agents' confidence regarding future outcomes (but not a measure of the uncertainty of future outcomes). Lower confidence requires higher interest rates to inveigle our agents to draw their capital from safe but unprofitable liquid deposits and invest it in volatile but potentially profitable illiquid assets. So although the agents themselves can behave pretty miserably, Keynes concludes that wise governmental moderation can, in principle at least, considerably stabilise an economic system by setting monetary policy to moderate the billow and bounce of otherwise outrageous market circumstance.[6] This proposition has been repeatedly echoed in subsequent experimental and theoretical studies concluding that well designed control policies can be very effective in moderating or eliminating asset bubble formation; some investigators report that dynamic policy control is superior in this regard to static controls [32].

These economic agents face a bimodal impetus, compelled to avoid loss on the one hand and to seek profit on the other, but they must undertake this activity under conditions of irreducible uncertainty. Interestingly, the source of macroeconomic volatility appears to lie less in the presence of fundamental uncertainty, which is unavoidable anyway, and more in the manner by which our agents attempt to avoid dealing with it. They retreat from it through the fallacious appeals that Keynes describes. Left unmanaged, the consequences of this behaviour are that numerous small failures accumulate unrecognised and unreconciled, eventually erupting in the sudden system-wide failure broadly known as a 'crash' when the edifice of false confidences in these measures can no longer be maintained under the accumulated burden of hidden errors.

Keynes' conventions seems to provide a concise summary of managerial methods, which makes sense from the point of view that management practices and investment behaviour are inextricably intertwined. A deeper connection is that one of the pillars of management theory is the inversion of the economics of externalities[7][9]. The

---

[6]This does not intend to imply that Keynesian economics is without assumptions nor subject to limitations, but merely to convey the component of the theory that pertains to uncertainty and its effects in relation to autonomy as an economic question. For instance, the details of how governments intervene really matter: the Keynesian interventions in the 1930s that directly supported broader populations deeply impacted by the Great Depression were clearly more successful than the bailouts after the GFC in which trillions were poured into the large failing financial enterprises whose activities had caused the bubble and ensuing crash, and which produced exploding deficits. The outcomes of the current downturn is leading many to question capitalist economic systems themselves, amidst a growing view that capitalism is inherently unstable, inefficient, antidemocratic, and not - as often assumed - synonymous with the presence of markets; such debates substantially question the premise as well as the limits of Keynesian intervention to stabilise capitalist systems.

[7]An externality, or transaction spill-over, is a cost or benefit that is not transmitted through the resource allocation mechanism and is instead incurred by a third party not involved in the transaction.

actions of the agents are reliant on strong assumptions about the measurable nature of future outcomes as a basis for supposedly justified action; the well documented over-reliance on risk models in finance is about precisely this kind of justification [26]. The conventions also capture the essence of the behavioural assumptions often built into automated systems, which brings us to what I regard as the fundamental question of autonomous systems development: if the uncertainty is not measurable, then how can we build systems that can effectively deal with it? The failures of machines and organisations that similarly incorrectly assume that the uncertainties of their operational environments are measurable will similarly see failures tend to accumulate into sporadic cascading systemic distress, and I warrant that it is largely in recognition of this unacceptable potential under extreme forms of uncertainty that military operational settings have proven largely unyielding to the best current technology has to offer.

How, then, should we conceive of autonomy? If we consider an autonomous agent as an economic agent self-allocating under fundamental uncertainty, then such agents - be they humans, organisations or machines - display a property I term plasticity: the ability to countenance unpredicted, and unpredictable, future states of the operating environment, in a social setting, within acceptable limits. The name is in reference to the implied need for the thing exhibiting the property to change itself in response to changing environmental conditions. Autonomy will then apply specifically to machines that satisfy this condition. This conception of autonomy has nothing to do with the inherent sophistication of algorithms, power supplies, sensors, actuators and circuits, but instead motivates technical finesse specifically to the extent and in the direction needed to fulfil the plasticity imperative adumbrated by the intended operational setting; this position reflects the primarily economic viewpoint of this chapter, in distinction to the algorithmic or robotics emphasis widely seen in the literature.

21.5 Black Swans and Universal Causality

Taleb famously coined the parable of black swans [48] to describe the occurrence of unforeseeable rare events having high consequences, and previously described the strong tendency of humans to find simple, though erroneous, explanations for their occurrence, after the fact [47]. It has since been established that this description of uncertainty draws the same basic distinction between stochastic risk and Knight's

---

Externalities may lead to inimical outcomes by upsetting the resource allocation mechanism, and there is typically a large magnification effect whereby a small benefit to one or both parties in the transaction generates a disproportionately large cost to the third party. Economics attempts to limit such effects, as represented most famously by Coase's Nobel-prize winning work on externality elimination cited in the text. In contrast, management theory contains a branch that aims specifically at generating externalities for the benefit of a specific party in the transaction (privatisation of profits) and the cost of other parties, which might include the second party in the transaction (socialisation of costs).

intractable epistemological uncertainty [13, 14, 50]. The black swan anecdote serves to illustrate Knight's epistemic uncertainty outside an obviously economic or financial setting. Taleb's description of the behaviour of humans in concocting reasons for the event after the fact mirrors Keynes' conventions describing economic agent behaviour, but does so in a manner that highlights the role in shaping expectations - whether economic or otherwise - of a widely discredited position usually known as universal causality in philosophy.

Taleb's observation that humans regard events as being much more attributable to determinable causes than they really are has a long history in philosophy as the notion of universal causation. Universal causation maintains that all events are the result of prior events, and this belief connects the construction of Taleb's post hoc explanations for rare events with Keynes' view that economic agents maintain undue reliance on the veracity of prediction. The post- economic meltdown criticisms of financial firms assuming that the relevant risks are all measurable is a modern manifestation of the same effect.

The intuitive appeal of this view lies in that whenever we ask simple questions after the fact about why a particular event occurred, we can obtain a plausible explanation for its occurrence in terms of some causal chain of earlier events. So it would seem on the face of it that every event is caused by something, albeit probabilistically, and hence that every event follows from prior events according to some governing logic. Yet the more deeply we dig, the more ostensibly antecedent sets of conditions look like reasons for deciding to act in a certain way, or, more pertinently, for not acting a certain way, and the less they look like the inevitable causes of an event that we first supposed. In other words, alleged causes are really only epistemic factors that influence the decisions we make from within a problem context, rather than immutable ontological features of an environment into which the agent peers from the outside.

Universal causality connects to predictability through causal determinism, which holds that every event is necessitated by some set of prior events. This claim is then the antecedent to so-called 'scientific' determinism, which concludes therefore that the world is basically predictable. To elaborate: 'scientific' determinism alleges that the structure of the world is such that future events can be predicted with precision depending on that of knowledge of the governing laws of the phenomena of interest and the accuracy of the account of past events. It is worth noting that 'scientific' determinism is poorly named, for it is not actually about determinism at all: determinism refers to the absence of arbitrary choices in the application of transition operators, with non-determinism then being the admission of arbitrary choices. Rather, 'scientific' determinism is an assertion about predictability, equivalent to assuming stability and logical completeness. Though perhaps intuitively appealing, the inference from all events having necessary causes to predictability is flatly wrong: it is well established that even fully deterministic systems in which the current state completely determines the transition to a unique subsequent state can be nonetheless savagely unpredictable [20, 22]. Conversely, non-deterministic systems can also be completely predictable. Even if we limit ourselves to completely deterministic systems and hold that this is a correct characterisation of the world, strong predictability remains the exception, not the rule.[8]

So the premise to the conclusion of predictability that we can rely on for, say, making investment decisions or for autonomously operating in a complex operational environment, is untenable in general, and we draw this conclusion without needing to reject the notion that some events may have causes, or even the stronger assertion that every possible event has causes. It collapses merely when we admit that causes may be only necessary but not sufficient for at least some events that matter to us in terms of our decision-making some of the time. Moreover, across a wide range of contemporary fields, including mathematics, computer science and physics, as well as economics, it appears increasingly clear that there are also events that just do not really have any cause at all [7, 8]. This ties in very closely with Keynes' ontological notion of uncertainty.

21.6 Ontological Uncertainty and Incompleteness

21.6.1 Uncertainty as Non-ergodicity

An ergodic system [20, 22] is one that tends towards a limiting form that is independent of its initial conditions; in a dynamical system sense, this means the phase-space average is the same as the infinite time average, for all Lebesgue-integrable functions almost everywhere (meaning except possibly in sets of measure zero). In other words, an ergodic system is one for which sampling - collecting more data - actually gives more information about the underlying system, so ergodicity characterises the precise conditions under which obtaining additional data provides additional information. The mechanisms governing the system are stationary, so they remain constant over time, and they satisfy some regularity conditions, essentially meaning that they are well behaved. Ergodicity effectively means that we can float detached from the world about which we make observations and predictions, thus avoiding the observation and prediction paradoxes that produce fundamental uncertainty.

Paul Davidson argues that the rational expectations hypothesis and efficient market hypotheses of textbook economics are worthless, and indeed positively dangerous, on the grounds that real economic systems are inherently non-ergodic: such systems are not regular or not stationary or both and consequently it is unreasonable to expect that they will converge to any equilibrium distribution, and they cannot be amenable to reliable forecast as a result [14]. Davidson holds that Keynes' ontological uncertainty pertains to the behaviour of such non-ergodic processes, in contrast to

---

[8]I will pick up on the question of exactly what are the conditions under which strong predictability in principle holds in the next section. In short, the conditions amount to the delineation of Keynes' ontological uncertainty, which is about absolute limits on what is knowable in principle. Knight's epistemic uncertainty and Taleb's black swans amount to further practical limits on the tractability of knowing.

Knight's epistemic uncertainty and Taleb's Black Swans, which still presume the presence of ergodic processes. In the latter case, the surprising outcome simply lies far out on the tail of a nonetheless fixed and well-behaved distribution, with apparent uniqueness given by the inordinate time between re-occurrences. In the new edition of Taleb's book, he concedes that there is a difference between non-ergodic processes and black swan events but dismisses the difference between the two as irrelevant.[9]

But there is a world of difference. The various species in the zoo comprising the ergodic hierarchy have very different properties, particularly in terms of the kind and degree of uncertainty they manifest, essentially in terms of the kinds of questions we might want to ask about the future behaviour of such systems and which of those questions can be answered in advance of simply waiting to see what happens. The most important distinction is between the class of ergodic processes and all the other classes of systems that fall somewhere in the much larger world of processes that are non-ergodic [20, 22].

To be slightly more formal: the ergodic hierarchy is a classification scheme for deterministic dynamical Hamiltonian systems, in terms of their relative unpredictability. Ergodic systems characterising certainty, stochastic risk and epistemic uncertainty are at the bottom of this hierarchy, being the most highly restrictive and correspondingly the lowest in terms of the level of uncertainty they can manifest. Weak mixings are next, then strong mixings, above which are K-systems, whose behaviour is already very strongly unpredictable, and the topmost currently recognised classification are Bernoulli systems, whose behaviour is the most deeply unpredictable in the hierarchy.[10] The criteria for strong mixings have been convincingly proposed as the demarcation of what is commonly regarded as deterministic chaos [55]. There are also interesting systems that straddle between K-systems and Bernoulli systems, known variously as C-systems or Anosov systems, but their relation to the other levels mentioned here in terms of unpredictability is more complicated and beyond the scope of the present discussion.

Note that non-ergodic systems do not undergo arbitrary change at any moment; the presence of fundamental uncertainty does not require or entail total disorder. To the contrary, systems that are non-ergodic will typically fall into transient states of apparent stability that dissipate as suddenly and unexpectedly as they start, never to repeat themselves. Strong prediction about future system behaviour is impossible, in the sense that the kinds of questions we might want to ask about the future behaviour of the system are not solvable, with higher classes in the hierarchy representing a situation of having fewer such problems for which there is a solution. Yet this does not mean that we cannot cope at with life in such a system - as individual economic agents, people and organisations certainly manage to do so - but rather that we cannot

---

[9]This dismissal of the significance of the difference is understandably not well received amongst researchers in ergodic theory and nonlinear dynamics, for reasons that will become clear.

[10]Interestingly, entropy is not sufficient to classify K-systems, meaning that there are uncountably many K-systems with the same entropy but that are not isomorphic; thus Ornstein's isomorphism theorem does not work for K-systems. All K-systems are also Bernoulli systems, but not vice versa; Bernoulli systems potentially manifest greater unpredictability. Yet Ornstein Theory is sufficient to classify Bernoulli systems.

expect to do so very successfully by using by relying on methods that presume that uncertainty is measurable, or that presume ergodicity, typically by using relatively strong predictions about what will happen in futures delineated by the time periods over which decisions will play out.[11]

The power of complex systems to produce long sequences of apparent predictability[12] is highly seductive. Self-reinforcing beliefs about predictability of future returns and thus future market behaviour that drive the formation of market bubbles is a highly visible example of this [1, 31, 32, 41, 46, 52]. The precision risk models heavily implicated in the mortgage-backed securities bubble preceding the Global Financial Crisis of 2007–2008 provides almost innumerable practical examples of the catastrophic failure of ergodic models in what are actually non-ergodic environments.[13] They will tend to fail suddenly and disastrously rather than smoothly, but often will do so only after mendaciously long periods of apparent positive success. Agents are more easily deceived because holding ergodic expectations about the world will naturally mean that they will also expect failures to be similarly be ergodic, and thus relatively benignly behaved and predictable. Yet there is simply no basis for this expectation. The distinctly irregular and non-stationary quality of the collapses of ergodic models in what turned out to be highly non-ergodic systems, came as something of a surprise to those invested in them, to say the least.[14]

The difference between Keynesian ontological uncertainty and Knightian epistemological uncertainty is that the former position holds that some things are simply not knowable, while the latter entails that with better information and greater ability to process it we could, in principle, calculate the probabilities for more kinds of events. The epistemic uncertainty notion ultimately sees the universe still as a collection of ergodic processes, and uncertainty as essentially a consequence of limitations that computer science calls tractability. Roughly speaking, tractability limits occur

---

[11]There is a growing general awareness in economics, as exemplified here by Davidson's work [13], that economic systems worth worrying about are inherently non-ergodic. The inescapable fact that real economic agents can, do and always have successfully operated under these conditions should suffice to refute the proposition that it is not possible to operate adequately in such an environment so we should not bother to do so in artificial intelligence research.

[12]Even a completely random sequence produces such sequences of lengths that are logarithmic with respect to the overall observed sequence length, which is deceptively long [35].

[13]The economic analyses typically describe this in terms of the catastrophic cascading failure of the application of precision risk models under conditions where the falsity of the strong underpinning assumptions to the effect that all relevant risks are measurable was never examined (nor were these assumptions even stated, so much were they taken for granted). Under Davidson's direct mapping from ontological uncertainty in economics to non-ergodicity, we have the stated interpretation.

[14]John Meriweather famously described the financial collapse of 2007 as a ten-sigma event, which means, according to the predominant economic models, that it should occur no more than about three times in the entire history of the universe. The models were designed from the outset to eschew the very possibility of catastrophic failure. Apparently it did not occur to the adherents of the orthodoxy even after the fact that their models might be flawed, despite the manifest empirical failure and the clear absurdity of many of their base assumptions. Even in 2010, Bernanke argued that the problem was not that the economic models failed to see the economic crash coming, but rather it was that the economic crisis was an event that was just not supposed to happen. Apparently, reality should consider itself refuted.

because although a problem may be formally solvable, the time and space requirements to solve it rapidly explode beyond the ability to meet them as the problem size increases. So the inability to sample a system for long enough in order to observe occurrences of ultra-rare events -black swan evens - are a tractability constraint of the type that characterises epistemic uncertainty.

21.7 Uncertainty and Incompleteness

The Keynesian position maintains that obtaining answers to some questions is just not ontologically possible, in exactly the same grain as the deep mathematical uncertainty expressed in results such as the incompleteness of every formal axiomatic system that contains arithmetic, the existence of recursively inseparable sets, and the unsolvable nature of most computing problems, most famously the Halting Problem. In other words, ontological uncertainty amounts to the fact that problems of determining future outcomes in non-ergodic economic systems are generally paradoxical, because such systems allow the possibility of self-reference. So the Keynesian uncertainty concept amounts to an economic manifestation of algorithmic randomness, which rests on computability theory and amounts to the modern study of incompleteness phenomena, by regarding effective procedures as compressions of potentially unbounded sequences of data generated by the system of interest and asking about what sequences have finite compressions.

Formal axiomatic systems are mathematical languages allowing us to talk formally about phenomena in which we are interested, including other axiomatic systems, so it's difficult to over-state their significance to autonomy. They each consist of a set of basic terms and a set of reasoning rules defined by axioms that describe, essentially, what we might conclude from what given conditions. Such a language allows us to formally state propositions, some of which might be provably true in the sense of being logically entailed by the axioms, such as "there is no largest prime number" in Peano Arithmetic (the basic theory of numbers, see for instance [24]). Other expressible propositions, such as "adding two positive numbers together yields a number smaller than either" in Peano Arithmetic, reduce to contradiction, which is a primitive term of a system that is false in all interpretations. The most basic question about whether we have a viable axiomatic system to use is whether or not the axioms themselves entail contradiction; if so then the system is said to be inconsistent, and it does not represent a viable basis for reasoning because in such a system it is possible to conclude absolutely anything. As famously shown by Kurt Gödel [19], this fundamental question of the consistency of formal reasoning systems turns out to be anything but trivial.

Some formal axiomatic systems, such as Turing machines [39, 45], describe computation and thus set the ultimate basis for machine intelligence. In this setting, the complexity of any other system is defined as the size of the smallest procedure, with respect to some reference machine model, that reproduces the data observed from that system [35]. The remarkable fact is that this complexity is asymptotically independent of the particular machine model, up to additive constants.[15] The conditional complexity of a sequence with respect to some information is the smallest effective procedure that takes the information as an input and produces the sequence as an output. A sequence is then said to be incompressible when the smallest size effective procedure is asymptotically comparable to the size of the sequence. For infinite sequences, the complexity is the limit as the length of an initial segment of the sequence approaches infinity of the size of the smallest effective procedure producing the segment, divided by the length of the initial segment of the sequence. If the complexity in the limit is non-zero then the sequence is incompressible, meaning that it represents a fixed individually random mathematical object, indistinguishable from the flips of a coin by any possible statistical test.

As indicated earlier, an irreducibly random sequence of this type provably contains surprisingly long sequences, of about a logarithmic function of the observed sequence length, that appear to be regular and stationary [35]. There is a deep connection between the non-linear dynamics view discussed earlier and the algorithmic information view of unpredictability: the trajectories of a non-linear dynamic system can be encoded as infinite sequences by dividing the state space of the system up into numbered cells and tracing the trajectories through these cells. A trajectory is random when there is a partitioning of the state space into cells such that the encoding of the trajectory is algorithmically random. Predictability means having a compression - in terms of some effective procedure - for anticipating the outcome in advance. Yet there are simply not enough compressions to go around. It is not even close: the shortfall is exponential, meaning vast majority of possible systems are left with no compression by which their trajectories can be predicted that is shorter than simply waiting around to see what eventually happens.

The underlying reason is that prediction in general is paradoxical: while a contradiction is both true and false, a paradox is neither true not false within the logical frame of reference in which it is stated. Though we usually think of paradoxes as obviously self-referential statements, they are usually not so obviously discernible because paradoxes actually do not need not be visibly self-referential. The famous Halting Problem for Turing Machines, the Busy Beaver Problem,[16] and their equivalents in other computational models, as well as the compression problem are all actually paradoxes that do not look like it on the face of it because their self-referential nature is hidden from immediate view.

The reason for this is that the self-referential expression is not as primitive a notion as it might at first seem: numerous systems of logic come with various kinds of implicit function theorem by which self-referential statements can be turned into equivalent statements, called "fixed-points", that lack obvious self-referentiality [5]. Paradoxes are normal, natural, and extremely common, and formal mathematical sys-

---

[15]This is one of those mathematical facts that seems remarkable upon first discovery, and entirely natural to the point of obviousness afterwards.

[16]Give me the largest natural number that can be generated by an effective procedure with respect to some model of computation - a program in your favourite programming language, if you prefer - limited to at most a given size.

tems are full of them[17] [7, 8]. Turing's original proof of the unsolvability of the Halting Problem for Turing Machines, by a Cantor Diagonalisation Argument,[18] further reveals the detailed nature of mathematical paradoxes, and hence of the irreducible nature of fundamental uncertainty: they are kind of folded-up infinite regresses. The proof in question is essentially just an infinite successive unfolding of the Halting Problem paradox to yield what amounts to an impenetrably unknowable number [8].

The basic lesson of Gödel's Incompleteness Theorems [19] is that any system that allows for the possibility of self-reference - any system containing basic arithmetic will do - will give rise to paradoxes, and this will manifest as uncertainty in the form of the presence of problems we might like to solve but to which there can be no solution from within the system. A bigger system might be able to provide a solution, but we don't in general have the luxury of stepping out to it and peering into the phenomena with which we are concerned from the outside. The basic example of this is that we are bound to compute things from within the limitations incumbent in the models of computation, which are all known to be equivalent and absolute, and under the Church-Turing thesis are not surmountable by any other realisable system either [39, 45].

Can the ergodic hierarchy provide a formal mathematical basis for characterising plasticity - the property of being able to survive in an unpredictable environment? I suggest so: if the observations that an agent makes of its environment meets the criteria of, say, a K-system, and yet that agent is able to survive in that system for better than a logarithmic function of time, where logically time would be taken in terms of successive observations the agent makes of its environment, then we know that the agent must be doing better than merely taking advantage of an appar-

---

[17] As an example of this, I recently played around with paradoxical statements about Peano Arithmetic - axioms about the behaviour of the natural numbers under the usual operators - using Provability Logic. Provability Logic system consists of familiar propositional logic with a modal operator $\Box$ meaning "it is provable that", its dual $\Diamond$ meaning "it is not disprovable that", and Löb's Theorem, which states that in any system containing Peano Arithmetic, any time we can prove that something implies its truth we may conclude that it is provable. We can use Provability Logic to explore and even to write computer programs to generate arbitrarily many generalisations of Gödel's Second Incompleteness Theorem [19] for us, by feeding it with paradoxical statements. Provability Logic's implicit function theorem guarantees that we have unique solutions to a large class of self-referential expressions. For instance, the solution $\neg(\neg\Box\bot) \to (\neg\Box\bot)$ to a paradoxical statement $p \leftrightarrow \neg\Box p$ happens to be direct restatement in Provability Logic of The Second Incompleteness Theorem, asserting that the system cannot prove that it is consistent, or equivalently, that if it can prove that it is consistent then it must be inconsistent. Here, the symbol $\leftrightarrow$ stands for if and only if, $\to$ is implication, and $\bot$ stands for contradiction. The statement $\neg\Box\bot$ says that the system is consistent.

[18] Cantor's Diagonalisation Argument was first used to prove that there are infinite sets that cannot be put into one-one correspondence with the natural numbers, and later to prove that the real numbers are uncountable, Russell's Paradox whereby attempted formulations of set theory prior to Zermelo set theory are inconsistent, and Gödel's First Incompleteness Theorem, as well as the unsolvability of Turing's Halting Problem [39, 45].

ently regular transient state, and thus must be dealing with some effectiveness with K-system uncertainty.[19]

21.8   Decision-Making Under Uncertainty

Keynes' ontological uncertainty corresponds to incompleteness phenomena in the same manner that Knight's epistemic uncertainty mirrors intractability, meaning that the root cause of ontological uncertainty is the possibility of self-reference and thus of logical paradox. Agents self-allocating in an environment where their actions affect the future states of that environment and that expectations about the future states of the environment impact on the agents' decision is logically self-referential. This is why I consider such self-allocation to be a feature delineating autonomy. The self-referential nature of self-allocation means that ultimately problems of maximising utility or efficiency in the customary sense must be formally unsolvable, so the question remains what we can do in terms of developing autonomous systems. In economics and finance, an examination of trading strategies is a good place to start for a solution, in light of the huge literature on both these kinds of strategies and their outcomes for both the agent and for the overall systems of which they are a part. To illustrate this: speculation trading specifically relies on the fact that asset prices are non-stationary, for there simply is no profit to be had for an asset speculator in a stationary price environment.

Finance economics identifies two basic types of trading behaviour: those who attempt to predict future price movements by looking for patterns in historical data are termed chartists, or sometimes technicians; those who trade on the basis of trying to determine the financial fundamentals of assets - their 'real' value - are termed fundamentalists (this terminology is common in the empirical studies of market dynamics; see for example [31, 32, 37, 46, 52]). Of course, real traders may represent a mix of trading strategies, and may alter this mix over time, so these types should be read as pertaining more to agent behaviours rather than to the agents themselves. The chartists are the speculators, and are willing to purchase assets at prices above their fundamental value, in the intention of making gains by selling those assets at still higher prices. Chartists thus operate essentially on the basis of scepticism about the rationality of other traders. They are traditionally held to be the bad guys insofar as asset bubble formation is concerned, because this speculative demand is well known to tend to build on itself in a self-reinforcing manner: speculative trading means higher demand, which pushes asset prices higher and higher above their fundamental values, resulting in bubble formation.

The fundamentalists no longer get off the hook so easily with respect to asset bubble either: the problem lies in the difficulty in assessing assets to determine their

---

[19]I have started to try to formalise this concept. The difficulty seems to lie in defining a general notion of what 'surviving' should mean in formulating a deterministic dynamical Hamiltonian model of agents interacting in an environment.

fundamental value and thus rational price. The crudeness of commonly used asset valuations is obvious from even a cursory examination of the kinds of models typically in use, such as the Gordon Growth Model, 2-stage model, 3-stage model and H-model [37], yet the broader issue facing the fundamentalist is that uncertainty must be an inevitable limitation with any asset value model, irrespective of its mathematically sophistication.[20] Asset valuation turns out to be effectively just a different mode of speculation (see for instance [31, 41]). The resulting systematic valuation errors may drive asset bubble formation, even in the absence of the speculation-driven demand of the chartists.

A market containing such traders allows the possibility of self-reference, because the behaviour of the market is a consequence of the behaviours of all the traders who operate within it, and whose behaviours are, in turn, deeply effected by the current state and history of the market. Consequently we should expect fundamental uncertainty here: the problems the agents are trying to solve are without complete solution, and indeed this manifests in real markets in the form of unpredictability, or what economists know as "volatility". In the final analysis, economically rational beliefs are harder to come by than it may appear at first blush. Maximising reward might make perfect sense when we consider an individual agent in isolation. Yet the collective consequences of agents effectively relying on strong assumptions about the independence of the future states with respect to their individual actions in the present means that reward maximisation becomes self-defeating, when, during the subsequent crash, almost everybody following such strategies loses.

21.9 Barbell Strategies

A barbell strategy[21] [37] is formed when a trader invests seeks to increase risk-adjusted returns by investing in a combination of safe long duration investments, and the small remaining portion in short duration securities, with nothing in intermediate duration investments. Closely related is laddering,[22] which avoids reinvesting assets in unfavourable economic environments, by investing in multiple instruments having

---

[20]Indeed, this is an example where increasing sophistication is actually dangerous, by creating false impressions about the reliability of the model. As explained elsewhere in the text, this factor of over-confidence in precision financial risk models is heavily implicated in the GFC of 2007–2008, and thus gives us a highly topical real-world example of the phenomenon.

[21]The term is very common in financial economics. See for instance http://www.forbes.com/forbes/2005/0509/144.html and http://www.investopedia.com/articles/investing/013114/barbell-investment-strategy.asp for brief overviews.

[22]Not to be confused with a type of insider trading known by the same name. Laddering is also used as a term to denote a process whereby insiders purchase stock at lower prices while artificially inflating the price to permit them to then sell at a higher price, by agreeing upon purchase at the lower price to also purchase additional shares at some higher price. This practice was a target of SEC investigations in the wake of the Global Financial Crisis of 2007–2008 [Fjesme, S.L, Initial Public Offering Allocations, Price Support and Secondary Investors, Journal of Financial and Quantitative Analysis (JFQA), Forthcoming].

different maturity dates; the difference is that laddering spreads investment across short, intermediate and long maturity instruments. Laddering can be seen as a kind of nesting of barbells; while real autonomous systems may in general have to ladder in this manner, I will focus here on barbells as the primitive basis of such a plan of attack.

The opposite of a barbell is a bullet strategy [37], where a trader invests in inter-mediate duration securities, to build a portfolio that has securities that mature con-sistently over time. Note that both fundamentalists and chartists typically employ a bullet strategy, just with respect to different choices of problem: the fundamentalist works on the basis of discounted average expected future dividend returns, while the chartist operates using expectations about patterns in asset price movements. The barbell strategy rests on a different base choice of problem: surviving the unexpected, rather than maximising expected returns. A barbell on only the shortest and longest bonds in a bond market is, under a simplistic forward rates assumption, known to be a maximiser of the modified excess return [10].

Taleb [49] invokes a version of the barbell strategy emphasising a large majority in extremely safe instruments that pay poor returns and the remaining in highly risky but potentially highly profitable instruments as a strategy to insulate against black swan events. Taleb presents barbells as applicable outside trading markets; this leap is not a large one having undertaken to regard decision-making in general as resource allocation. The strategy works best in periods of high inflation: put options are cheaper under high interest rates in accordance with the Black Scholes Option Pricing Formula [37], and market crashes tend to coincide with periods of high interest rates. In other words, the strategy works well under conditions of high volatility, or when viewed over the long term where such periods will manifest (usually quickly and unexpectedly), which is precisely the conditions of Taleb's claim.

Indeed, a formalisation of the uncertainty of asset distributions as entropy maximi-sation, without any utility assumption - most of the mathematical finance literature dealing with entropy assumes entropy minimisation as the optimisation goal - yields the barbell portfolio as the optimal solution [17]. In this sense, the barbell strategy seems to constitute a kind of fixed-point solution to an entropic formulation of sur-vival in high-uncertainty environments, where utility assumptions cannot properly apply; recall that a fixed-point is an invariant that resolves a self-referencing question by removing the direct self-reference. Their apparent success across a wide vari-ety of environments and conditions seems to support this interpretation, and makes them a viable starting point for defining the kinds of behaviour that self-allocating autonomous agents might use.

Keynes [25] provided a solid macroeconomic basis for barbell strategies with his explicit separation of the impetus for profit seeking from that for failure exposure, which sets up a difficult trade-off with which agents in such a system must grapple, and which provided the basis for the necessity of governmental moderating control, particularly through monetary policy. Barbells in investment amount to replaying this split at a microeconomic level: instead of retreating into prophesies, going along with majority views, and trying to optimise future returns by riding the middles of economic waves, agents that employ a barbell (or to a lesser extent laddering) are effectively attempting to match the bimodal nature of their investment problem with a bimodal solution aimed specifically at producing a favourably asymmetric effect. To put it more in terms suited to autonomous decision-making, such agents manage their exposure to disastrous outcomes they can envisage but that they cannot brook on the one hand, while investing whatever they can thereafter reasonably afford to lose on a selection of bets that they expect will usually bomb but that will sporadically and unpredictably return disproportionately large rewards. The first component is about hedging against unacceptable outcomes, and it only requires that agents determine their sensitivity failure and be able to determine hedging actions against it, not that they predict the future states of their environment.

The second component is about taking advantage of possible opportunities but doing so only with what the agent can bear to lose, and again this requires only that agents be able to recognise potentially propitious junctures, not that be able to predict what will happen with them in the future. This reasoning about affordable bets on sporadic high return investments is not the same as the more frequently notion of high risk and high reward, which refers to a symmetric situation in which there is high chance of an unacceptable outcome and a low chance of a very high reward. Such barbell agents will not be interested in situations of high risk and high reward in the usual sense, which would at least implicitly presuppose that they are able to reason about known or at least in principle knowable distributions over known or at least knowable sample sets of outcomes. The basic mechanism is about making asymmetric opportunity bets: the rewards are potentially high, the risk of failure is unquantifiable, but losing the bet is affordable.

Under this approach, the largest part of available resources, both intellectual and physical, are typically devoted to simply avoiding exposure to decisive failure; it would be a rare circumstance where this concern did not dominate the allocation of limited resources in a military operation. Modern defence forces arguably already operate along the lines of a barbell investment, though it seems to have not previously been formulated in these terms. Military forces plan and then plan again. They reconnoitre, looking for exposure to their vulnerabilities as they best conceive of them at the time. They inefficiently keep a third in reserve, without knowing in advance how or if it will be required, and position assets to have them available to respond quickly to the unexpected, arguably given more determination of their critical vulnerabilities than positive predictions about the future will hold. They constantly review plans in light of sudden unpredicted experiences of subordinate organisations or shifts in the strategic context, and they unabashedly change our whole approach, in principle, at least, at a moment's notice, if the available evidence compels them to refute their plans. They do everything they can to cope with the reality of being constantly disrupted.

Having so hedged against unacceptable outcomes, to the extent possible under the circumstances, with whatever resources remain, military command and control will put a little something into trying to create and exploit opportunities that just might reap disproportionately large benefits. The central point in barbell strategies is that such opportunity investment is restricted to that which the agent can afford to lose -

noting that just what it might have to be prepared to lose is a highly context-specific determination. Note that nowhere in a barbell strategy are we concerned with directly considering efficiency and maximising expected utility, nor does it heavily rely on knowledge of distributions over known outcomes. Although Taleb's version of the barbell is aimed at addressing the problem of rare events on the tails of stationary distributions, the barbell strategy appears to potentially applicable to Keynes' onto-logical uncertainty as well: it remains feasible in principle even within a non-ergodic system, even when the agent cannot determine the sample sets of the relevant possible outcomes.[23] It seems that our would-be autonomous machines capable of handing high uncertainty environments including those of military operations could follow the same approach.

21.10 Theory of Self

There is a further complication to consider. In economics, capital rationing refers to the process by which limits are imposed on the ability of economic agents to invest of resources [11]. An economically efficient market implies access to capital markets to obtain resources, whereby an agents could, in principle at least, access virtually any amount of capital at market rates in order to pursue any and all investment opportunities that promise a positive return, allowing for the cost of the capital and other expenses and some margin dependent on the perceived risk of the project. In contrast, an agent operating under capital rationing faces potentially high decision-making complexity because of the investment options can no longer be considered in isolation.

Soft capital rationing occurs when the agent itself exercises an internal policy restricting the size of investments, which can be understood as an attempt to manage exposure to uncertainty. Examples of soft capital rationing include internal budget allocations, setting aside capital for unforeseen contingencies, and setting a hurdle rate, which is a minimum rate of return as a required compensation for the perceived risk of the option. A hurdle rate can also be viewed as an opportunity cost, which might be evaluated as the rate of return from the next best investment opportunity having a similar perceived risk profile. Hard capital rationing is externally imposed, where the agent cannot raise capital through equity or debt. Regulatory capital requirements on banks, an inability to raise capital because of previous low performance, and legal prohibition on national defence organisations from accessing capital markets are all examples of hard capital rationing.

In addition to forcing the agent to simultaneously consider multiple options against one another, the presence of capital rationing violates the conditions of market effi-

---

[23]The dismissal mentioned earlier that Taleb makes towards the distinction between rare events in an ergodic system and events in a non-ergodic system being inconsequential might be generously re-interpreted as a recognition of the potential applicability of barbell strategies to classes of situations of non-ergodic uncertainty, and hence to the weaker ergodic rare events of Knightian uncertainty with which Taleb is concerned as well.

ciency, so we should not expect to have an economically efficient allocation of resources. This is a central point in this chapter: conventional notions of utility maximisation break down rapidly under complexity. Moreover, the quality of decision-making of agents under capital rationing is especially sensitive to unforeseen changes in the future cost of capital, which means for autonomous agents potentially exquisite sensitivity to failure under uncertainty.

Given hard and soft constraints under ontological uncertainty, decisions regarding how much resource to put into each side of the strategy amounts to judgements about own tolerance to loss, more than it is about the potential for the environment to dish up favourable or unfavourable outcomes. So at the core of the strategy is the requirement for autonomous machines to be equipped with a theory of self, specifically for the purpose of evaluating exposure to unacceptable failure and deciding hedging plans, and for recognising feasible opportunity for reward and determining investment plans given a range of such opportunities. Theory of mind in the usual sense is then really an extension of theory of own mind as the more fundamental concept for autonomous systems research. The strategy amounts, therefore, to a mechanism for substituting the unapproachable problems of prediction and knowledge acquisition in uncontrolled unstructured environments in general with the much more manageable problem of self-knowledge.

Note that soft capital rationing is self-imposed, and amounts to judgements and this condition appears to very directly imply a requirement for an agent theory of self. It also seems that a theory of self would imply that, in a sense, such agents would effectively talk to themselves, much as humans do [28, 38], constructing a kind of narrative of self as they debate with themselves about different investment options, and moderate and alter their own beliefs and expectations.

There is a deep issue lurking here that motivates and underpins this proposition about agency. Incompleteness means that self-knowledge, and thus knowledge of one's own sensitivity to failure, is inherently limited; after all, we all observe ourselves from inside ourselves. A theory of mind supporting the development and application barbell-type strategies accommodates this in two ways simultaneously. Firstly, the judgement caveat on plasticity and thus on autonomy concerning limits we consider to be operationally acceptable is about making visible to the agent itself the consequences of the limits of its own self-knowledge in terms of managing the effects of the limited ability of anything - or anyone - to determine its own failure modes. Secondly, with respect to determining potential exposure to unacceptable failure, I am advocating a defensive kind of posture: exposure to decisive failure is a matter of choosing boundaries beyond which unacceptable failure is a potential rather than a certainty. We cannot determine failure sensitivity completely or with exactitude, but we can choose those boundaries conservatively or optimistically and in priority order depending on our faith in the broader social enterprise to absorb the possible consequent failures. It seems that this problem of self-knowledge is much more manageable, however, both by virtue the fact that the system we then have to deal with is much smaller than that of the entire environment, which, after all, includes the agent itself, and by virtue of the fact that we humans are a testament to how successful in uncertain worlds agents armed with self-knowledge can be.

21.11 Conclusion

There is a deep connection between economics and autonomous systems research, for the simple reason that the two fields have at their core questions about the nature of agency as autonomous decision-making. The difference is roughly that economics is mainly concerned with descriptive models of agency, while artificial intelligence is squarely focussed on engineering prescriptive models. The interface has been, and must be, permeable. Both fields face the same basic issues about the nature of agency and, in particular, suffer from the inadequacy of current approaches with respect to decision-making under conditions of fundamental uncertainty. Previous work [44] in the economic literature has sought to exert influence primarily on economics audiences about poor representations of human agency, has noted the role in artificial intelligence of wishful mnemonics in masking the severe limitations incumbent in standard assumptions, and has concluded that artificial intelligence has not even begun to replicate the abilities of real humans to cope with fundamental uncertainty as a result.

This chapter is firstly about raising awareness among machine learning, automated reasoning and robotics communities about the relevance of the economic literature on decision-making under uncertainty. In particular, economic theory distinguishes between stochastic risk and unmeasurable epistemological uncertainty, and between epistemological uncertainty and a deeper notion of ontological uncertainty. Secondly, it is about relaying and extending the mathematical underpinnings of this economics literature. The connection from the basic economic notions of uncertainty to non-linear dynamics and ergodic theory that has been relatively recently established in economics [13, 50], with epistemological uncertainty and ontological uncertainty formally distinguishable on this basis. I have also sought here to extend this viewpoint with reference to the well established mathematical practices of formulating questions about future behaviours of non-linear systems as computational problems, whereby ontological uncertainty then manifests as formally unsolvability and incompressibility [35, 39].

Yet the formal unsolvability of computational problems is a particularly deep extension to the slightly earlier results establishing the incompleteness of non-trivial systems of formal reasoning, which means that within such systems there are always questions that cannot be answered, even with infinite resources. Ontological uncertainty can perhaps be best understood on this basis: the possibility of self-reference in any system means that some problems we might like to solve, such as predicting what will happen in the future, are paradoxical in the sense of being unresolvable within the system as either true or false.

Far from constituting an abstruse irrelevance, the practical consequences in economics of failures to handle this kind of uncertainty are difficult to overstate. The sophisticated risk models heavily implicated in the 2007–2008 GFC [3] are now widely acknowledged as having failed so spectacularly for the precise reason that they fail to address epistemological - let alone ontological - uncertainty. There is abounding circumstantial evidence that the same kinds of failures have been felt with respect to applications of artificial intelligence for a very long time, particularly in the form of the manifest dearth of genuinely autonomous military operational systems. The failure of ergodic models in non-ergodic environments will be non-ergodic, meaning that questions about the future occurrences of failure will be unanswerable, and thus failure will be observed as unpredictable and sporadic after seductively long periods of apparent success.

In addition to the prime example of this type of failure, financial economics provides a window into the kinds of strategies that have already been applied successfully in securities markets to cope with uncertainty. Such environments display ontological uncertainty because they allow the possibility of self-reference, because future outcomes are dependent on agent behaviour, which depends on expectations about future outcomes. The consequence is that supposedly rational reward-maximising behaviour for an individual agent may easily be ultimately self-defeating; what makes apparent sense for the individual does not necessarily make sense for the system as a whole.

Barbell trading strategies and their relatives aim to change the problem choice from expected reward maximisation to survival in face of intrinsically unknowable futures. The idea is to divide resource allocation into two logical steps, with the first step being allocation of resources to avoid exposure to unacceptable outcomes. In the second step, remaining resources can be utilised to pursue opportunities, which will usually amount to affordable failures yet will sporadically reap large returns. While not additive, both concerns have to be considered together under conditions that amount economically to hard capital rationing, which make decision making much more difficult; moreover, the resource limits concerned may not be fully determinable in advance. At the centre of this picture is a requirement for self-knowledge: a theory of self seems necessary to recognising failure sensitivity and opportunity in high uncertainty environments under partially observable hard resource limits. The limits of self-knowledge together with the complexity of decision-making under hard capital rationing with the possibility of unexpected budget changes appears to imply that autonomous problem problem solving must be intrinsically social.

Autonomy boils down to developing decision processes for machines for solving problems in complex environments; ontological uncertainty, which I have emphasised in importance over epistemic uncertainty, boils down to the common occurrence in complex environments of seemingly straightforward decision-making problems for which there can be no solution. The future of autonomous research - as with economic theory - will be about changing the problem choices of the past, and, in doing so, effectively altering what it means for the problem to be acceptably solved, than it will be about advancing the technical development of most currently established methods. The technical developments matter, but are subordinate to wiser choices about how they are applied. Fundamental uncertainty has to feature as the central concern of robotics, machine learning and automated reasoning, for otherwise the account of genuinely autonomous operationally usable systems will surely remain at zero.

References

1. K. Adam, A. Marcet et al., *Booms and busts in asset prices* (Technical report, Institute for Monetary and Economic Studies, Bank of Japan, 2010)
2. M. Allais, Le comportement de l'homme rationnel devant le risque: critique des postulats et axiomes de lîêcole amèricaine. Econometrica **21**(4), 503–546 (1953)
3. M. Auerback. Risk vs uncertainty: the cause of the current financial crisis. Technical Report Occassional Paper No 37, Japan Policy Research Institute, 2007
4. M. Bar-Hillel, A. Margalit, Newcomb's paradox revisited. Br. J. Philos. Sci. **23**(4), 295–304 (1972)
5. G. Boolos, *The Logic of Provability* (Cambridge University Press, Cambridge, 1995)
6. C. Carr, *The Book of War: Sun-Tzu's "The Art of War" & Karl Von Clausewitz's "On War"* (Random House Inc., New York, NY, USA, 2000)
7. G.J. Chaitin, *The Unknowable* (Springer Science & Business Media, New York, 1999)
8. G.J. Chaitin, *The Limits of Mathematics: A Course on Information Theory and the Limits of Formal Reasoning (Discrete Mathematics and Theoretical Computer Science)* (Springer, New York, 2003)
9. R.H. Coase, *The problem of social cost, Classic Papers in Natural Resource Economics* (Springer, New York, 1960), pp. 87–137
10. P. Cotton. When a Barbell Bond Portfolio Optimises Modified Excess Return (2012)
11. T. Cowen, *Modern Principles: Microeconomics* (Worth Publishers, Basingstoke, 2011)
12. P. Davidson, *Post Keynesian Macroeconomic Theory* (Edward Elgar Publishing, Aldershot, UK, 1994)
13. P. Davidson, Black swans and knight's epistemological uncertainty: are these concepts also underlying behavioral and post-walrasian theory? J. Post Keynesian Econ. **32**(4), 567–570 (2010)
14. P. Davidson, Is economics a science? should economics be rigorous? Real-World Econ. Rev. **59**, 58–66 (2012)
15. I. Erev, A.E. Roth, Predicting how people play games: reinforcement learning in experimental games with unique, mixed strategy equilibria. Am. Econ. Rev. **88**(4), 848–881 (1998)
16. R. Frydman, G.P. O'Driscoll, A. Schotter, Rational expectations of government policy. South. Econ. J. **49**, 311–319 (1982)
17. D. Geman, H. Geman, N.N. Taleb, Tail risk constraints and maximum entropy. Entropy **17**(6), 3724–3737 (2015)
18. I. Gilboa, A. Postlewaite, D. Schmeidler, Rationality of belief or why bayesianism is neither necessary nor sufficient for rationality. Technical Report Cowles Foundation Discussion Papers 1484, Cowles Foundation for Research in Economics, Yale University (2004)
19. K. Gödel, *On formally undecidable propositions of principia mathematica and related systems* (Dover Publications, New York, 1992)
20. R.M. Gray, R.M. Gray, *Probability, Random Processes, and Ergodic Properties* (Springer, New York, 1988)
21. F. Hahn, R.M. Solow, *A Critical Essay on Modern Macroeconomic Theory* (MIT Press, Cambridge, MA, 1997)
22. P.R. Halmos, *Lectures on Ergodic Theory* (Martino Fine Books, Eastford, 2013)
23. D. Kahneman, A. Tversky, Prospect theory: an analysis of decision under risk. Econometrica. J. Econ. Soc. **47**, 263–291 (1979)

24. R. Kaye, *Models of Peano Arithmetic* (Cndon Press, Oxford, 1991)
25. J.M. Keynes, *The General Theory of Employment, Interest and Money* (Macmillan, London, 1973)
26. M. King, *The End of Alchemy: Money, Banking, and the Future of the Global Economy* (WW Norton & Company, New York, 2016)
27. F.H. Knight, *Risk, Uncertainty and Profit* (Beard Books, Washington DC, 2002)
28. E. Kross, E. Bruehlman-Senecal, J. Park, A. Burson, A. Dougherty, H. Shablack, R. Bremner, J. Moser, O. Ayduk, Self-talk as a regulatory mechanism: how you do it matters. J. Pers. Soc. Psychol. **106**(2), 304 (2014)
29. P. Krugman, R. Wells, Microeconomics (2012)
30. A.D. Lane, R.R. Maxfield, Ontological Uncertainty and Innovation. J. Evol. Econ. **15**(1), 3–50 (2005)
31. V. Lugovskyy, D. Puzzello, S. Tucker, An experimental study of bubble formation in asset markets using the tâtonnement trading institution. Technical Report Working Papers in Economics 11/07, University of Waikato, Department of Economics (2011)
32. V. Lugovskyy, D. Puzzello, S. Tucker, A. Williams, et al., Can concentration control policies eliminate bubbles? Technical Report Working Papers in Economics 12/13, University of Waikato, Department of Economics (2012)
33. M.J. Machina, Choice under uncertainty: problems solved and unsolved. J. Econ. Perspect. **1**(1), 121–154 (1987)
34. D. McDermott, Artificial intelligence meets natural stupidity. ACM SIGART Bull. **57**, 4–9 (1976)
35. L. Ming, P. Vitányi, *An introduction to Kolmogorov complexity and its applications* (Springer, New York, 2008)
36. P. Mirowski, *Machine Dreams: Economics Becomes a Cyborg Science* (Cambridge University Press, Cambridge, 2002)
37. P. Moles, N. Terry, *The Handbook of International Financial Terms* (OUP, Oxford, 1997)
38. A. Morin, Self-talk and self-awareness: on the nature of the relation. J. Mind Behav. **14**(3), 223–234 (1993)
39. A. Nies, *Computability and Randomness*, vol. 51 (Oxford University Press, Oxford, 2009)
40. P. Roberts, *The Failure of Laissez Faire Capitalism* (Clarity Press, Atlanta, 2013)
41. J.R. Shiller, Speculative asset prices. Am. Econ. Rev. **104**(6), 1486 (2014)
42. O. Shy, *The Economics of Network Industries* (Cambridge University Press, Cambridge, 2001)
43. A.H. Simon, Rationality as process and as product of thought. Am. Econ. Rev. **68**(2), 1–16 (1978)
44. R.E. Smith, Idealizations of uncertainty, and lessons from artificial intelligence. Econ.: Open-Access Open-Assess. E-J. **10**, 1 (2016)
45. R.I. Soare, *Turing Computability: Theory and Applications* (Springer, New York, 2016)
46. K. Steiglitz, D. Shapiro, Simulating the madness of crowds: price bubbles in an auction-mediated robot market. Comput. Econ. **12**(1), 35–59 (1998)
47. N. Taleb, *Fooled by Randomness: The Hidden Role of Chance in Life and in the Markets* (Random House, New York, 2005)
48. N.N. Taleb, *The Black Swan: The Impact of the Highly Improbable*, vol. 2 (Random House, New York, 2010)
49. N.N. Taleb, *Antifragile: Things that Gain from Disorder* (Random House, New York, 2012)
50. A. Terzi, Keynes's uncertainty is not about white or black swans. J. Post Keynesian Econ. **32**(4), 559–566 (2010)
51. A. Teversky, D. Hahneman, Judgement under uncertainty: Heuristic biases. Science **185**(4157), 1124–1131 (1974)
52. G.A. Timmermann, How learning in financial markets generates excess volatility and predictability in stock prices. Q. J. Econ. **108**(4), 1135–1145 (1993)
53. T. von Ghyczy, B. von Oetinger, C. Bassford, *Clausewitz on strategy: Inspiration and insight from a master strategist* (Wiley, New York, 2002)

54. T.A. Weber, A robust resolution of newcomb's paradox. Theory Decision **81**(3), 339–356 (2016)
55. C. Werndl, What are the new implications of chaos for unpredictability? Br. J. Philos. Sci. **60**(1), 195–220 (2009)

Index

© The Editor(s) (if applicable) and The Author(s) 2018

393

H. A. Abbass et al. (eds.), *Foundations of Trusted Autonomy*, Studies in Systems,
Decision and Control 117, https://doi.org/10.1007/978-3-319-64816-3

Printed in the United States
By Bookmasters